Advances in Spatial Science

Springer
Berlin
Heidelberg
New York
Barcelona
Hong Kong
London
Milan
Paris
Singapore
Tokyo

Institutet för regionalforskning
Swedish Institute for Regional Research

UNIVERSITY
TROLLHÄTTAN·UDDEVALLA
DEPARTMENT OF ECONOMICS
AND COMPUTER SCIENCE

Börje Johansson · Charlie Karlsson
Roger R. Stough
Editors

Theories of Endogenous Regional Growth

Lessons for Regional Policies

With 85 Figures
and 48 Tables

Springer

Prof. Dr. Börje Johansson
Prof. Dr. Charlie Karlsson
Jönköping International Business School
Economics Department
Box 1026
SE-55111 Jönköping
Sweden

Prof. Dr. Roger R. Stough
George Mason University
The Institute of Public Policy, M5-3C6
Fairfax VA, 22030-4444
USA

ISBN-13:978-3-642-64030-8 Springer-Verlag Berlin Heidelberg New York

Library of Congress Cataloging-in-Publication Data applied for
Die Deutsche Bibliothek - CIP-Einheitsaufnahme
Theories of endogenous regional growth: lessons for regional policies; with 48 tables / [Institutet för Regionalforskning, University Trollhättan Uddevalla]. Börje Johansson ... ed. - Berlin; Heidelberg; New York; Barcelona; Hong Kong; London; Milan; Paris; Singapore; Tokyo: Springer, 2001
(Advances in spatial science)
ISBN-13:978-3-642-64030-8 e-ISBN-13:978-3-642-59570-7
DOI:10.1007/978-3-642-59570-7

Springer-Verlag Berlin Heidelberg New York
a member of BertelsmannSpringer Science+Business Media GmbH

Softcover reprint of the hardcover 1st edition 2001

Cover design: Erich Kirchner, Heidelberg

SPIN 10733998 42/2202 - 5 4 3 2 1 0 - Printed on acid-free paper

Preface

The "Fyrstad region" (Region of four towns) is located on the west coast of Sweden about 100 kilometers north of Göteborg, the second largest metropolitan region of Sweden. In this region a series of development activities have been initiated in the second half of the 1990s, to a large extent supported by EU's structural fund program for regional revitalization. In 1997 a supporting research activity was established in the form of a collaboration between the University College of Trollhättan-Uddevalla (HTU) and Jönköping International Business School (JIBS). A major approach in this project, called "New Economic Life in the Fyrstad Region", is a systematic effort to learn from comparisons of the development patterns and policies in the Swedish region and similar regions in a global outlook. This is manifested by an annual international workshop.

The first of these meetings took place in the town of Uddevalla 1998 and the outcome is documented in this book Endogenous Regional Development and Policy. In four parts the book presents a contemporary view of theories and models to analyze and guide policies for regional renewal and development in a new economic regime of interacting regions in a global world economy. The workshop was organized by E. Bergman, B. Johansson, C. Karlsson and R. Stough.

The workshop has received financial as well as other forms of support from the Swedish Institute for Regional Research (SIR), Sparbanksstiftelsen Väst, the European Union, Länsstyrelsen i Västra Götalands Län, the Institute of Public Policy and the Center for Regional Analysis at George Mason University, JIBS and HTU. These contributions are gratefully acknowledged.

Uddevalla and Jönköping, November 2000

Börje Johansson, Charlie Karlsson and Roger Stough

Contents

Part I

Introduction

1 Introduction: Endogenous Regional Growth and Policies

Charlie Karlsson[a)], Börje Johansson[a)] and Roger Stough[b)]

[a)]Jönköping International Business School, Box 1026, SE-551 11 Sweden

[b)]The Institute of Public Policy, The Mason Enterprise Center for Regional Analyis and Entrepreneurship, and The Transport Policy and Logistics Center, George Mason University, Fairfax VA, 22030-4444 USA

A rich literature has grown up around the concept of the rise in the importance of regions in the global economic system. According to this view global trade has the form of interaction between functional regions, rather than between countries. As this vision has evolved, empirical observations suggest that global and national economic change should be understood as a process, which is dependent on local dynamics operating on the regional level. Perceptions and models of such change can be underpinned with various theoretical perspectives generally referred to as the new endogenous growth theory. The term endogenous implies that economic growth is influenced by the use of "investment resources" generated by the economy itself – in contradistinction to the reference made to exogenous factors in the Solow type growth models.

In aggregate macro models this form of endogeneity implies that investments in production capital, infrastructure, education (or human capital) and R&D affect the growth rate of the economy. Today, an increasing number of scholars claim that this perception of the growth mechanism is relevant for understanding economic change since the beginning of the industrial revolution. At the same time, it may be argued that models of endogenous growth and dynamically increasing returns have an even more prominent role to play in the analyses of contemporary economies. Moreover, endogenous growth models imply that policy matters – as regards both the supply of public services and the investment in tangible and intangible infrastructure.

Obviously, there are endogenous macro mechanisms that affect the growth of productivity and income per capita at the country level as well as for regions. In this volume the contributions focus on endogenous processes in regional economies and the pertinent role of national and regional policies. This focus reflects an insight that vital change processes are initiated and materialize at the regional level. From this position we derive an associated set of questions, what are the regional policy initiatives that can support and revitalize regional economies and under which conditions are they effective?

Endogenous Specialization and Economic Change in Regions

According to the new growth theory, cumulative processes will obtain with either self-reinforcing decline or continuing growth, which may last over long periods and which may transform location patterns considerably. In the individual region such dynamics are recognized as change processes including entrepreneurship, learning, education, learning by doing, acquiring institutional capability, but also migration of firms and households embodying business visions and knowledge, technology and skills. Such processes are orchestrated in varying degree by regional and higher-level government, as well as by regional leadership in general.

Thus, a significant part of economic growth can be modeled as induced by technical and organizational change, which is related to local forces such as education, learning, regional leadership and institutions, and government action. As a consequence, the rationale and guide for economic development has moved away from a pure neo-classical economic thinking, according to which static factor cost-minimization can explain comparative advantages of regions. This view has been an intellectual cornerstone of regional development policy-making in the industrial age. In contrast, the papers presented in this book examine the application of the endogenous perspective and the related driving forces in a regional context.

In models of endogenous growth knowledge capital is treated as an independent factor of production. However, it is also recognized that knowledge capital has the characteristics of a public good. Hence, the new knowledge that is generated within a certain firm in a given sector and region will over time spill-over to other firms, to other sectors and to other regions. These diffusion effects give rise to dynamically increasing returns in the total economy and, thus, stimulate economic growth. This phenomenon is all the more important in cases when the investment is financed by private sources. Under such circumstances the likelihood of under-investment increases, and the overall economy will fail to benefit from the total growth options that are embedded in the potential of increasing returns.

When the human capital of households and other capital resources are located together in the same region, a self-reinforcing concentration may obtain in certain regions, and such a concentration process has the power to further stimulate economic growth in these regions. This type of agglomeration has been associated with both positive productivity-enhancing effects and negative counteracting impacts such as crowding, noise and environmental decay. The endogenous growth models and analyses presented in this book stress that agglomeration and localization phenomena generate positive external effects that outweigh the negative effects, especially if these phenomena are accompanied by appropriate regional infrastructure investments. This may result in a process of increasing spatial concentration of economic resources. At the same time one must ask: what are the limits of cumulative growth, and how do the counteracting forces develop?

In a spatial context endogenous growth demands a consideration of trade flows and location dynamics. Interregional trade has basically the character of exports and imports between functional regions. Such trade between industrialized regions is mainly determined by internal and external scale economies and by generalized spatial interaction costs, i.e. the sum of generalized transportation and transaction costs. The demand from other regions creates together with the demand from the internal regional market the total market potential of a region. The interplay between market potential and scale economies generates agglomeration economies that function as centripetal, concentrating forces. The advantages for firms to locate in regions with a large market potential increase with the degree of product differentiation.

Next we can observe that cumulative growth processes which generate large functional regions are counterbalanced by high land prices and crowding costs. Low spatial interaction costs together with well-functioning markets tend to increase the prices for land and premises, while increased crowding tends to increase the spatial interaction costs, i.e. spatial friction.

It is an inherent characteristic of a world with endogenous agglomerations that factors of production will not become evenly distributed but will instead tend to concentrate spatially. However, due to the existence of inertia the migration of factors of production over the borders of functional regions will normally be a slow process. This is, in particular, true for households, whose geographical mobility is limited by the fact that their assets (housing, social networks, information, and so on) are tied to the locality and to the functional region. When a household moves to another region, part of these assets is lost.

The development trajectories described above are such that the small and medium-sized regions seem to have a bleak future. Is that so? Several of the contributions in this book examine this issue. One option for non-metropolitan regions is the development of narrow localization economies that may provide these regions with sustainable economic development. This conclusion may be underpinned with insights from the so-called "new trade theory" with reference to monopolistic competition, localization economies and cluster formation. One may also investigate under what conditions these regions can develop resource-based advantages in a world with multiregional product cycles and endogenous growth. Moreover, can small regions enhance their market potentials through improved interaction links to neighboring regions? And what is the role of local policy in this context?

According to the endogenous approach, regional economic growth and renewal is a market phenomenon mainly driven by a self-organized, interdependent and cumulative relationship between demand and supply. Demand creates geographical market potentials that stimulate production at the same time as the supply of resources and products creates demand at the market. How does policy-making become a part of such self-organization processes?

The Role of Policy in Endogenous Growth Processes

As regional economic dynamics have become more endogenous the role of policy and its meaning has been questioned. For example, in a world where change is increasingly driven by endogenous forces and where regions are becoming relatively more important, what is the role of national policy? Is there a role for national policy? Are top down policies capable of producing desired regional dynamics or facilitating the achievement of national goals? Are bottom-up policies capable of impacting national goals positively or negatively? Further, are there combinations of top-down, inside and bottom-up policies that offer the best guide to development regionally and nationally? Finally, as with all policy it is ultimately dependent upon culture and values, which vary significantly from region to region and country to country. So another major issue this book deals with is the nature and relevance of regional economic development policy in a comparative context and in the new reality of endogenously driven economic growth.

There is another important aspect of the role of policy in regional economic development that is emphasized. In the old economy of the industrial age, policy from higher levels of government was used to drive development but with mixed results. In the new economy of the late 20th Century development is often driven most efficiently by endogenous self-organization and self-adjustment processes. Thus, there is an important policy question regarding the degree to which it is possible to create and implement policies that will efficiently and effectively induce self-organizing and adjustment processes.

Information technology has emerged as a new core technology during the latest three decades. The recent adoption and spread of this technology into all forms of societal endeavor has transformed the meaning of competitiveness, and it can be expected to do so even more in the future. In the earlier industrial era, competition was local (or national). It was relatively easy to get a comprehensive overview of markets, work was task oriented, factor inputs (land, labor and capital) and their cost were fundamental determinants, products were mass produced, hard infrastructure was critical to regional competitiveness and the dominant role of government was to provide services (Table 1.1). Moreover, national governments could exercise an influence on macro performance and capital markets. Most of this has changed. In the new economy, competition is global, markets are volatile and require continuous adjustment on the part of producers. Consequently, it should be no surprise that production processes are becoming increasingly flexible. Further, life-long learning has become a survival technique for the workforce as work is continuously changing. Today innovation, quality, speed and agility in the value delivery chain tend to be more important than minimizing traditional factor costs.

While hard infrastructure like roads, airports, water and waste water systems physically underpin regional economic systems, soft infrastructure in the form of education and institutional structure and flexibility has become relatively more important. Finally, in the new economy, government is seen more as an enabling agent than as a provider of services.

In the evolution of the "new economy", the influence of information technology on the entire economy is accelerating. This technology is rapidly becoming embedded in many economic sectors including technical ones and increasingly so in non-technical ones such as law, institutional arrangements, retail, wholesale, trans-

Table 1.1 Attributes of the old and new economies

Issue	Old economy (Fordist)	New economy (Neo-Fordist)
Organizational form	Vertically integrated	Horizontal networks
Scope of competition	National	Global
Markets	Stable	Volatile
Competition among sub-national	Medium	High
Geographic mobility of business	Low	High
Role of government	Provider	Steer/row/end
Labor and Workforce Characteristics		
Labor-Management relations	Adversarial	Collaborative
Skills	Job-specific skills	Global learning skills and cross-training
Requisted education	Task specialization	Lifelong learning and learning by doing
Policy goal	Jobs	Higher wages and incomes (productivity)
Production Characteristics		
Resource orientation	Material resources	Information and knowledge resources
Relation with other firms	Independent ventures	Alliance and collaboration
Source of competitive advantage	Agglomeration economies	Innovation, quality, time to market and cost
Primary source of productivity	Mechanization	Digitization
Growth driver	Capital/labor/land	Innovation, invention and knowledge
Role of research and innov. in the economy	Low moderate	High
Production methodology	Mass production	Flexible production
Role of government	Infrastructure provider	Privatization
Infrastructure Characteristics		
Form	Hard (physical)	Soft (information and organizations)
Transport	Miles of highway	Travel time reduction via application of information techn.
Power	Standard generation plant	Linked power grid (co-generation)
Organizational flow	Highly regulated	Deregulation
Telecommunications	Miles of copper wire	Wireless and fiber
Learning	Talking head	Distance learning

* For a more extended discussion see: Jin D. and R.R. Stough (1998) "Learning and learning capability in the Fordist and Post-Fordist age: an integrative framework", Environmental and Planning A, V 40, pp. 1255-1278.

portation, etc. While this technology is a dominant global vehicle that generates change in regional economic systems it is also locally driven by know-how, leadership, institutions (government and intermediate) that extend and transform it. Such local influences lead to trajectories of success of some regions relative to others. The purpose of the book is to explore and examine from theoretical, applied and policy perspectives how one can formulate models of and draw conclusions about this endogenously driver regional economic development. The following part of this introduction outlines the organization of the book and describes some of its central features as well as the various parts and the individual contributions.

Organization and Contents of the Book

The book is organized into five parts with the Introduction serving as Part I. Part II focuses on endogenous growth theory as applied in a regional context. Part III addresses issues tied to interregional processes, scale economies and agglomeration. Part IV deals with functional regions, clustering and local economic development. Part V focuses on endogenous regional economic policy issues. These include questions about how policy can influence structural renewal, innovations, specialization, competitiveness and trade.

Part II: Endogenous Growth Theory Applied in a Regional Context

Five chapters form the content of Part II. The first paper, by R.R. Stough, focuses on an often forgotten institution in the regional economic development equation, leadership. This paper argues first that strategies for economic development cannot be well targeted even with the best of leadership if gleaned only from traditional quantitative analytical methods such as input-output analysis will be imperfect unless it is supported and augmented by techniques that generate expert, primary information about these factors. A methodology called Multi-Sector Analysis is outlined and illustrated with a brief case study. This work is followed by the formulation of a regional leadership model and a test using data from U.S. metropolitan regions. The paper concludes that leadership is a central amplifying variable in the economic growth and development process.

The second paper by J.W. Harrington and D. Ferguson examines the role of institutions in labor and human capital formation processes. This opening paper is especially focusing the new institutional economics. Successful endogenous regional development increasingly depends on institutions to mobilize, direct and execute development strategy. This chapter illustrates the nature of institutions and their role in endogenous led development.

The next paper by W.B. Zhang focuses on more traditional and established modeling approaches. Using a dynamic two-region, multi-sector model with endogenous knowledge accumulation, the author examines how differences in knowledge utilization and creativity affect the economic geography of the regions. The paper shows that the regional economic system may have either a unique solution or multiple equilibria and that each possibility may be either stable or unstable. The outcomes depend in an intrinsic way on the level of knowledge utilization and the creativity of the production sectors in the two regions.

J. Rees in the next paper focuses directly on technology based theories of regional development and emphasizes the view that technology is the prime "motor" for understanding the process of regional development. In short, technology is the primary determinant of productivity and, therefore, regional economic growth and change. He then emphasizes that the primacy of technology is significantly affected by agglomeration, learning and leadership processes. In so doing the theme returns to institutional considerations with the final part of the paper providing case study evidence of the role and importance of leadership in technology-led economic development that further support the Stough thesis.

The last paper in Part II by G. Maier reviews the recent developments, known as the new growth theory, that have modified the predominate view and understanding of economic growth. These developments are viewed as fundamental in the sense that they demonstrate that some of the assumptions of traditional growth theory and most of neo-classical economic theory in general, are inconsistent with the basic phenomena of the modern economy. The author argues that these recent advances make it possible to investigate more thoroughly the consequences of policies embedded in traditional economic thinking. The paper begins by looking at the implications that the allowance of scale effects and minor change in the traditional assumptions has major consequences on policy and brings into importance in the traditional view. Maier develops and applies a model to test and demonstrate the argument that externalities and scale effects must be considered in regional economic development policymaking.

Part III: Interregional Processes, Scale Economies and Agglomeration

The third part of the book brings together several papers that consider the importance of interregional processes and scale economies from an endogenous regional innovation and development process perspective. Emphasis is also placed on agglomeration and network effects on development and growth. The first paper by I. Gordon and P. Cheshire considers the importance of location advantage in territorial competition drawing upon lessons from Europe. Through an interpretive analysis of the literature the authors first introduce the hypothesis that the shift to more flexible, market-mediated forms of economic co-ordination, as well as the shift to smaller production units characteristic of post-industrial activities, imply an enhanced role for highly agglomerated economies like those found in major metropolitan regions. Next they present a counter argument. Suggesting that an emphasis on the role of networks in urban economies, supported by relations of trust rather than the haphazard connections enabled by pure agglomeration, provides a contrary perspective in which the territorial economy may operate as a "club" promoting its own (endogenous) development. The paper evaluates these hypotheses and offers some prescriptive conclusions about outcomes associated with the adoption of the two alternatives.

B. Johansson and C. Karlsson investigate the relationship between scale economies and product-specific home market borders in the next paper. In particular this chapter introduces a theoretical framework based on the concepts of a functional region and its product-specific market potentials. A basic conclusion is that large urban regions have a fundamental competitive advantage, because their much larger market potentials allow them to host many more scale-dependent activities than smaller regions can. Recognizing this, their approach is developed to improve the analysis of small and medium-sized regions, with an emphasis on the size of their internal market and the extension of their external markets. The authors show that regions with a large internal market potential have an absolute advantage in finding a diversified specialization. Moreover, when a region has a large internal as well as external market-potential the competitive advantage increases even further, with increased possibilities of becoming a host of a wide range of sectors, many of which will export to other regions. The major conclusion as regards small and medium-sized regions is that their main opportunity is to develop a narrow specialization based on external economies of scale with specific cluster formations.

E. Haddad and G. Hewings create and use an interregional CGE model of the Brazilian economy to examine the open door policies of the 1990s in Brazil and the national strategies for increasing international competitiveness. The CGE model is used to examine the short-run and long-run effects of trade liberalization policies, represented by tariff cut simulations. The general equilibrium nature of economic interdependence and the fact that the policy impacts in various regional markets differ are considered. Results of the analysis suggest that the interplay of market forces in the Brazilian economy favor the more developed region of the country (Center-South). In the short-run all regions are positively impacted. In the long-run only the less developed Northeast region is negatively impacted. However, in both the long and short-run the tariff reduction worsens the Northeast's relative position in the country. The important contribution of this paper is the use of a robust CGE model to simulate national policy impacts at the regional level.

S. Mun and K. Sasaki next examine empirically and numerically the economic systems of small-to-medium sized regions in Japan. The analysis focuses on the comparative examination of agglomeration economies at the level of the small and medium-sized region and on the effects of transport network change on the whole system of regions. The impact of the attraction of the central area (Tokyo) on the performance of small and medium sized regions is also considered. Again simulation modeling for testing policy impacts is a central contribution of the paper. The consistency requirements in the model design imply that a set of multiregional interdependencies can be taken into account.

The next paper in this part of the book examines the relationships among productivity, agglomeration and enterprise size. In particular, E. Feser investigates the relationship between firm size and local (business) external economies, and focuses first on the issue of measurement. However, early in the paper he reviews and examines a collection of relevant theoretical and empirical approaches to this problem from which he develops a new and improved test for examining the link between enterprise proximity and productivity. In particular, he models a set of commonly postulated sources of agglomeration economies (proximity to input suppliers, proximity to purchasing sectors, labor pools, and knowledge or information spillovers) as technical efficiency parameters in a cross-sectional, establishment level production function. The results of the research suggest the presence of spatial economies for both large and small enterprises, with the strongest differences between firms being manifested in terms of knowledge spillovers.

Part IV: Functional Regions, Clustering and Local Economic Development

Part IV dealing with functional regions, clustering and local economic development starts with a paper by D. Maillat and L. Kebir, which examines the concept of learning embedded in a regional economic or territorial production system. The learning process is viewed as central to regional competitiveness in that non-material or constructed resources have become relatively more important to economic success than physical ones. Learning and learning process are critical resources for regional development in an era where constructed resources are a fundamental attribute of competitiveness. After an examination of several foundation concepts – including interactive learning, institutional learning, organizational learning and regional learning – the authors introduce a two-dimensional regional typology composed of evolutionary and dynamic axes. Learning and innovation scenarios are associated with each quadrant of the typology thereby enabling the classification of regions according to these attributes. A second typology for classifying territorial production systems (firms) is also offered in the learning region context. This enables the investigator to classify regions and firms in terms of their learning and evolutionary dynamics.

The next paper is by M. Steiner where relationships between industrial clustering and economic change are examined using data from the Styria region in Austria. Steiner begins with an assessment of the change in industrial cluster structure between the old and the new emerging economy of Styria. He then examines the efficacy of alternative methodologies for the construction and analysis of clusters and for constructing policy strategies and instruments. The final part of the paper examines the strengths and weaknesses of adopting a cluster-oriented approach to regional economic development policy.

D. Martellato next introduces a paper that examines various features of the so-called anomaly of the preponderance and maintenance of a large proportion of small and medium sized firms in the Italian economy. Yet, he argues the alleged inefficiency and non-sustainability of this small-scale dominated production system is in patent contrast to the operation of these systems in the Italian context where there is a history of sustained strong economic performance. Further, these systems have high innovation rates, strong export performance and a high level of internationalization. His analysis focuses on the coexistence of (i) firm-size structural stability with turbulent market dynamics and (ii) persistence of high rates of firm entry and exit flows. A model from biology is adapted to depict and examine the conditions under which firms with heterogeneous behaviour patterns, rather than size, coexist.

The paper by N. Hansen presents a variety of theories that are concerned with how the geographic clustering of firms jointly promotes the competitive advantage of firms and the regions within which they are located. All of the theories examined imply that firms in geographically peripheral locations are disadvantaged by not participating in urban scale or agglomeration economies. Hansen then presents evidence that firms can benefit from participation in national and global innovation networks, whether or not they participate in local or regional networks. It is then argued that regardless of the nature of the spatial networking process, much of the knowledge required for innovation is gained from sources outside of a firm or sector, and that innovations are often made as a result of user driven needs. The final part of this paper examines the role of information and communications technologies (ICT) as a means for enterprises to not only communicate their innovations to potential users, but also to communicate their needs for innovation to potential suppliers of innovative activities. The author concludes that those advanced firms and regions have benefited from the network externalities of ICT, but ICT has not been sufficient to generate economic development in regions that lack knowledge workers and the necessary organizational and institutional capacity. He also finds that ICT cannot create local innovation networks where none existed before but it can contribute greatly to the capacity of peripheral areas to engage in external networking on a global scale and to innovate, if the appropriate educational and civic environment has been established. This again stresses the importance of institutional infrastructure in the new economy.

In the last paper in this part B.L. Weinstein examines the applicability of alternative regional growth theories in an examination of two U.S. case studies of the local economic development process. Both endogenous and exogenous factors are seen to have influenced the development trajectories of the Dallas-Fort Worth, Texas metropolitan region and the much smaller Dalton, Georgia region. A combination of 12 different regional economic theories is used to explain ways in which a strong leading technical sector transformed the Dallas-Fort Worth region. A similar analysis and interpretation is used to explain how the Dalton region was able to diversify its economy away from a heavy dependence on one sector and thereby become more sustainable.

Part V: Endogenous Regional Economic Policy Analyses

Part V of the book, which includes four papers, focuses on endogenous regional economic development policy. The first paper in this part is by A. Varga and examines the role of universities in regional economic development. This paper uses metropolitan level data from the U.S. to examine the relationship between agglomeration and local academic technology transfers in the electronics and instruments sectors. The results show that agglomeration must surpass a threshold level before substantial local economic effects of academic research are achieved. Further, simulations of university knowledge-effects suggest that pure university-based regional economic development policies are not effective enough to "upgrade" localities to a higher level of innovative activities. In short, institutions are critical to the success of local development strategies and, therefore, in capitalizing on university research activities and infrastructure. The result of the Varga paper again emphasize the often, unrecognized importance of institutions as variables that amplify the role of more traditional economic variables in the growth and development process.

The next paper by K. Haynes and M. Dinc investigate the foundations of economic performance and employment change in manufacturing sectors for a sample of U.S. states using a new extension of the shift-share methodology building on their earlier work. The extension that is introduced enables them to take account of industrial and regional differences in labor and capital (non-labor) productivity and the impacts of productivity differences on regional manufacturing employment change. The findings suggest that the differential effects of output growth (decline) drove the change in manufacturing employment during the study period. New investments (under-investment) in physical capital, and improvement in technology, though labor productivity also played an important role. The research suggests that the methodology and the extension developed in this paper can be used to both evaluate and guide regional growth policy.

The paper by H. Westlund examines spatial policy in Sweden and its effects on the peripheral, sparsely populated areas. The empirical analysis shows that the international tendencies of counter-urbanization are reflected in the sparsely populated periphery of Sweden. Westlund argues that the center-oriented spatial policy adopted by the central government has been in part misdirected and that comparative advantages of the sparsely populated periphery have been missed or neglected. This suggests that economic renewal lies more in attracting migrants via quality of life attributes, rather than attempting to attract enterprises that bring with tem new residents.

The editors of this book have written the final paper in the book. The aim is to present a synthesis of the various aspects introduced in the earlier chapters. It starts by asking, who are the policy makers and what are the objectives? It then investigates what policies that according to an endogenous perspective might support inter- and intra-regional change processes. In particular, it discusses the role of knowledge, technology, infrastructure, institutions, and clusters. The role of territorial competition and the dynamic formation of location advantages are also highlighted. The overview ends with the rethoric question as whether regional policy itself is endogenous?

Part II

Endogenous Growth Theory Applied in a Regional Context

2 Endogenous Growth Theory and the Role of Institutions in Regional Economic Development

Roger R. Stough
The Institute of Public Policy, The Mason Enterprise Center for Regional Analyis and Entrepreneurship, and The Transport Policy and Logistics Center, George Mason University, Fairfax VA, 22030-4444 USA

2.1 Introduction

Regional economic development involves a wide array of factors whose importance varies across regions and time. Thus, an economy that is resource dependent will have different defining characteristics than one that is service based. Likewise, an economy embedded in a transition to a knowledge based structure will exhibit different attributes than at an earlier time when it was manufacturing based. This understanding is not new. However, two types of factors are essential to the understanding of regional economic systems: endogenous and exogenous. In all regional economic systems these factors are entwined and intermingled, and functionally inseparable.

Recent work in the area of endogenous growth theory (Romer 1986 and 1990) separates endogenous and exogenous factors for analytical purposes. Such an exercise leads to a conclusion that it is possible for closed economic systems, e.g., regional or national economies, to grow and develop. That is, growth and development are not, in a fundamental sense, dependent on external conditions. Thus, as the knowledge base of a regional economic system is enhanced from within, for example, through learning (Arrow, 1962), it becomes a continuous and internally created source of competitive advantage and monopoly power (Romer, 1986 and 1990; Lucas 1988 and 1993). Such internal learning may, at the same time, lead to a decision to build new infrastructure, e.g., a road or improved waste water treatment, to enhance development. Nonetheless, knowledge creation, through learning, appears to be the central proposition of the endogenous growth formulation. Through learning it is possible to envision how a closed regional economic system could survive develop and sustain itself.

It is not the position of this paper to argue that exogenous factors such as trade, labor mobility and migration, knowledge or innovation diffusion, foreign exchange, business cycles and capital mobility are unimportant to a region's economic performance. Clearly these are important but, for the most part, are not substantially under the control of local development efforts. What is significant about endogenous growth theory is that it emphasizes the importance of local factors in creating and maintaining sustained development. Such a cohesive argument for the importance of local factors has been lacking. Perhaps that is why the fields of community development on the one hand and regional science and economics on the other have minimally overlapped in the past. Endogenous growth theory provides a way to see a broad array of community and institutional and non-traditional economic variables, e.g., leadership, learning, and social capital, as major inputs for economic development.

Endogenous growth theory provides a rationale for the notion that it is possible for regional economic growth and development to be sustained by local internal forces. While these forces include a wide array of factors some of the more important are learning, leadership, institutions, physical infrastructure and human capital. Through these it is possible for endogenous closed economic systems with feedback to become self-sustaining and experience the phenomenon of dynamically increasing returns (Kilpatrick, 1998; Arrow, 1962 and 1994; Arthur, 1994). This in turn fuels and sustains growth and development.

This paper has several purposes. First, it examines ways to identify and target critical competitiveness factors in local regional economies that can be used to inform endogenous development policy. Second, it views local leadership as a far more important variable than has been the case in traditional regional science and economics. Consequently, the role of leadership in regional economic development is examined conceptually and empirically. Finally, the notion of regional endogenous growth policy is considered. While the role of infrastructure in endogenous growth is important it is not examined in this paper at length, however, see Haynes, 1991; Seitz and Licht, 1992; Boarnet, 1997; and Stough, et al., 1998.

Institutions: Institutions in their broadest definition are social rule structures with associated standing patterns of behavior and procedures (Buchanan, 1991; Lavoie, 1994). For example, at a given time societies have rules (formal/statutory and informal) that define accepted behavior and action patterns for institutions such as property rights, provision of infrastructure (private vs. public), management practices, governance, the role of markets, and so on. Institutions are sometimes confused with organizations, e.g., universities vs. higher education; government rather than governance; associations vs. influence circles or structures; and companies vs. markets or competition. In this paper institutions are defined in terms of the broader definition. Concern thus rests with institutions defined as rule structures not necessarily as specific organizations.

As noted, learning is critical to endogenous economic growth and development. Further, social capital and local institutions are relatively immobile yet they help to induce development in some regions but not others (Malecki, 1998). This is important given that regions and localities are increasingly attempting to create regional systems of innovation through local and, therefore, decentralized policies, i.e., endogenous policies (see Camagni, 1995). Thus, local regional leadership to guide and direct economic development in a time of rapid change will become increasingly important. The second part of this paper offers and tests a leadership theory for endogenous led economic growth.

2.2 Identifying and Targeting Endogenous Growth Factors

A variety of approaches have been developed to identify strengths and weaknesses in local regional economies. SWAT analysis is one broad and familiar approach. However, SWAT in its many different forms often suffers in application with a less than systematic effort. Other more analytical based approaches include input-output, shift-share and location quotient analyses. These, while much more systematic and replicable share a common shortcoming tied to the fact that they rely on archival data that is dated. Thus they fail to capture the influences of current trends and thinking. Finally, few if any of the approaches used in practice collect information and data on a full array of potentially important development factors (e.g., stock, flow, endogenous, exogenous, soft vs. hard infrastructure). Below these issues are examined and new approaches are suggested.

The endogenous objective in economic development strategic planning and management is to identify those factors in a region's economy that if improved somewhat will have a disproportionately large impact on performance, i.e., will induce an increasing returns to scale outcome. This set is here called the *auto-catalytic set.* Below several approaches are described for identifying auto-catalytic variables. All approaches utilize a list of factors and ratings of the region by local public and private sector officials on these factors.

Smart infrastructure model: One interesting approach is by Smilor and Wakelin (1990) who have developed a nested hierarchical system of factors that they call smart infrastructure. They argue that in the late 20th century "[I]nnovation and economic growth are the outcomes of successful promotion of technology-oriented economic activities" (p.53). This approach begins with an assumption that technology based economic activities are themselves catalytic, which is both a strength and a weakness. Smart infrastructure – talent, technology, capital and know-how – are the center pieces of their formulation as presented in Figure 2.1. Each of these in turn is erected, they argue, on environmental or context conditions as shown in Figure 2.2. Finally, each of the context conditions rests upon various public policy positions or lack thereof, Figure 2.3. While this system is seemingly over represented in the areas of institutional infrastructure it does provide a window to more traditional infrastructural components and provides a clear link to public policy. It could thus serve as a template for identifying critical strengths and weaknesses which could be effectively linked to policy. A chief weakness of the smart infrastructure model is that it is probably too narrowly conceived to serve as a broad screening and identification tool.

2.3 Multisector Analysis (MSA)

An alternative, first developed by Roberts and Stimson (1998) and subsequently extended (Stough, Stimson and Roberts, 1999) is entitled Multisector analysis (MSA).

Roberts and Stimson outline how their approach is derived from sectoral analysis techniques (e.g., input-output analysis) that reveal related and unrelated sub-systems to display known variables having a key role in influencing the structure of a system being investigated (Tenière-Bouchot, 1973, cited in Godet 1994; Lefebvre 1982; Barrand and Guigou, 1984). MSA builds on structural analysis by measuring regional

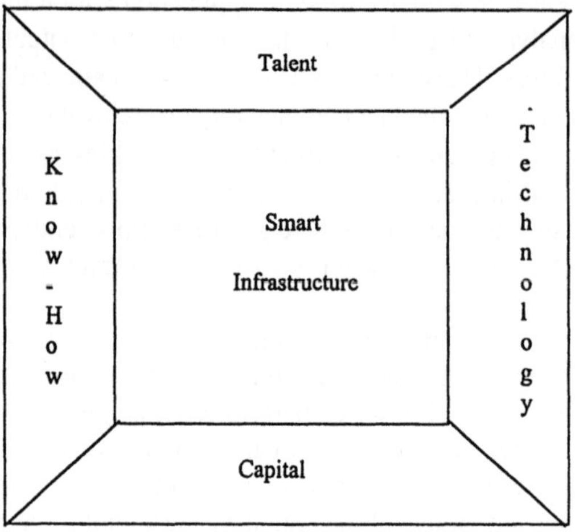

Figure 2.1 Key factors in the development of a smart infrastructure

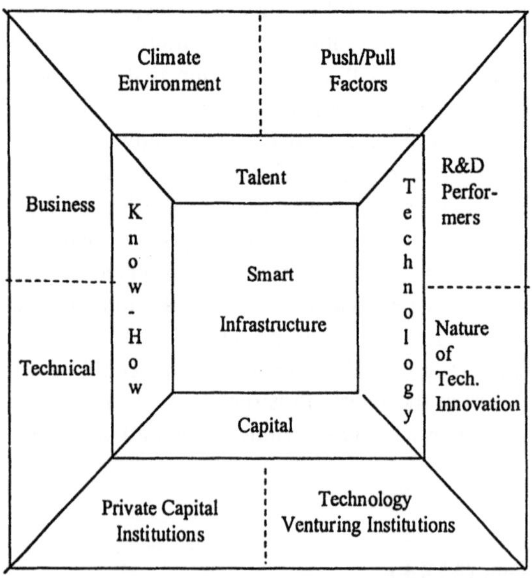

Figure 2.2 Environmental conditions

structural economic variables on a sectoral industry basis. This enables a more detailed qualitative assessment to be made of the performance and potential of industry sectors by deriving a MSA matrix as illustrated in Figure 2.4. It is a technique that can help identify key elements for enhancing regional competitiveness and can be used as a platform for fostering future economic development. MSA draws upon the industry linkages framework identified through the traditional quantitative analytic tool, input-output analysis, but it applies qualitative analysis to identify relative strengths and weaknesses of the core competencies of a region, and to assess the risks and economic development potential of different industry sectors, and in particular, to identify existing and potential new industry clusters that might drive the future development of a region.

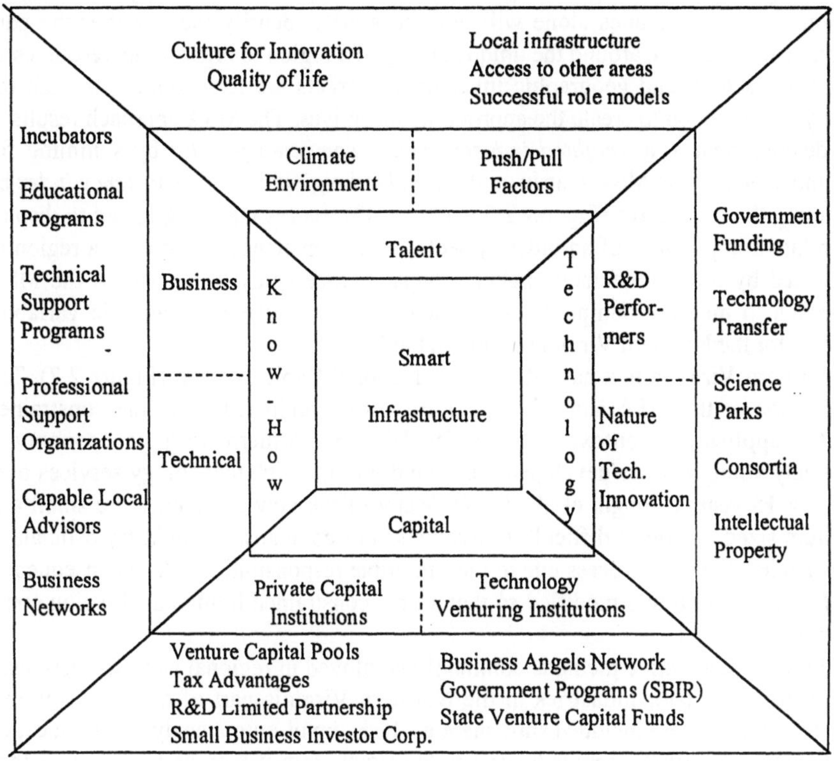

Figure 2.3 Policy implications

		Industry sectors			
		A	B	C	Σ
Evaluation	X	0	1	0	1
criteria	Y	1	3	0	4
	Z	0	0	2	2
	Σ	1	4	2	

Figure 2.4 The MSA matrix

The MSA approach uses major decision maker groups to derive a series of indices to measure the strengths and deficiencies (both existing and potential) of a region. The initial focus is on the region's *core competencies*. These core competencies of a region may be seen as a bundle of skills and technologies within a physical resource base that can be synthesized to produce distinctive products and sets of skills, and technology and knowledge that give a region competitive advantage. Traditional regional audit techniques alone will not necessarily identify them; rather the core competencies revolve around the unique ability of a region to organize its resources so that new sectoral and geographic investment markets are forthcoming through the ability of the region to create the appropriate incentives. The MSA approach results in the development of a *weighted industry sector competency index* by summing for column scores in the MSA matrix, and a *weighted regional core competency index* by summing the row scores (Figures 2.5 and 2.6). The scores may be graphed to display the relative importance of an industry sector or a competency criterion in a region as evaluated by the key decision maker groups. A case study using part of the MSA approach to measure competitiveness factors and identify auto-catalytic variables follows for the Northern Virginia region in the U.S.

Northern Virginia is a part of the U.S. National Capital region (Figure 2.7). The application of the MSA analysis in this case was conducted in a time constrained limited application context (Stough, 1997). The Northern Virginia economy is relatively young but has developed very rapidly in the high technology services over the past 15 years (Stough, et al., 1997). Many of the new companies are small and medium sized making it difficult to involve senior company and industry officials in an extended planning process due to their multiple responsibilities. As a consequence, the MSA approach was modified so that a time constrained limited application could be tested.

Traditional quantitative methods commonly employed in regional analysis were used to identify the major industries in the Northern Virginia part of the U.S. National Capital region. These included shift share analysis, input-output analysis, and location quotients to identify the core industries in the region. Industry group leaders were then organized to participate in a series of structured sessions in which an objective was to undertake an assessment of the region and the performance of its industry sectors using the MSA methodology (Stough, 1997). This included the use of a competitiveness survey instrument (i.e., factor rating list) designed to elicit evaluation information on a variety of regional competitiveness factors including the quality of the region's hard, soft and institutional infrastructure. Participants were selected from industrial directories and from economic development agency information bases to ensure that they represented senior officials from the region's major industries. The sampling methodology was a non-random expert sample. The survey was faxed and/or sent by E-Mail to the respondents.

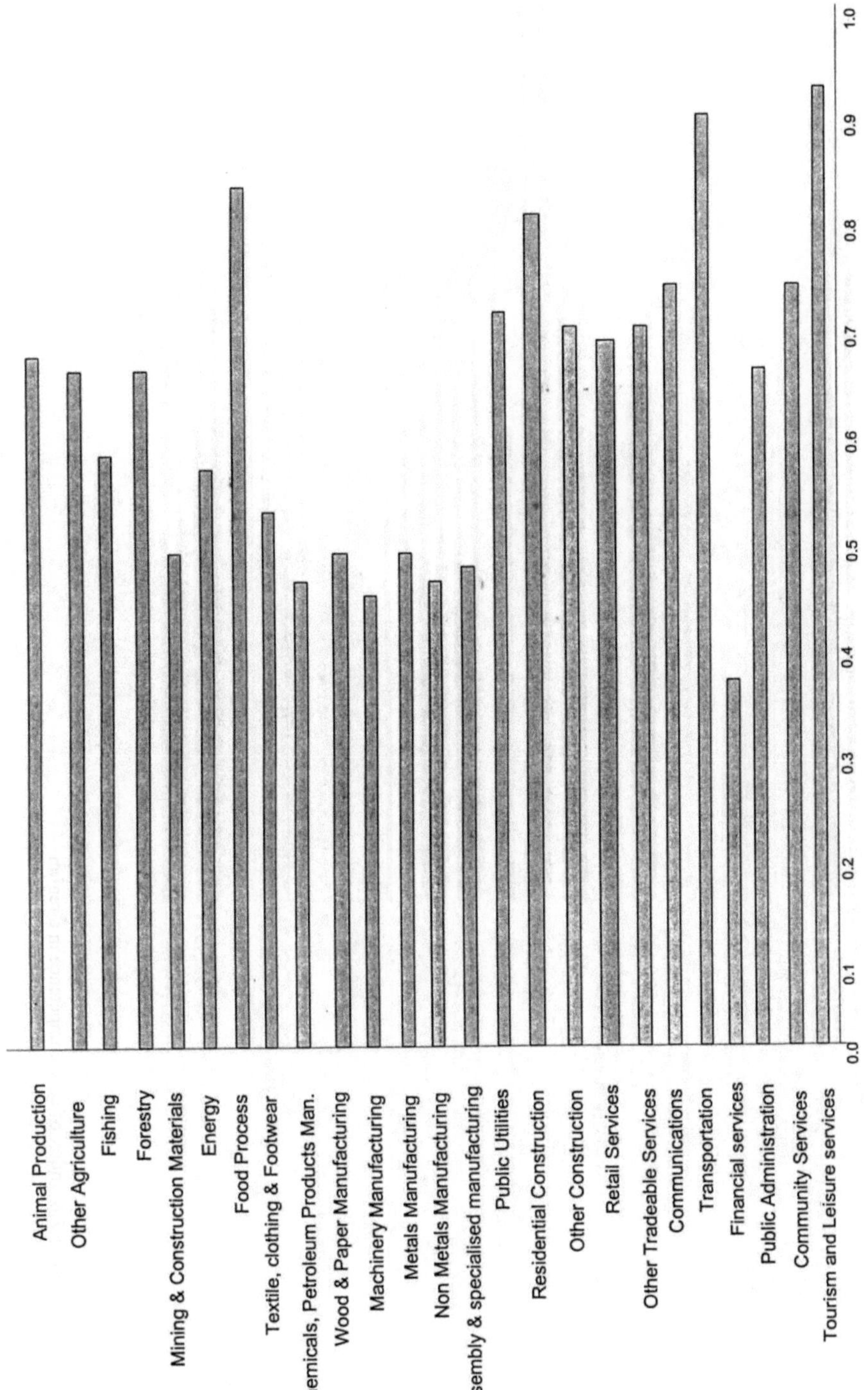

Figure 2.5 Index of sector industry competence in a region

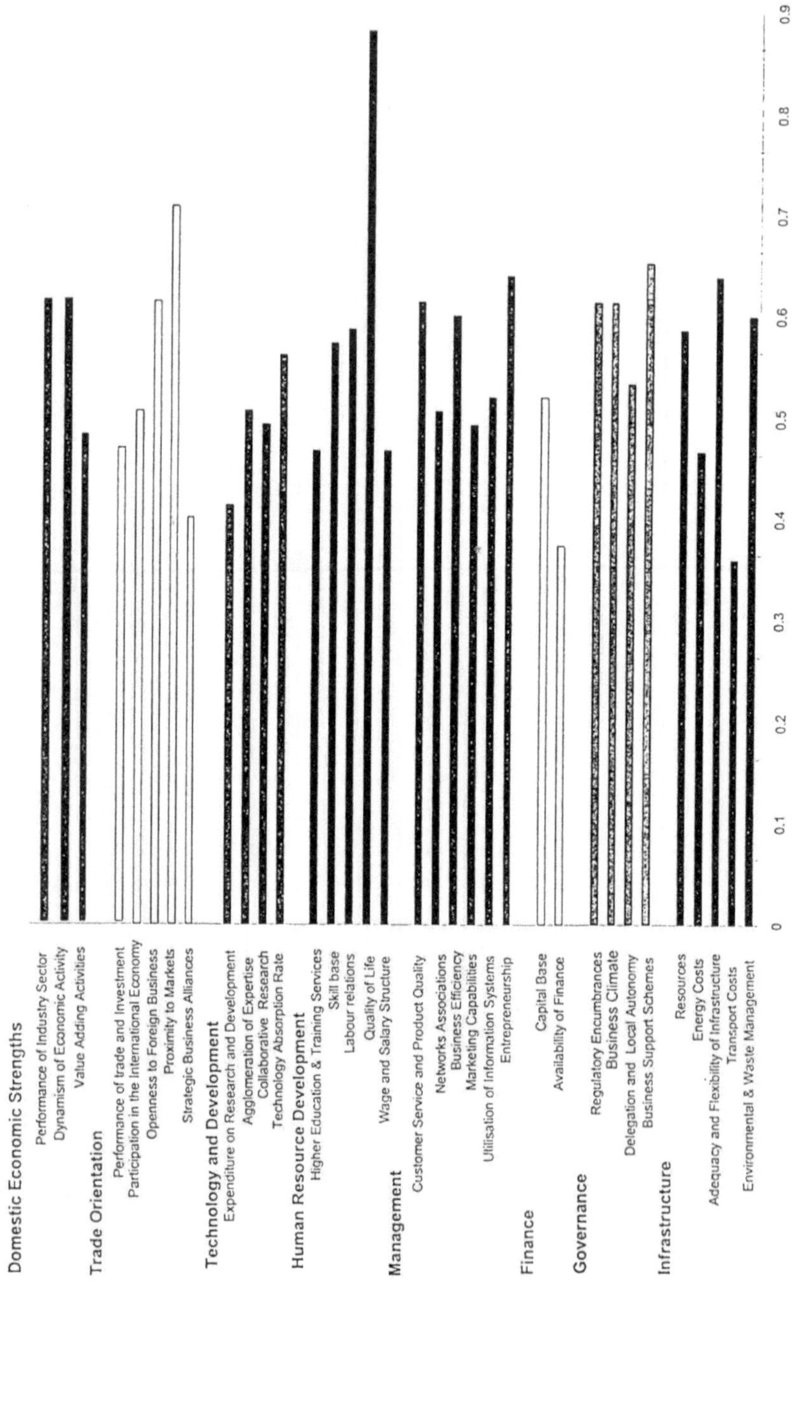

Figure 2.6 Index of core competence in a region

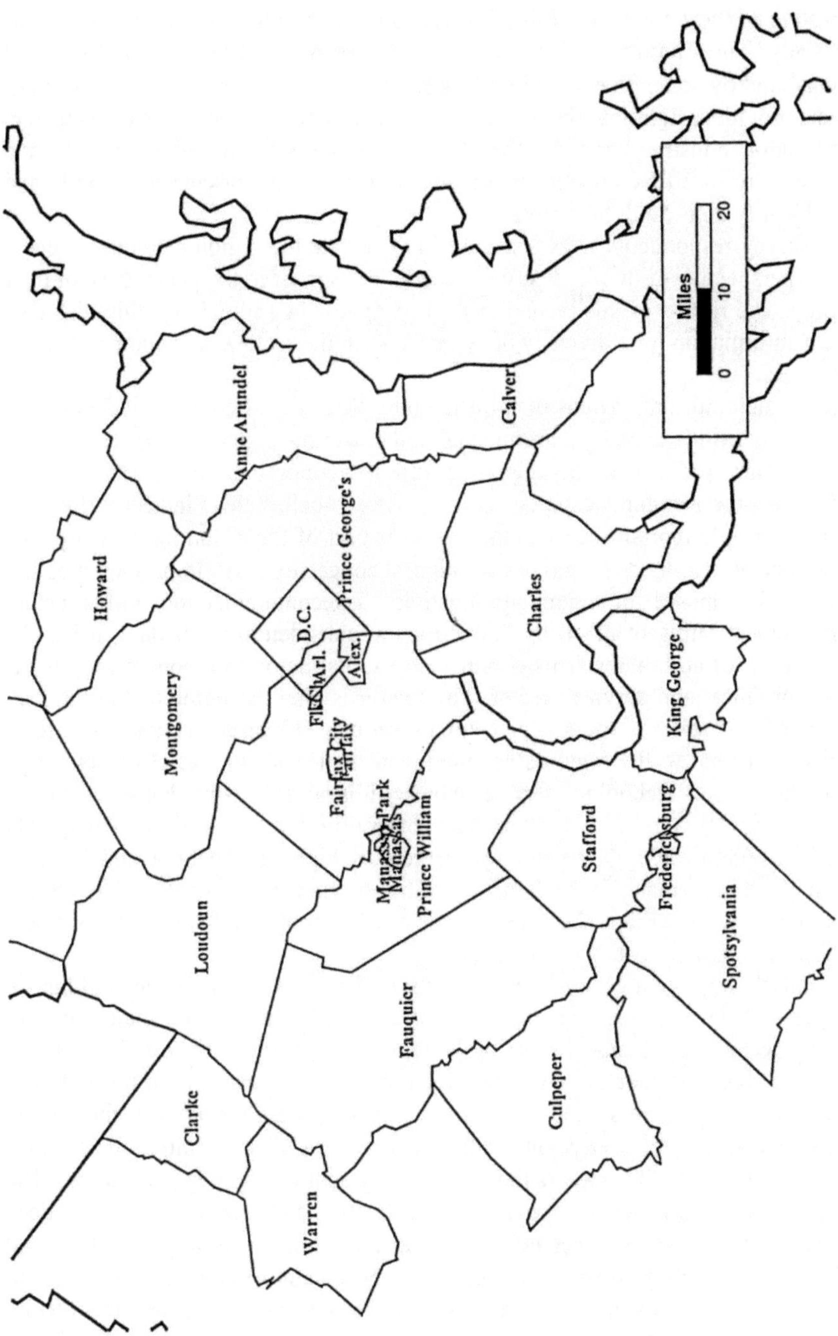

Figure 2.7 Washington D.C-MD-VA MSA

Respondents were first asked to evaluate the region's competitiveness from the perspective of their company. Table 2.1 shows these results from the Northern Virginia study. The findings are organized by 11 primary economic sectors (column headings) and by some 35 competitiveness and infrastructure factors (row headings). The data provide insight into the overall assessment by the 35 competitiveness factors and by major industry. They also may be used to assess the region's strengths and weaknesses by specific competitiveness factors or industries (or clusters of both) and thus to identify auto-catalytic factors.

The survey respondents were also asked to evaluate the region's competitiveness from the perspective of their industry in general, in contrast to the perspective of their firm only. The results of this second evaluation appear in Table 2.2. Again, this data provides information for a diversity of assessments at the sectoral and competitiveness factor levels.

For example, all dimensions of infrastructure were identified as critical competitiveness dimensions except environmental and waste management. Waste management is a minor problem in this region because its economic base is primarily composed of the producer and technology services, i.e., so-called clean industry. Highway infrastructure is important because the region is part of the second most congested metropolitan area in the U.S. and has the highest congestion cost. Improving mobility would clearly enhance the region's productivity. Telecommunications infrastructure has the highest rating of all of the infrastructure considerations. In fact, it has the highest rating of any factor. This obtains because its major core competency is the network products and services sector. The region is also the home of the internet. Maintenance of state of the art telecommunications infrastructure and high level capacity is critical to the region's economic and industrial future. Given the large scale of the region's technology services industry it is not surprising that scheduled air service is considered to be a very important competitiveness factor along with quality of life. The region's 300,000 technology workers are highly competitive and thus produce an effective demand for high quality of life services and amenities. Each of the above factors with perhaps the exception of environmental and waste management may be viewed as auto-catalytic variables.

A similarly important general competitiveness dimension lies in the area of human resources and development. The survey results showed there was a major concern across the technology intense sectors of the regional economy with workforce issues. Namely the ability of the region to attract and train workers qualified for this industry. Subsequently, this issue surged to the top competitiveness problem in the region, when a survey was released showing that in Northern Virginia alone there was a shortfall of nearly 20,000 workers (later verified by another independent survey). The workforce issue defines another catalytic consideration and it continues to be a deep concern to the region. In becoming better at attracting technology workers and developing them through training, the region can become more competitive relative to other technology regions in the U.S. and abroad. In short, by addressing infrastructure and workforce issues the region can leverage its investments to improve if not optimize returns from regional investments.

In the continuing efforts to address the above referenced competitiveness issues (Table 2.2) the region found that it has a weak institutional infrastructure for addressing such problems. While these issues are driving the development of a more cohesive leadership consciousness, the region has learned that its ability to marshal consensus is relatively weak. Rapid growth has characterized the region over the last 25 years resulting in relatively uncoordinated regionwide activities. The rules or structure required to produce cohesive leadership on such issues requires a relatively close coupling between formal rules and informal norms. This is what promotes high performance, i.e., effective performance, in organizations and regional economies (Nee, 1998). In the U.S. intermediate leadership group(s) are what drive and maintain a relatively close coupling among formal and informal contexts. Regions with loosely coupled formal and informal institutions are, therefore, likely to be less successful than those with tight coupling. A leadership model is developed and tested below.

The next step in the Northern Virginia application of the MSA process was the presentation of the findings of the analysis of the survey results to the groups drawn from the survey respondents in meetings lasting about 3 hours. These presentations included data showing demographic and economic trends (including sector specific) derived from archival sources. With the modified approach suggested in this paper the results of the survey were also presented in the context of a modified Nominal Group Technique. This data is essentially expert data as it is derived from senior officials (experts) representing the region's core industry sectors. Each group was provided with the summary evaluation data from its sector, (e.g., aerospace, information technology, professional services, tourism). A major objective of the group meetings was for participants to verify and validate and, as necessary, modify and reinterpret the survey results. For example, local economic development efforts were viewed as relatively weak by the information technology sector group which indicated that the development community was making a great effort to reduce or eliminate the personal property tax levied on their industry (unfairly in their view) but these efforts had not been successful, i.e., the tax was still in place.

Following the verification and reinterpretation of the survey findings the groups then identified new business opportunities for the future of their sectors, and assessed the risk associated with developing these options. Out of this exercise it was possible to create alternative proposals for deepening, and stretching and leveraging the sectors. An illustration of the deepening opportunities for the information technology and telecommunications (IT&T) sector appear in Figure 2.8. Stretching and leveraging possibilities are illustrated for the information, technology and telecom industry clusters in Figure 2.9 and for the tourism industry in Figure 2.10. A fully integrated vision of future industry clusters for Northern Virginia is summarized in Figure 2.11.

Table 2.1 Responses to how factors affect the performance of your firm in the Northern Virginia region (cont. next page)

Competitiveness factors	Sectors										
	Aero-space	Bio-tech	Inform techn.	Sys-tem inte-gra-tion	Tele-comm	Trans-port	Asso-ciation hdqts.	Real estate	Finan-ce	Pro-fes-sional servi-ces	Tou-rism	Mean across factors
Adequate highway system	3,429	4,000	2,143	3,542	3,625	3,091	3,556	4,500	3,444	3,750	3,667	3,552
Scheduled air service	4,000	4,375	2,571	3,292	3,500	3,182	4,000	3,250	2,889	4,250	4,667	3,510
Telecommunications	4,571	4,625	4,429	4,375	4,375	3,546	4,000	3,250	4,111	4,250	3,667	4,095
Environmental and waste mgmt.	2,570	3,000	1,857	2,250	2,625	1,818	2,667	3,167	2,000	2,000	2,333	2,400
Regional quality of life	4,143	4,125	3,571	3,958	4,500	3,273	3,444	4,417	3,889	3,500	3,833	3,905
Finance	**3,00**	**4,31**	**2,64**	**2,81**	**3,19**	**2,09**	**2,28**	**2,96**	**3,17**	**1,38**	**1,92**	**2,78**
Availability of financing	3,143	4,375	3,000	3,125	3,625	2,455	2,556	3,417	3,667	1,750	2,000	3,095
Venture capital	2,857	4,250	2,286	2,500	2,750	1,727	2,000	2,500	2,667	1,000	1,833	2,457
Human resource development	**4,00**	**3,45**	**3,49**	**3,94**	**4,18**	**2,45**	**2,98**	**3,25**	**3,62**	**2,80**	**3,87**	**3,50**
Higher education /training services	4,000	3,375	3,286	3,958	4,125	2,455	2,778	2,333	3,667	2,250	2,833	3,286
Availability of skilled labor	3,571	3,500	4,000	4,375	4,500	2,182	3,222	3,833	3,556	2,750	4,167	3,705
Availability of prof. employees	4,429	4,000	4,429	4,500	4,375	2,909	3,667	3,250	4,000	3,500	3,500	3,924
Flexible labor-mgmt. relations	4,000	2,750	2,429	2,833	4,000	2,091	2,222	3,250	3,333	2,750	4,333	3,010
Competitive Wage/Salary Structure	4,000	3,625	3,289	4,042	3,875	2,636	3,000	3,583	3,556	2,750	4,500	3,591
Technology and development	**2,14**	**3,15**	**2,11**	**2,06**	**2,63**	**2,60**	**1,69**	**1,70**	**1,60**	**0,95**	**1,83**	**2,06**
University research programs	2,000	2,875	2,429	2,125	2,625	2,273	2,000	1,917	2,000	1,750	1,667	2,162
University-industry partnerships	2,286	3,625	2,429	2,208	3,125	2,636	1,889	1,750	2,333	1,000	2,500	2,352
Federal research lab programs	2,286	3,000	2,429	2,000	2,375	2,182	1,667	1,667	1,111	1,000	2,000	1,971
State research initiatives	1,857	3,125	1,571	1,833	2,250	3,364	1,333	1,583	1,111	0,500	1,667	1,810
Private research efforts	2,286	3,125	1,714	2,125	2,750	2,546	1,556	1,583	1,444	0,500	1,333	2,000
International trade orientation	**3,23**	**2,83**	**2,11**	**2,05**	**2,03**	**1,90**	**1,16**	**2,25**	**2,09**	**0,25**	**2,60**	**2,09**
Current overseas trade activities	3,286	3,375	2,143	2,417	1,750	1,800	1,222	2,333	2,222	0,000	2,667	2,212
Foreign investment into this region	2,429	2,125	1,286	1,625	1,500	1,400	1,222	2,333	2,444	0,500	3,500	1,846
Overseas investment of your firm	2,571	2,250	1,143	0,917	1,750	1,300	0,889	1,833	1,556	0,000	0,500	1,346
Business alliances (with U.S. firms)	3,857	3,250	3,714	3,250	3,250	3,000	1,333	2,500	2,444	0,500	3,667	2,894

Table 2.1 continued

Competitiveness factors	Sectors											
	Aerospace	Biotech	Inform techn.	System integration	Telecomm	Transport	Association hdqts.	Real estate	Finance	Professional services	Tourism	Mean across factors
Business alliances (foreign firms)	4,000	3,125	2,286	2,042	1,875	2,000	1,111	2,250	1,778	0,250	2,667	2,144
Government	**3,64**	**3,63**	**3,61**	**3,92**	**3,81**	**2,84**	**3,34**	**4,23**	**3,89**	**3,13**	**4,58**	**3,73**
Local regulation of business	3,429	3,375	3,571	3,917	4,000	3,000	3,444	4,167	3,556	4,000	3,833	3,686
General business climate	3,714	3,875	3,714	4,250	3,875	2,727	3,375	4,417	4,444	2,750	5,000	3,914
Local econ. development efforts	3,429	3,625	3,143	3,250	3,500	2,636	3,000	4,250	3,889	1,750	4,667	3,410
Local tax structure	4,000	3,625	4,000	4,250	3,857	3,000	3,556	4,083	3,667	4,000	4,833	3,904
Regional economic strengths	**3,46**	**2,83**	**3,76**	**3,54**	**3,54**	**3,06**	**2,07**	**3,94**	**3,67**	**2,17**	**4,22**	**3,36**
Performance of your industry sector	3,667	2,875	3,857	4,000	3,625	3,182	2,000	4,000	3,778	2,500	4,833	3,567
Strength of No.VA regional economy	3,143	2,875	3,857	3,542	3,714	3,091	2,444	4,500	4,222	2,500	4,833	3,558
Cross-industry information flow	3,571	2,750	3,571	3,083	3,286	2,909	1,778	3,333	3,000	1,500	3,000	2,962
Your firm's management characteristics	**3,79**	**3,25**	**3,98**	**3,94**	**3,92**	**3,32**	**3,04**	**3,60**	**4,04**	**2,92**	**3,61**	**3,65**
Customer service/product quality	3,857	3,375	4,571	4,833	4,625	3,818	4,000	4,667	4,889	3,500	4,833	4,381
Inter-business networking	4,143	3,000	4,286	3,708	3,625	3,091	3,000	3,333	4,444	2,500	4,333	3,600
Available management Ccnsultants	2,714	2,625	2,857	2,609	2,250	2,818	2,222	2,417	3,000	2,000	1,500	2,519
Marketing capabilities	4,000	3,250	4,000	4,000	4,250	3,455	2,778	4,167	3,889	2,250	4,333	3,762
Entrepreneurship	3,429	3,500	3,857	4,125	4,250	3,182	2,667	4,000	3,889	3,000	3,500	3,686
Info/telecommunication systems	4,571	3,750	4,286	4,375	4,500	3,546	3,556	3,000	4,111	4,250	3,167	3,933
Mean through firm sectors	**3,393**	**3,377**	**3,088**	**3,226**	**3,375**	**2,681**	**2,517**	**3,161**	**3,145**	**2,157**	**3,305**	

Values are average scores of responses to 5 point importance ratings (5 is most important)

Table 2.2 Responses to how factors affect the performance of industry sectors in the Northern Virginia region

Competitiveness factors	Sectors											
	Aero-space	Bio-tech	Inform techn.	Sys-tem inte-gra-tion	Tele-comm	Trans-port	Asso-ciation hdqts.	Real estate	Finan-ce	Pro-fes-sional servi-ces	Tou-rism	Mean across factors
Infrastructure	**4,54**	**4,35**	**3,60**	**4,09**	**4,06**	**3,89**	**4,00**	**4,18**	**4,07**	**3,95**	**4,33**	**4,10**
Adequate highway system	4,714	4,625	4,000	4,333	4,143	4,273	4,111	4,727	4,333	4,000	4,500	4,364
Scheduled air service	4,714	4,625	3,500	3,952	4,286	4,273	4,222	4,273	4,111	4,500	4,667	4,232
Telecommunications	4,714	4,750	4,000	4,524	4,429	4,182	4,444	4,000	4,556	4,250	4,667	4,414
Environmental and waste mgmt.	4,000	3,625	2,167	3,476	3,429	2,909	3,556	3,364	3,222	3,250	3,500	3,343
Regional quality of life	4,571	4,125	4,333	4,143	4,000	3,818	3,667	4,546	4,111	3,750	4,333	4,131
Finance	**4,29**	**4,25**	**3,50**	**3,12**	**3,86**	**3,64**	**3,61**	**3,73**	**4,06**	**3,50**	**3,75**	**3,68**
Availability of financing	4,286	4,250	3,500	3,571	4,429	4,000	3,667	3,818	4,222	3,750	3,667	3,889
Venture capital	4,286	4,250	3,500	2,667	3,286	3,273	3,556	3,636	3,889	3,250	3,833	3,465
Human resource development	**4,31**	**4,08**	**4,03**	**3,93**	**4,09**	**3,78**	**3,44**	**3,89**	**4,36**	**3,85**	**4,60**	**4,00**
Higher education/training services	4,143	4,375	4,000	4,143	4,143	3,909	3,556	3,909	4,556	3,500	4,167	4,061
Availability of skilled labor	4,286	3,875	4,167	4,286	4,571	4,000	3,444	4,364	4,778	3,750	5,000	4,232
Availability of prof. employees	5,000	4,375	4,500	4,429	4,429	4,182	3,889	3,909	4,667	4,500	4,500	4,364
Flexible labor-mgmt. relations	3,857	3,750	3,167	2,810	3,143	3,182	3,000	3,455	3,667	3,750	4,500	3,354
Competitive Wage/Salary Structure	4,286	4,000	4,333	4,000	4,143	3,636	3,333	3,818	4,111	3,750	4,833	3,980
Technology and development	**3,77**	**3,73**	**2,62**	**2,66**	**2,97**	**3,00**	**2,90**	**2,64**	**3,11**	**2,85**	**3,20**	**2,98**
University research programs	3,714	3,875	2,500	2,619	3,000	2,909	3,125	2,546	3,444	2,750	3,167	3,000
University-industry partnerships	4,286	4,125	2,833	2,810	3,000	3,182	3,000	2,818	3,333	2,750	3,667	3,192
Federal research lab programs	4,000	3,500	2,600	2,619	2,857	2,818	2,875	2,727	2,889	3,500	3,167	2,959
State research initiatives	3,429	3,625	2,333	2,524	3,000	3,000	2,625	2,636	2,778	2,250	2,833	2,806
Private research efforts	3,429	3,500	2,833	2,714	3,000	3,091	2,875	2,455	3,111	3,000	3,167	2,959
International trade orientation	**3,14**	**2,98**	**2,63**	**2,43**	**3,29**	**2,13**	**2,38**	**2,75**	**2,95**	**1,90**	**3,17**	**2,66**
Current overseas trade activities	3,571	3,375	2,833	2,476	3,429	2,364	2,556	2,818	3,000	2,500	3,833	2,879
Foreign investment into this region	3,143	3,125	2,167	2,400	2,857	2,091	2,111	3,273	3,111	2,250	3,167	2,674

Table 2.2 continued

Competitiveness factors	Sectors											
	Aero-space	Bio-tech	Inform techn.	Sys-tem inte-gra-tion	Tele-comm	Trans-port	Asso-ciation hdqts.	Real estate	Finan-ce	Pro-fes-sional servi-ces	Tou-rism	Mean across factors
Overseas investment of your firm	2,000	2,375	2,167	1,619	2,714	1,000	2,111	1,818	2,143	1,000	1,333	1,814
Business alliances (with U.S. firms)	3,571	3,125	3,167	3,381	4,000	3,000	2,667	3,000	3,500	2,000	4,500	3,276
Business alliances (foreign firms)	3,429	2,875	2,833	2,286	3,429	2,182	2,444	2,818	3,000	1,750	3,000	2,674
Government	**4,18**	**4,28**	**4,08**	**3,95**	**3,88**	**3,31**	**3,96**	**4,09**	**3,64**	**3,94**	**4,67**	**3,97**
Local regulation of business	4,167	4,000	4,000	3,944	3,667	2,857	4,286	4,000	3,444	4,000	4,667	3,895
General business climate	4,286	4,250	4,000	4,238	3,857	3,600	4,111	4,000	3,778	4,500	4,333	4,071
Local econ. development efforts	4,286	4,500	4,333	4,238	3,857	3,600	3,778	4,182	3,778	3,750	4,833	4,102
Local tax structure	4,000	4,375	4,000	3,381	4,143	3,200	3,667	4,182	3,556	3,500	4,833	3,806
Regional economic strengths	**3,90**	**3,75**	**3,39**	**3,68**	**3,76**	**3,18**	**3,11**	**3,67**	**3,73**	**3,25**	**3,89**	**3,58**
Performance of your industry sector	4,143	3,500	4,000	4,238	4,143	3,273	3,222	3,636	3,625	2,250	4,833	3,784
Strength of No.VA regional economy	3,714	4,250	3,333	3,571	3,857	3,182	3,000	3,909	4,000	4,250	3,833	3,667
Cross-industry information flow	3,857	3,500	2,833	3,238	3,286	3,091	3,111	3,455	3,556	3,250	3,000	3,293
Your firm's management characteristics	**3,81**	**3,96**	**3,89**	**3,83**	**3,76**	**3,50**	**3,59**	**3,55**	**3,84**	**3,71**	**3,69**	**3,74**
Customer service/product quality	3,571	3,875	4,333	4,524	4,571	3,727	3,667	4,091	4,111	3,750	5,000	4,141
Inter-business networking	4,000	3,750	4,500	3,714	3,429	3,364	3,778	3,546	3,889	3,250	4,333	3,748
Available management consultants	2,857	3,500	2,333	2,757	2,571	2,818	3,333	2,727	3,222	3,250	2,000	2,879
Marketing capabilities	4,143	4,000	3,833	3,810	4,000	3,636	3,556	3,636	3,667	3,750	4,000	3,796
Entrepreneurship	3,857	4,000	3,667	3,810	3,857	3,636	3,222	3,727	3.888	4,250	3,167	3,727
Info/telecommunication systems	4,429	4,625	4,667	4,381	4,143	3,818	4,000	3,546	4,333	4,000	3,667	4,152
Mean through firm sectors	**3,956**	**3,887**	**3,462**	**3,458**	**3,683**	**3,277**	**3,347**	**3,516**	**3,687**	**3,345**	**3,892**	

Values are average scores of responses to 5 point importance ratings (5 is most important)

2.4 Leadership and Regional Economic Development

Stough (1990) on the basis of a number of regional economic development case studies in the U.S. identified five characteristics of successful initiatives. These are:

1. Local initiative is critical for initiating and sustaining community economic development.
2. Local initiative is consistently undertaken by nongovemment community (intermediate) organizations.
3. Community organizations are effective economic development planning organizations.
4. Economic development plans are a basis for cross-sector collaboration.
5. Successful communities have access or create access to a broad range of local and extra local national resources.

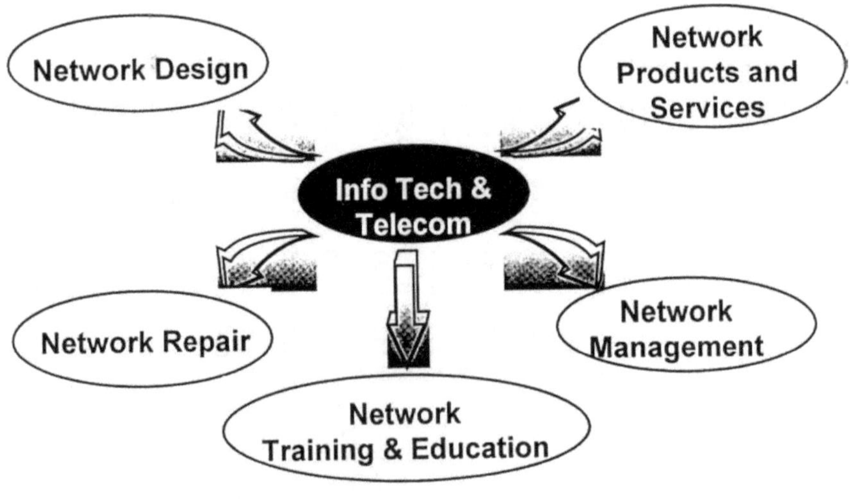

Figure 2.8 Sectoral deepening (specialization) development for the Northern Virginia region

DeSantis (1993) in a further review of this topic identifies several factors that effect local regional economic development effectiveness. These are: 1) weak and fragmented public authorities, 2) degree of cooperation between the local stakeholders, 3) the level of locally available resources for economic development, and 4) the tendency of a community to engage in local problem-solving. The last factor is a distillation of the five characteristics described by Stough. DeSantis concludes, however, that more fundamental forces underlie these attributes: economic development effectiveness is influenced by leadership and resource endowments, where the basic elements of the leadership construct are included in factors (1), (2), and (4) and the resource endowment concept is accounted for in factor (3).

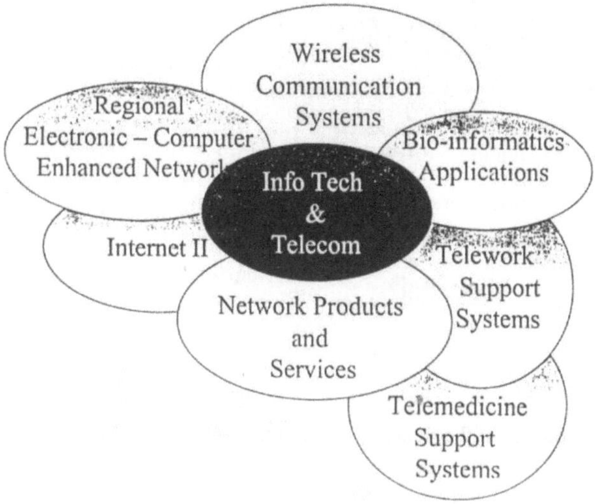

Figure 2.9 Information technology and telecom cluster in the Northern Virginia region

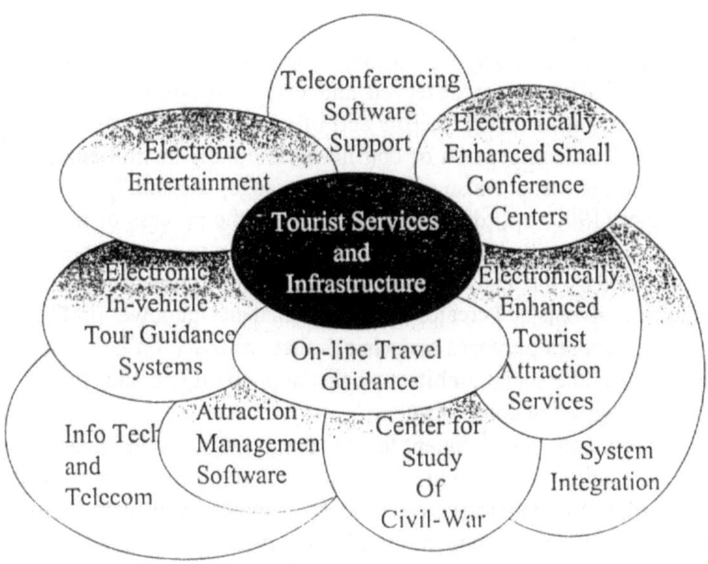

Figure 2.10 Tourism cluster in the Northern Virginia region

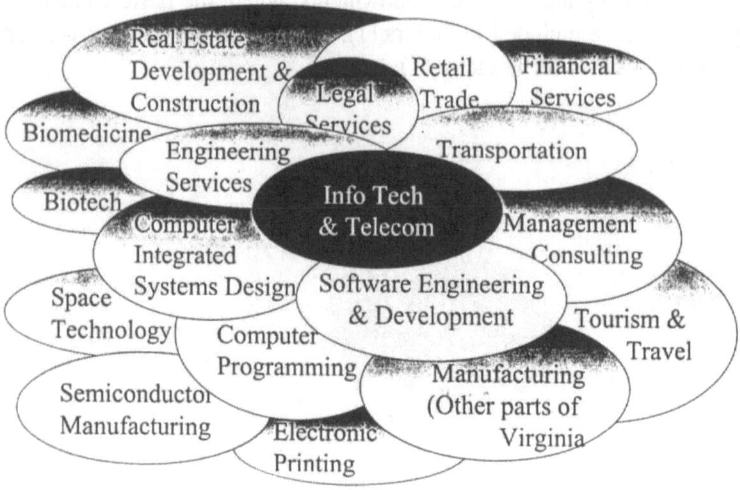

Figure 2.11 Integrated economy of the future of the Northern Virginia region

In the following part of this discussion the basic constructs of this formulation are explicated and then used to define a conceptual model of local regional economic development.

***Leadership*:** Leadership has many definitions. For example, Burns (1978) asserts leadership is "...one of the most observed and least understood phenomenon on earth..." and he defines it as (p.19) "...[the act] of persons with certain motives and purposes to mobilize, in competition or conflict with others, institutional, political, psychological, and other resources so as to arouse, engage, and satisfy the motives of followers... Gardner (1990, p.1) defines leadership as "the process of persuasion or example by which an individual (or group) induces a group to pursue objectives held by the leader or shared by the leader and his followers." For Bennis and Nanus (1991, p.218), leadership "...invents and creates institutions that can empower [individuals] to satisfy their needs, chooses purposes and visions that are based on key values of the work force and creates the social architecture that supports them, and, finally moves followers to higher degrees of consciousness."

Leadership for economic development that is community-wide in impact necessarily is not based on traditional hierarchical authority relationships between leader and follower, but is rather a collaborative relationship between local institutional actors and is based on mutual trust and cooperation (Bower, 1982; Bryson and Crosby, 1992; Fosler, 1992; Gray, 1989; Judd and Parkinson, 1990; Osborne, 1988). That is, no single local individual or institution is in a position of authority to undertake fully effective community-wide economic development. As a result, local leaders must inspire and motivate followers through persuasion, examples, data informed arguments and empowerment, not through command and control (Burns, 1978; Kouzes and Posner, 1987; Bunch, 1987; Neustadt and May, 1986).

Political theorist Bower (1983) maintains that "when problems or opportunities... exceed the capabilities of individual institutions, we must look for plans from associations of the institutions [at which]... there follows a series of meetings among leaders of the various institutions that might be-involved in the problem (or opportunity)... formally or not, an intermediate-level institution gets created." Bryson and Crosby (1992) perceive a new world in which no single institution is "in charge" or has the legitimacy, power, authority, and knowledge required to tackle any major public policy issues, and institutions must "join forces" in a "shared-power world." For Bryson and Crosby (1992, p.13), shared-power is "... shared capabilities exercised in interaction between or among actors to the further achievement of their separate and joint aims." The concepts of "shared-power" or "intermediate institutions" are basic elements of local leadership for economic development.

Leadership for economic development requires a multidisciplinary approach to understanding and defining this phenomenon and must therefore include leadership theory, community leadership, and regional economic development. Rost (1991) provided a summary of definitions of leadership which is appropriate as a starting point for defining leadership for economic development. Rost (1991, p.102) defines leadership as "an influence relationship between leaders and followers who intend real changes that reflect their mutual purposes." Theories of community leadership compliment Rost's definition and, as a consequence, build on the leadership construct by pointing out that local groups or coalitions and their interactions affect local public policy (Friedland and Bielby, 1982). Further, economic development theory and practice reveals that these identifiable local leadership groups cooperate to influence the economic future of the community. A definition of leadership for local regional economic development is thus: *the tendency of a community to collaborate across sectors in a sustained, purposeful manner to enhance the economic performance of its region.*

Resource endowments: One of the most original and fundamental concepts in economic analysis is that economic growth and performance are related or tied to resources. The more well endowed a region is in terms of resources the better it should perform, *ceteris paribis*. However, physical resources are only a minor part of the value of finished products in the economies of the late 20th Century.

A contemporary expression of a resource dependent model may be erected on the concept of slack institutional resources which is defined as the difference between "the resources available to a firm ... [organization] ... and the total necessary to maintain [it]" (Cyert and March, 1963, p.36). Organizational slack exists at varying levels and times in all organizations and represents an organization's "excess" resources. This excess or slack is the source of voluntary contributions to "civic activities ", or locally based and focused community efforts by local public, private and nonprofit organizations. Therefore, resource endowments are defined as: *the aggregate concentration or mass of public, private and non-profit organization's capacity to voluntarily commit resources to economic development.*

The model: Specifying a leadership model for local regional economic development is a straight forward extension of the above discussion (see Stough and DeSantis, 1999). In short, regional economic development effectiveness is directly related to a region's resource endowments and its propensity for leadership *ceteris paribus.* That is, once exogenous factors are controlled for, regional economic performance depends on leadership and resources as shown in Figure 2.12. In this formulation leadership is modeled as a variable that amplifies the independent effect of resources.

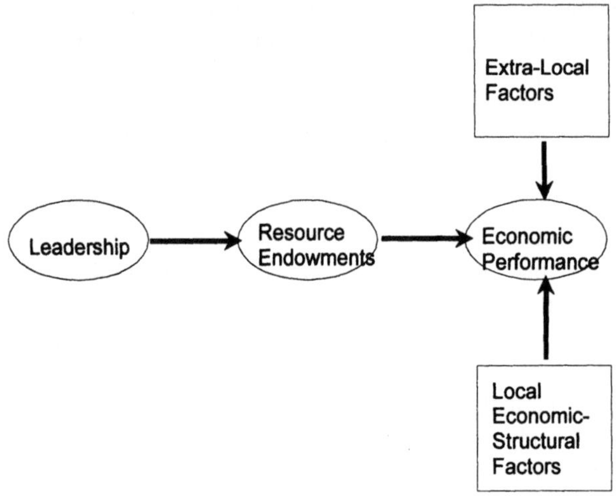

Figure 2.12 Leadership path model

Testing the model: Sample Selection: Metropolitan regions in the U.S. were selected as the basic unit of analysis for testing the theory. The sample includes 35 of the approximately 350 Metropolitan Areas (MAs) in the U.S. (Office of Management and Budget, 1990; Federal Committee on Standard Metropolitan Statistical Areas, 1979). A MA must contain either a place with a population of 50,000 or a Census bureau-defined urbanized area and a total MA population of at least 100,000 (75,000 for New England) . An MA comprises one or more central counties and it may include one or more outlying counties that have close economic and social relationships with the central county. An outlying county must have a specified level of commuting to the central counties and also must meet certain standards regarding metropolitan character, such as population density, urban population, and population growth.

Conceptually, a metropolitan "region" is a functional economic area constituting an integrated labor market surrounding an urban core (Wolman, 1987, p.1-2). Metropolitan Statistical Areas (MSAS) and Primary Metropolitan Statistical Areas (PMSAS) or regions, have several major advantages as units of analysis. First, they are discrete economic entities that allow for analytic comparability (Wolpert, 1988). Second, a great deal of data describing the economic and social aspects of regions is readily available. Third, regions typically encompass multiple political jurisdictions, resulting in a likely higher occurrence of leadership in local economic development (Giarrantani, 1991).

A list of the sample metropolitan areas appears in Table 2.3. The sample reflects data availability rather than a specific sampling strategy. Although recent and comprehensive profiles of local economic development organization activity are scarce, a Council for Urban Economic Development (CUED) study (1992) provided a 1990 survey of economic development organizations in 35 metropolitan regions in the U.S. The CUED survey sample includes only those metropolitan areas that had response levels sufficient to make an accurate assessment of the aggregate funding levels of regionally focused economic development.

Table 2.3 Sample metropolitan regions

Albuquerque, NM (MSA)	Atlanta, GA (MSA)
Austin, TX (MSA)	Baltimore, MD (MSA)
Boston, MA (PMSA)	Charlotte, NC (MSA)
Chicago, IL (PMSA)	Cleveland, OH (PMSA)
Dallas, TX (PMSA)	Denver, CO (PMSA)
Detroit MI (PMSA)	Eugene, OR (MSA)
Houston, TX (PMSA)	Indianapolis, IN (MSA)
Jackson, FL (MSA)	Kansas City, KA (MSA)
Las Vegas, NV (PMSA)	Los Angeles, CA (PMSA)
Louisville, KY (MSA)	Miami, FL (PMSA)
Mobile, AL (MSA)	New York, NY (PMSA)
Orlando, FL (MSA)	Philadelphia, PA (PMSA)
Phoenix, AZ (MSA)	Portland, OR (MSA)
Sacramento, CA (MSA)	St. Louis, MO (MSA)
San Antonio, TX (MSA)	San Diego, CA (MSA)
San Francisco, CA (PMSA)	Seattle, WA (PMSA)
Tampa, FL (MSA)	Washington, D.C. (MSA)

Two sample biases may exist in this data. First, the sample regions are "self selected" in that only those regions with significant development efforts responded to the CUED (1992) survey. Second, regions that did not have region-wide economic development efforts but, nevertheless had significant regional aggregate development budgets (suggesting a highly regionally fragmented approach) are excluded from the survey. For example, regions with large numbers of sub-regional development organizations such as neighborhood development organizations, were excluded from the survey if there were no development organizations whose geographic area of concern was the entire region.

The sample may be defended on two grounds. First, the self selection inherentin the sample was essential in order to obtain a sample of regions with significant, but varying levels of regionally focused economic development. Second, leadership for regional economic development is by definition regionally focused and thus inclusion of regions with highly fragmented neighborhood level development efforts would misrepresent the actual level of economic development effort of such regions. The sample regions ranged in size from Eugene, Oregon with a total population of 280,000 to New York City with over 8.9 million. The median population was 2.9 million.

2.4.1 Variable Specification and Measurement

The three theoretical variables: leadership, resources and economic performance, are operationalized in multiple ways given the exploratory nature of this paper (Table 2.4). First, and most straight forward, is economic performance which is specified in terms of local employment and earnings (Table 2.5). Throughout, all measures are standardized for population size.

The theoretical formulation implies that the independent variables are either coincident with (static relationship) or precede in time the occurrence of the dependent variable (dynamic relationship). Thus, the variables are specified in terms of both coincident (static) and lagged (dynamic) frames of reference (Table 2.4). Variable specifications for the static formulation are in1990 values. The lagged (dynamic) specifications are the actual changes in the mean data values from 1980 to 1985 for the independent variables and actual changes in the mean data values for the dependent variables from 1985-1990.

Ideally the leadership variable should be measured on the basis of substantive local information and knowledge. Unfortunately, resources to achieve this were not available to the authors. Consequently, surrogate measures of leadership were developed. That is, it was assumed that leadership would be manifested in various outcomes. Four outcome (surrogate) leadership variables were specified: 1) voluntary Community Effort; 2) Number of Voluntary Community Organizations; 3) Expenditures of Voluntary Community Organizations, and 4) Economic Development Effort (see Tables 2.4, 2.5 and 2.6).

The Voluntary Community Effort variable measures the local propensity to commit resources to community problem-solving. United Way fund raising is a useful measure of this variable because its bylaws require funds to be spent where they are raised. Further, Whitt and Lammers (1991) found, in a case study of Louisville, Kentucky, that of 47 non-profit organizations that local metropolitan United Way fund raising was most strongly and positively correlated with the presence of local regional economic development organizations.

The Number of Voluntary Community Organizations is a measure of the propensity for local voluntary extra-governmental community problem-solving. Non-profit social service organizations and selected membership organizations define this variable. Non-profit social service organizations include the following: individual and family social service establishments; job training and vocational rehabilitation services; child day care services; residential care; and community and neighborhood improvement organizations. Selected membership organizations include: business associations; professional membership organizations; and, civic, social and fraternal associations.

These three classes of organizations represent membership organizations most active in local community affairs for which comprehensive data is readily available.

Table 2.4 Dependent variable descriptions

DEPENDENT VARIABLES

Economic performance (P)

Static variables (1990):

LEMP	=	Local employment
LERNP	=	Local earnings dynamic

Variables (change 1985 to 1990):

ΔLEMP	=	Change in local employment
ΔLERNP	=	Change in local earnings

Leadership variables (L)

Static variables (1990):

VCEL	=	Voluntary community effort
NVCOL	=	Number of voluntary community orgs
EVCOL	=	Voluntary community org. expenditures
EDEL	=	Economic development effort

Dynamic variables (actual change 1980 to 1985):

ΔVCEL	=	Change in voluntary community effort
ΔNVCOL	=	Change in number of voluntary community orgs.
ΔEVCOL	=	Change in expenditures of voluntary community organizations
ΔEDEL	=	Change in economic development effort

Resource endowment variables (R):

Static variables (1990):

NCHR	Number of corporate headquarters
ECHR	Employment at corporate headquarters
NFIR	Number of financial institutions
MDFIR	Marker deposits of financial institutions
NCUR	Number of colleges and universities
ECUR	Enrollment of colleges and universities
PIR	Personal income

Dynamic variables (change from 1980 to 1985):

ΔNCHR	Change in number of corporate headquarters
ΔECHR	Change in corporate headquarters' employment
ΔNFIR	Change in the number of financial institutions
ΔMDFIR	Change in market deposits of institutions
ΔNCUR	Change in number of colleges and universities
ΔECUR	Change in enrollment of colleges and universities
ΔPIR	Change in personal income

Table 2.5 Operational economic performance variables

$\frac{\text{Total region 1990 employment}}{\text{Total region 1990 population}}$	=	Local employment
$\frac{\text{Total region 1990 earnings}}{\text{Total region 1990 employment}}$	=	Local earnings

Table 2.6 Operational leadership variables

$\frac{\text{Total region 1990 united way dollar contributions}}{\text{Total region 1990 population}}$	=	Voluntary community effort
$\frac{\text{1990 number of social service organizations based in the region} + \text{1987 number of selected service organization based in the region}}{\text{Total region 1990 population}}$	=	Number of voluntary community orgs.
$\frac{\text{1990 expenditure of social sservice orgs. based in the region} + \text{1990 expenditures of selected service orgs. based in the region.}}{\text{Total region 1990 population}}$	=	Expenditures of voluntary community orgs.
$\frac{\text{Total region 1990 united way dollar contributions}}{\text{Total region 1990 population}}$	=	Economic development effort

The Expenditures of Voluntary Community Organizations are defined by the annual expenditures of non-profit social service organizations and selected membership organizations included in the Number of Voluntary Organizations variable.

Economic Development Effort is measured by the local commitment of resources to economic development. This variable is defined by the region's total public, private and non-profit economic development organization budgets. Only the budgets of economic development organizations whose area of focus is most or all of the region were included in the measurement of this variable.

The resource endowment variable is easier to measure than leadership in that a variety of archival data sources provide direct measures for this variable. The resource endowment variable is specified in seven ways: 1) number of Corporate Headquarters; 2) Corporate-wide Employment of Corporate Headquarters; 3) Number of Financial Institutions; 4) Market Deposits of Financial Institutions; 5) Number of Colleges and

Universities; 6) Enrollment of Colleges and Universities; and 7) Level of Personal Income.

The variables are defined in Table 2.4 and specifications given in Table 2.7. Corporate Headquarters are places where the activities of multi-location businesses are planned and monitored, including staff functions, such as finance, general counsel, planning, employee relations, marketing, advertising, public relations and are the location for top operating executives (Boyle, 1988). All manufacturing and nonmanufacturing corporations with the following attributes: 1) maintain a net worth of at least \$500,000; 2) conduct business from more than one location; and 3) retain a controlling interest (51% or more) in at least one subsidiary company are included in the sample.

Employment of Corporate Headquarters includes the total corporate-wide employment of all of the corporate headquarters in the previously described corporate variable. Including total world-wide employment in this measure is important because this represents a measure of scale of influence that is controlled or directed from headquarters.

The Financial Institution variable is defined by the central offices of banks and financial services organizations, exclusive of holding companies within a region. The functions included in central offices are analogous to those included in the definition of corporate headquarters presented above (see, Boyle, 1988). Market Deposits of Financial Institutions is again a measure of the scale of influence of the locally based central office.

Colleges and Universities comprise all two and four year post secondary education institutions based in a region. Vocational and technical institutions were not included in the analysis. Enrollment is, again, a measure of the scale of influence of the set of colleges and/or universities. Personal Income is defined as income from all sources to all individuals that reside in a region.

2.4.2 Multiple Regression and other Analyses

The first step in the analysis was to conduct an unconstrained multiple regression analysis for both the static and dynamic specifications of the model, i.e., each dependent variable (employment and earnings) was regressed on all four leadership and all seven resource variables. The results are presented in Table 2.8. The multiple adjusted R^2 is near or above 0.7 for all four models. There was considerable multicollinearity among the leadership and resources variables. An attempt to manage this was made using factor analysis but with speculative results (see, DeSantis, 1993). Similar results obtain for the dynamic formulation (Table 2.9).

Table 2.7 Operational resource endowment or "slack" variables

Total region 1990 number of corporate headquarters / (Total region 1990 population) / 1,000,000 population	=	Number of corporate headquarters
Total region 1990 corporate-wide employment of all corporations headquartered in the region / (Total region 1990 population) / 10,000 population	=	Employment of corporate headquarters
Total region 1990 number of all central offices of financial institutions / (Total region 1990 population) / 1,000,000 population	=	Number of financial institutions
Total region 1990 number of colleges and universities / (Total region 1990 population) / 10,000 population	=	Number of colleges and universities
Total region 1990 enrollment of all colleges and universities based in the region / Total region 1990 population	=	Enrollment of colleges and universities
Total region 1990 number of corporate headquarters / Total region 1990 population	=	Level of personal income

Table 2.8 Multiple regression analysis of the static variables

Dependent variable	Correlation results
LEMP	Adjusted $R^2 = 0.7329$, $p <= 0.0002$
LERNP	Adjusted $R^2 = 0.6925$, $p <= 0.0009$

Table 2.9 Multiple regression analysis of the dynamic variables

Dependent variable	Correlation results
ΔLEMP	Adjusted $R^2 = 0.7554$, $p <= 0.0001$
ΔLERNP	Adjusted $R^2 = 0.7182$, $p <= 0.0001$

Static stepwise regression analyses: Stepwise regression was adopted for a second part of the analysis. Stepwise regression creates the most efficient regression model (i.e., the one employing the fewest independent variables as noted in Feldman. et al., 1987, p.152). A stepwise regression for the static formulation using employment variable LEMP) as the dependent variable generated a relatively strong statistically significant model (adjusted $R^2 = 0.54$) with one resource endowment variable PIR and one leadership variable VCEL. Similarly, a stepwise regression for an alternative model using the dependent earnings variable LERNP produced a statistically significant model (adjusted $R^2 = 0.59$) with one leadership variable EVCOL and two resource endowment variables MDFIR and PIRR. In this case PIR is negative despite the expectation that it would be positively related to the dependent variable. However, it is not uncommon to find sign reversals occurring in stepwise regression when there is multi-collinearity among the independent variables. Further, collinearity is not a problem unless one is trying to interpret the coefficients. Here, the concern is with identifying the most efficient model.

It is important to note that two statistically significant models were identified for the static formulation of the model. Further, the first two steps in each model included one leadership and one resource endowment variable. Thus, the static analysis suggests that there are two underlying explanatory dimensions corresponding to the theoretical concepts in the hypothesized model.

Dynamic stepwise regression analysis. The dynamic formulation and test of the model parallels the static approach. Dynamic analyses are conducted by lagging the mean values of the explanatory variables five years behind the mean values of the dependent variables. The five year lag has no rationale other than data availability.

The stepwise regression model using the dependent variable ΔLEMP was statistically significant (adjusted $R^2 = 0.67$) and was composed of two leadership variables ΔNVCOL and ΔEDEL, and one resource endowment variable ΔNFIR. This model has a strong bias toward leadership in that the first two steps entered leadership variables and is, therefore, a notable deviation from the structure of the static models.

The stepwise regression model using the dependent variable ΔLERNP defines a statistically significant model with one leadership variable ΔEDEL) and three resource endowment variables ΔPIR, ΔMDFIR and ΔNCHR. The adjusted R^2 for this model is 0.67. While somewhat complex, this model is consistent with the static models that identified leadership and resource endowment variables at the first two steps of the analysis.

Path analysis. Results of selected static and dynamic path analytic model tests are presented in Figures 2.12 and 2.13. Three of the models (two static and one dynamic) support the argument that leadership amplifies the effect of resource endowments on economic performance. The dynamic model with ΔLEMPP as the dependent economic performance variable, ΔNFIR as the resource variable and ΔANVCOL as the leadership variable shows leadership as the dominant factor. Generally however, these models may be viewed as supporting the leadership amplification hypothesis.

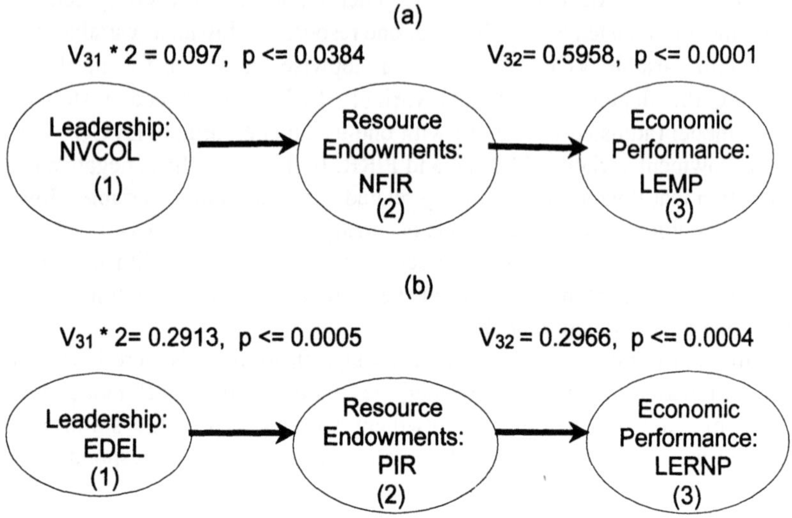

Figure 2.13 Static path analysis

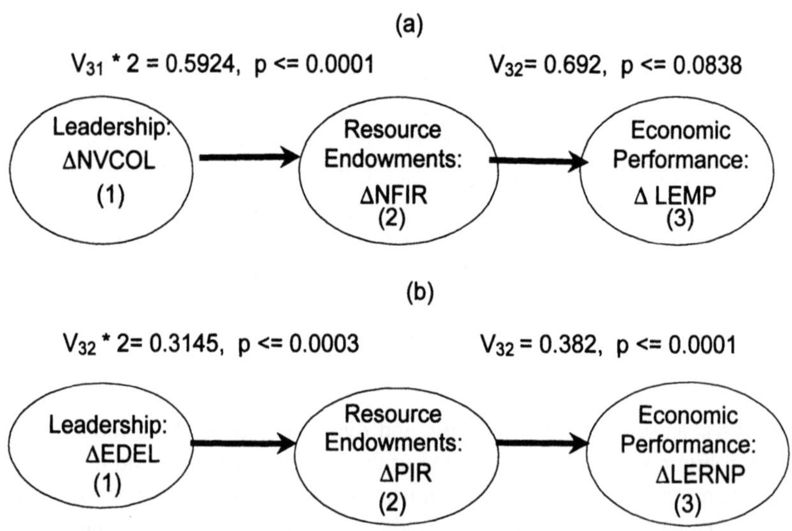

Figure 2.14 Dynamic path analysis

Summary and conclusions. Multiple and stepwise regression were used to examine the general hypothesis that regional economic performance depends on a region's resource endowments and its leadership. Causal models were formulated and path analysis (2.9 and 2.10) was used to test the hypothesis that leadership amplified the effect of resource endowments on economic performance. The results of the analyses of static and dynamic specifications support the general and interactive (leadership amplifies) hypotheses.

A new theoretical model of the relationship between regional economic performance and leadership was developed through a review of the literature. The model posits leadership as a factor that amplifies the effect of resource endowments on economic performance. The empirical tests presented in this paper provide some support for this model. While the results are encouraging there are a number of contingencies.

First, measures used for the independent leadership variables were all surrogates. Other surrogate measures could be developed that might better operationalize this theoretical construct. More direct measurement of this variable would be ideal but it is difficult-to see how to accomplish this without the kind of insight that is acquired through in-depth on site information and analysis.

Second, there is some multi-collinearity among the independent variables. In a separate study DeSantis (1993) applied factor analysis to the set of independent surrogate variables used in this study and then regressed the dependent economic performance variables on the factors. The results were inconclusive and the factors were difficult to interpret.

Third, the five year mean value lag used in this paper may not be an appropriate time period for the dynamic analysis. However, the multiple regressions for both the static and dynamic analyses showed strong relationships between the dependent and independent variables. An examination of different lagged values may yield a more robust relationship.

Fourth, there appears to be some interaction (correlation) between the leadership and resources constructs. Thus, it is possible that leadership in its amplifying role levers the effect of a region's resources. A multiperiod analysis could provide deeper insight into this possibility.

2.5 Summary, Conclusions and Policy Considerations

The paper has illustrated ways to identify potential auto-catalytic institutional and non-institutional variables in the context of a regional endogenous growth perspective. An illustration using data from the Northern Virginia region in the U.S. was provided. The conceptual analysis led to a conclusion that local regional leadership is critical for sustained regional economic development, especially at a time in history when technological change is driving and demanding rapid adjustment in nearly all institutions. A leadership model was developed and tested in an effort to build both theoretical and operational understanding in an area that is of broad and deep importance to endogenously led regional economic development Leadership is fundamental to not only the identification of auto-catalytic variables but particularly for acting upon such knowledge.

There are a few policy relevant results. Clearly, it is easier to direct policy once a set of catalytic factors have been identified. However, action may not always flow from such knowledge unless local leadership is capable and willing to induce such action. The weight of evidence in the literature suggests that the group and cooperative model of leadership is probably more applicable. Yet groups that are broad and inclusive are often unable to act suggesting an interesting dilemma. The way out of the dilemma, as many regions have found, is through leadership groups that are dominated by those who have considerable influence over discretionary resources (e.g., the Greater Baltimore Committee or the Greater Indianapolis Progress Committee in the U.S.). Despite the seemingly elite nature of such organizations, strategies that are developed by such organizations have been fruitful when other constituencies are provided a voice and final action plans are marketable to those who may not have fully participated. The analysis suggests that appropriate policy is stimulated by such intermediate organizations. In conclusion, regions without leadership organizations will suffer non-optimal outcomes in general and experience difficulty adjusting to relatively rapid change.

References

Arrow, K.J., 1962, "The Economic Implications of Learning by Doing", *Review of Economic Studies,* 29:155-173.

Arthur, W.B., 1994, *Increasing Returns and Path Dependence in the Economy,* University of Michigan Press, Ann Arbor.

Barrand, J. and C. Guigou, 1994, *Anaauze Structurelle*, University Thesis, Dauphine, Paris.

Bennis, W.G. and B. Nanus, 1991, *Leaders: Strategies for Taking Charge,* Harper and Row Publishers, New York.

Boarnet, M.G., 1997, *Spillovers and the Location Effects of Public Infrastructure*, Department of Urban and Regional Planning and the Institute of Transport Studies, Irvine, University of California.

Bower, 1.L., 1983, *The Two Faces of Management,* Mentor Books, New York.

Boyle, M.R., 1988, "Corporate Headquarters as Economic Development Targets", *Economic Development Review,* Vol. 6, 1:50-56.

Bryson, J.M. and B.C. Crosby, 1992, *Leadership for the Common Good & Tackling Public Problems in a Shared-Power World,* Jossey Bass Publishers, San Francisco, California.

Buchanan, J., 1991, *Constitutional Economics*, Blackwell, New York.

Bunch, C., 1987, *Passionate Politics,* Harper Collins, New York.

Burns, J.M., 1978, *Leadership,* Harper and Row Publishers, New York.

Camagni, R., 1995, "The Concept of Innovative Milieu and its Relevance for Public Policies in European Lagging Regions", *Papers in Regional Science*, 74:317-340.

Committee for Urban Economic Development (CUED), 1992, *Census of Economic Development Organizations,* Committee for Urban Economic Development, Washington D.C.

Cyert, R.M. and J. G. March, 1963, *A Behavioral Theory of the Firm*. Prentice Hall Inc., Englewood, New Jersey.

DeSantis, M., 1993, *Leadership, Resources Endowments and Regional Economic Development,* Ph.D. Dissertation, George Mason University, Fairfax, VA.

Fosler, R.S., 1992, "State Economic Development Policy: The Emerging Paradigm", *Economic Development Quarterly,* Vol. 6, 1:3-13.

Friedland, R., 1981, "The Power of Business in the City", in *Urban Policy Analysis*, Sage Publications, Beverly Hills, CA.

Gardner, J.W., 1990, *On Leadership,* The Free Press, New York.

Giarrantani, F. 1991, "A Note on Public-Private Partnerships and the Restructuring of a Regional Economy", Paper from the Department of Economics, University of Pittsburgh, Pittsburgh, Pennsylvania.

Godet, M., 1994, *From Adoption to Action: A Handbook of Strategic Prospective,* UNESCO Publishing.

Gray, B., 1989, *Collaborating: Finding Common Ground for Multiparty Problems,* Jossey-Bass Publishers, San Francisco, California.

Haynes, K.E., 1991, "The Role of Infrastructure in Regional System Dynamics", Presidential address, Western Regional Science Association, Monterrey, CA, Feb. 26.

Judd, D. and M. Parkinson, 1990, *Leadership and Urban Regeneration: Cities in North America and Europe*, Sage Publications, London.

Kilpatrick, H. E. Jr., 1998, "Empirical Complexity: A Study of Dynamic Increasing Returns in the Semiconductor Industry and its Policy Implications", Doctoral Dissertation, The Institute of Public Policy, George Mason University, Fairfax, Virginia

Kouzes, J.M. and B.Z. Posner, 1987, *The Leadership Challenge: How to Get Extra-ordinary Things Done in Organizations,* Jossey-Bass, San Francisco.

Lavoie, D.E., (ed.), 1994, *Expectations and the Meaning of Institutions: Essays in Economics by Ludwig Lachman,* Rutledge, New York.

Lucas, R.E., 1988, "On the Mechanics of Economic Development", *Journal of Monetary Economics,* Vol. 22, 1:3-42.

Lucas, R.E., 1993, "Making a Miracle", *Econometrica,* Vol. 61, 2:251-272.

Malecki, E.J., 1998, "How Development Occurs: Local Knowledge, Social Capital, and Institutional Embeddedness", Paper presented at the Meeting of the Southern Regional Science Association, Savannah, Georgia, April, 1998.

Nee, V. 1998, "Norms and Networks in Economic and Organizational Performance", *The American Economic Review*, Vol. 88, 2:85-89.

Neustadt, R.E. and E.R. May, 1990, *Thinking in Time: The Uses of History for Decision Makers,* Free Press, New York.

Office of Management and Budget, 1990, *Revised Standards for Defining Metropolitan Areas 1990's Notice,* Federal Register, Washington, D.C. (March, 30).

Osborne, D., 1988, *Laboratories of Democracy,* Harvard Business School Press, Cambridge, Massachusetts.

Roberts, B.H. and R.J. Stimson, 1998, "Multi-sectoral Qualitative Analysis: Assessing the Competitiveness of Regions and Developing Strategies for Economic Development", *Annals of Regional Science.*

Romer, P., 1986, "Increasing Returns and Long-run Growth", *Journal of Political Economy,* Vol. 94, 5:1002-1037.

Romer, P., 1990, "Endogenous Technological Change", *Journal of Political Economy,* Vol. 98, 5:71-102.

Rost, J.C., 1991, *Leadership for the Twenty-First Century,* Praeger, New York.

Seitz, H. and G. Licht, 1992, "The Impact of the Provision of Public Infrastructures on Regional Development in Germany", Zentrum for Europaische Wirtschafts-forschung Gmbh, Discussion Paper #93-13.

Smilor, R.W. and M. Wakelin, 1990, "Smart Infrastructure and Economic Development: The Role of Technology and Global Networks", in *The Technopolis Phenomenon,* IC^2 Institute, University of Texas at Austin.

Stough, R.R., 1990, "Potentially Irreversible Global Trends and Changes: Local and Regional Strategies for Survival" Paper prepared for presentation at the meetings of the American Association for the Advancement of Science, New Orleans, Louisiana, February 17-20, 1990.

Stough, R.R., K.E. Haynes and H.S. Campbell, Jr. 1997, "Small Business Entrepreneurship in the High Technology Services Sector: An Assessment for the Edge Cities of the U.S. National Capital Region", *Small Business Economics,* 9:1-14.

Stough, R.R., 1997, "Linking Technology and Traditional Sectors of the Northern Virginia Economy", in Stough, R.R., (ed.), *Proceedings of the Fifth Annual Conference on the Future of the Northern Virginia Economy,* Center for Regional Analysis, George Mason University, Fairfax, VA.

Stough, R.R. and M. DeSantis, 1999, "Fast Adjusting Urban Regions, Leadership and Regional Economic Development", *Region et Developpement,* November, 1999.

Stough, R.R., R. Stimson and B. Roberts, 1999, "Merging Quantitative and Expert Response Data in Setting Regional Economic Development Policy: Methodology and Application", Working Paper, Mason Enterprise Center, The Institute of Public Policy, George Mason University, Fairfax, Virginia.

Stough, R.R., 1998, "Infrastructure and Technology in U.S. Metropolitan Regions", Paper presented at the Workshop on Infrastructure Policy, The Tinbergen Institute, Amsterdam, The Netherlands, February.

Whitt, J.A. and J. C. Lammers, 1991, "The Art of Growth: Ties Between Development Organizations and Performing Arts", *Urban Affairs Quarterly,* 26:376-393.

Wolman, H., 1987, "U.S. Urban Economic Performance: What Accounts for Success and Failure?", *Journal of Urban Affairs,* 9:117.

Wolpert, J., 1988, "The Geography of Generosity: Metropolitan Disparities in Donations and Support for Amenities", *Annals of American Geographers,* 78:665-679.

3 Social Processes, and Regional Economic Development

James W. Harrington, Jr., and Deron Ferguson, Department of Geography, University of Washington, Box 353550, Seattle WA 98195-3550, USA

Recognizing the salience of waged labor and labor-force characteristics for regional development, this chapter proposes an institutional framework for analyzing local or regional labor processes and underscores the potential utility of that framework for effecting change in regional labor (re)production, "quality"" allocation, and wages.

Labor figures prominently in all explanations of and prescriptions for regional growth and development, traditionally as a key "factor" of production and industrial location. The heterogeneity of labor has been recognized, and increasing the "quality" of labor through education and training has become an important tool of national and regional development policy. However, most regional-development writing lacks explicit attention to the reproduction of labor, the structuring of labor markets, the process of employment search, labor control, and the design of work – referred to collectively as "labor processes". In a world whose local and national economies are becoming globally integrated, these labor processes define and distinguish places. Labor processes are central to the relationship between global economic change and local development. In addition, explicit conceptualization of labor should allow better informed interregional comparisons, leading eventually to structural assessment of economic development policy with respect to region-specific labor processes. This chapter provides an overview of labor processes with three purposes: understanding what is local and what is global in *fin-de-siècle* economies, establishing a basis for regional assessment and comparison, and pointing to potential interventions that may affect the outcomes of regional labor processes. Out of recognition of the importance of formal and informal institutions in the structure of labor processes, we use concepts and terms from institutionalist study in economics and sociology.

3.1 Rationales

3.1.1 Labor Income is Important

Wage, salary, and proprietors' income are the major sources of personal income (79 percent of total personal and corporate income in the U.S.). In modern economies that are increasingly dominated by formal markets for goods and services including health care and child care, personal income is an important indicator of well-being. Therefore, the ability of people to find gainful employment (including self-employment in the formal economy) in their local areas, and the wage levels thereby achieved, are crucial standards by which regional economic development can be measured. The structure and operation of labor markets, including the availability of people for employment and employment for people, the job-search and -hiring processes, and the organization of workers and work, are important structural features of regional economies.

3.1.2 Labor Processes Define Regions in the Midst of Economic Globalization

The advent of electronic communications and computing, and the transport and logistics advances they engendered, have rendered the other factors of regional development – capital, resources, and technology – increasingly mobile. In a world where financial capital is highly mobile, resources are easily transported, and much technical capability is readily diffused, what remains local? For one thing, much production remains local or national, especially the provision and delivery of services.[1] More importantly, even "global" corporations and inter-corporate networks seek out, not just *any* places, but the best places in the world for particular functions, based on some combination of: (1) transitory localized characteristics and (2) long-term, hard-to-replicate localized characteristics.

The transitory characteristics are the simple cost factors of low wages, low taxes, and limited government regulation. These hardly *anchor* productive activities, and unless they are matched with developmental policies of infrastructure and educational investment, they represent a dead end of competition for the lowest of the low, a competition that provinces, regions, or localities in wealthy countries are not likely to win. To add to the downward spiral, the substantial proportion of local economic activity that *is* local, that is not directly subject to international competition, suffers in direct proportion to the reduction in local real wages and public services. The longer-term, hard-to-replicate characteristics are the supply factors of communications and transport infrastructures, educated and innovative workforces, and environmental amenities, and the demand factors of growing household, commercial, and government markets for high-quality goods and services. These characteristics are not only hard to replicate, and thus earn high economic rents, but are hard to sustain, and require large economic investments (Storper 1995; 1997). Labor processes and labor-related institutions are important components of these truly localized characteristics.

---

[1] While total trade flows have grown from 14% to 20% of the U.S. economy between 1986 and 1996, that leaves 80% of the U.S. formal economy not accounted for by imported supply or export demand.

3.1.3 Labor Processes and Endogenous Growth

Information embodied in workers, combined with inter-organizational mobility of workers, provides an important basis for the localization of technical and market learning and consequent regional economic growth. In addition, publicly implemented arrangements for training, employment security, employment mobility, and entrepreneurialism affect the rate of employment creation and labor productivity within regions.

"Endogenous" growth theory provides a neoclassical basis for modeling these influences on growth. It incorporates technological advance, long seen as an important contributor to economic growth. Romer (1986) provided an explicit function for the production of new knowledge by application of capital and labor inputs, and making existing knowledge an argument in the aggregate production function. His primary postulates were: (1) there are increasing returns in the use of knowledge inputs in production; (2) increased knowledge is produced directly through the allocation of inputs, but that there are diminishing returns to the allocation of resources to knowledge production; and (3) some portion of the returns to new knowledge can be appropriated by the actors that invest in its production. The interpretation of these stylized facts is flexible. "Knowledge" can be interpreted as disembodied, embodied in capital inputs, in labor inputs, or as all three (see Lucas, 1988). Institutional arrangements such as patents and licensing, and the creation of firm-specific applications, allow the appropriation of returns to knowledge (Teece, 1986). The externalities of knowledge production reflect the possibilities of reverse engineering, industrial intelligence gathering, and the mobility of skilled and technical workers among employers and self-employment. The labor component of these externalities is often partially localized, because of the greater job mobility within the same local area. The resultant model allows for endogenous modeling of knowledge creation, allows knowledge inputs to serve as a fixed cost in production of goods and services (i.e., the quantity of output is not a function of the quantity of information), allows unconstrained increases in per capita output, and maintains a competitive equilibrium (via the decreasing returns to allocating resources to knowledge creation and via the inability of actors to appropriate all the returns from increased knowledge). The model also generates welfare implications based on the likely private under-investment in new knowledge. Romer thus explained persistent or increasing inter-locational disparities in growth rates not by differences in the externalized knowledge available (which is presumed to flow freely) but by differences in cumulative investment in knowledge inputs (disembodied or embodied) that allow full use of available knowledge. While Romer did not recognize locational stickiness in the externalities of knowledge creation, this theme continues in the literature.

The phrase "labor process" has been used to refer to the institutionally derived relationship among workers, material, and machinery: the process of transforming human labor into marketable products (Braverman, 1974). However, additional processes determine labor characteristics and employment trends in regional economies: the reproduction of a labor force, the creation of employment demand, and the allocation of people to jobs (by occupation, wage level, and employment security). We will use the phrase "labor processes" to refer to all of the social, political, and economic processes by which people become employees and the source of productive capability.

3.2 Institutional Analysis in Regional Development

Before discussing the major conceptualizations of institutions, it is important to distinguish "institutions" from "organizations". Institutions are collectively held beliefs, values, mores, and rules which condition or constrain individual action. Organizations are tangible groups or entities. Organizations have particular interests and supporting institutions, including a collectively held belief in their *raison d'être*. Organizations are often referred to as "institutions," which generates considerable confusion in the literature. A labor union, for example, itself is an organization insofar as one is referring to the "brick-and-mortar" facilities, the assemblage of people and equipment, and its everyday operation. The labor union as an organization, however, reflects a number of underlying institutions, chief among them a particular philosophy behind employment and employee representation. It should be noted that these beliefs about the purpose and legitimacy of the organization's purpose distribute (and limit) the politico-economic power of that organization. The way in which social and economic power is distributed (often unequally) in a community is, essentially, a characteristic outcome of its complex of institutions.

3.2.1 Economic Approaches to Institutions

In the current state of institutional economics, one may distinguish two general approaches: Original (or "Old") Institutional Economics (OIE), and New Institutional (or "Neo-institutional") Economics (NIE). Both OIE and NIE are grounded in the recognition that institutions condition economic behavior. The OIE approach is highly critical of the approach taken by conventional or "mainstream" economics in explaining economic behavior, while NIE is more of a broadening of mainstream economics, through a reliance on the notion of transactions costs, to include considerations of the role of institutions in that pursuit. While both OIE and NIE are grounded in the explicit recognition of the importance of institutions, they differ considerably in their scope and method (Stanfield, 1998). Before offering a framework for analyzing the institutional underpinnings of labor processes, it will be useful to trace the scope and method of each approach to institutional analysis.

Original institutional economics. The theoretical roots of OIE are varied. Important roots of the school can be found in the work of Karl Polanyi, Thorstein Veblen, John R. Commons, and Clarence Ayres. The scope of OIE is perhaps best articulated through a brief retracing of Polanyi's work in economic anthropology, and can be summed up in his phrase "the economy as instituted process" (Polanyi et al., 1957, Ch. 13). To sustain and reproduce itself materially, any society must develop ways to organize the provisioning of its material needs and wants. The cohesion and continuity of a community requires some degree of durability of this economic organization, which amounts largely to an organization of the division of labor or work, of planning ahead, and dealing with uncertainties or unforeseen events. Communities inherit and develop systems of moral beliefs, values, and ideologies – in short, "cultures" – in which this organization of material provisioning – the "economy" – is integrated. It is important to note here that the possibilities for different configurations for this integration are numerous, if not infinite, which is captured in the notion of cultural relativity (Mayhew 1987) and/or institutional indeterminacy (Peck, 1996) and further discussed below.

The study of how the economy is instituted is the subject of what Polanyi called the "substantive" meaning of economics; this is clearly an anthropological approach to setting the scope of economic inquiry. This is a major departure from the scope defined by conventional economic theory, which Polanyi argued is grounded in the "formal" meaning of the word "economic". The scope of this formal meaning of economic is limited to the study of individual choice subject to the logics of scarcity, hedonic calculation, and maximizing behavior (Polanyi et al., 1957). The substantive meaning recognizes that economies and individual economic actions are embedded within a cultural context, where that context articulates what kinds of actions are socially allowed, expected, encouraged, or prohibited – that is, the social context is articulated by institutions.

Much has been made of the notion of "embeddedness" in recent scholarship in regional and economic geography. While the notion itself was first developed by Polanyi in his distinction between pre-capitalist and capitalist economies, its use has become more narrowly specified in attempts to distinguish the more subtle regionally and culturally specific variations in the interdependencies between "non-market" and "market" transactions in what are basically advanced market economies (Granovetter, 1985). These applications of the concept argue that local "embeddedness" of economic transactions can produce quasi-rents to the extent that the subordination of market transactions to socialized reciprocal interdependencies (e.g., those founded on "trust" (Sabel, 1992) result in greater flexibility, efficiency, and hedges against uncertainty. This use of the concept clearly represents an evolution beyond Polanyi's application.

Equally important to the origins of OIE is the work of Thorstein Veblen (1899), who articulated the central role of institutions in economic life. Veblen observed that institutions derive from two broad bases of legitimation or valuation: "ceremonial" and "instrumental" (an observation referred to as the "Veblenian dichotomy"). *Ceremonial* institutions are inherited beliefs which socially legitimate status, class, and distribution of power, via a naturalized logic of invidious distinction. *Instrumental* institutions are the beliefs and systems which legitimate and motivate problem-solving action or skills acquisition – the source of technical progress, industriousness, and inquiry. Many behaviors are results of a mixture of ceremonial and instrumental validation, and the question of what roles particular institutions play in patterning action is largely a question of identifying the degree of ceremonial or instrumental "dominance" (Bush, 1987). Veblen proposed that institutional evolution is continual, cumulative, indeterminate (non-teleological), and conflictual. The conflict arises from the resistance of those accorded power through ceremonial validation, operating through ceremonial institutions, to changes which are brought about by the unfolding of problem-solving which may challenge the very legitimacy of ceremonial values. An important part of institutional analysis is uncovering the relative power of institutional actors to establish, to change, or to ignore conventions. This uncovering can be accomplished by investigating the reaction of ceremonial institutions in the face of exogenous change, or by comparing the pervasiveness and longevity of identifiable components of institutions.

The scope of OIE does not simply "widen" the net cast by conventional economics to include "culture" and "society." It begins with placing institutions at the center of economic behavior and stresses evolution, cumulative change, and indeterminacy in the functioning of economic systems. Economic institutions which regulate provisioning, once identified, are not evaluated according to *a priori* notions of rationality. This provides the basis for OIE's rejection of conventional economics as a science of choice under assumptions of insatiable, exogenously-determined wants and the rule of a maximizing mentality. While it would be redundant to repeat the critique of conventional economics which is by now well known amongst social scientists (e.g., see Hodgson and Screpanti, 1991), it is worth briefly reviewing the difference in method which corresponds to OIE's difference in scope.

Much of the past rejection of OIE by conventional economics has been, presumably, partly due to the non-quantifiable nature of OIE inquiry. Conventional economics has developed a tradition of using positivist, quantitative inferential methods for hypothesis testing. In contrast, the methodology of OIE has consisted mainly of qualitative, context-sensitive empiricism, with a greater reliance on the descriptive rather than inferential use of quantitative information. More profound than just this empirical difference in methodology, however, is a philosophically disparate approach to research. By virtue of the questions it sets out to answer – what is the nature and implication of the interdependence between economic institutions and economic processes in a place and time? – OIE method is comparative and inductive, proposing what appear to be general rules from collected information, such as ethnographies, from groups. This comparative method has relied on case studies and field studies of "the activities, the rules, and the applicable understandings or cultural underpinnings that comprise human behavior unfolding in an institutional context. The summation of these field studies is the ethnographic record, the codified assemblage or what we know of human behavior in place and time" (Stanfield, 1998).

Neo-institutional economics. The "beginning point" of NIE is an *a priori* acceptance of conventional economic theory as the science of individual choice. As proposed by North (1990:3), the project of NIE is then to develop an "analytical framework to integrate institutional analysis into economics and economic history." This approach accepts the methodological individualism of conventional economic theory.

> Defining institutions as the constraints that human beings impose on themselves makes the definition complementary to the choice theoretic approach of neoclassical economic theory. Building a theory of institutions on the foundation of individual choices is a step toward reconciling differences between economics and the other social sciences. The choice theoretic approach is essential because a logically consistent, potentially testable set of hypotheses must be built on a theory of human behavior ... hence, our theory must begin with the individual (p.5).

For NIE, institutions function as constraints on the "choice sets" of individuals. But the real reason for institutions is that they reduce uncertainty about the individual actions that require cooperation from others to be successful, which we can think of as "transactions." The effect of institutions on economic behavior (including the nature and costs of interpersonal transactions) is the subject of institutionalist analysis. North (1990:16) noted that institutions are developed, not necessarily to be efficient (to reduce transactions costs), but to serve the interests of those with the power to devise and enforce formal and informal constraints. As economies grow in scope and in the diversity of exchanges, that growth must be matched by appropriate institutional change to assure adequate economic performance.

3.2.2 Key Dimensions of Institutions: Toward a Framework for Regional Analysis

Regionalists and economic geographers have invoked the institution concept in a variety of ways and for a number of different purposes. Drawing from original and neo-institutionalist economics, as well as the variety of sociological and political insights encapsulated in regional-science and economic-geographic writing, we suggest a framework from which to conduct institutionalist analysis of labor processes in sub-national regions.

Our framework defines the attributes of institutions along five dimensions. First, *formality* refers to the continuum of the social codification of institutions, that is, whether they evolve informally or are the result of deliberate creation. Second, *domain* refers to the discursive origin of institutions, that is, whether they represent an attitude for intervention (or "philosophy of intervention") derived through a conscious "public" deliberation among organizations (especially within the state), or whether they derive from traditional habit or custom. These two dimensions affect the extent over which institutions hold sway: capacity, spatial scale, and temporality. *Capacity*, the third dimension, measures the numbers of actors and social groups that can be effectively influenced by a given institution. Institutions exhibit *spatial scale* to the degree to which they "reach" across distances or territorial boundaries. For example, the administration of economic security or welfare programs can vary locally or regionally while the establishment of workplace safety rules may occur on a national scale. Finally, *temporality* recognizes that institutions can be of a long-standing or transitory quality. Racial or gender discrimination in hiring or firing may stem from long-lasting (e.g., intergenerational) social attitudes, but may wax or wane in practice, depending on the circumstances governing the restructuring of labor demand. Thus, "undercurrents" of racism or sexism may be long-standing institutions within a society, while the incidence of discrimination will vary temporally. All institutions fall within these dimensional bounds, which can now be discussed in greater detail.

Formality. Institutionalists often describe institutions as being "formal" or "informal" (e.g., North, 1990, Ch. 5-6). The formality of an institution refers to its place within a continuum of codification, or "explicitness". For example, in a job interview, both the interviewer and interviewee will most likely be aware of informal rules of behavior and etiquette, including exchanges of greetings, dress, allowable topics of conversation, and so on. These informal rules will vary by situation and context – by size and type of firm, type of work and occupation, etc. – and play a part in "signaling" to the interviewer the possible fit of the interviewee in the company. At the same time, the interview may be regulated by formalized rules, whether established within the firm or externally, for example by public anti-discrimination laws specifying the bounds for the kinds of questions that may be asked. Both of these examples of formal and informal institutions flow from the broader society's values toward privacy, (prospective) workers' rights, and attitudes toward the relative bargaining strength of employers and workers.

The significance of this dimension is made clear by Nee (1998:87):

> When the formal rules of an organization are perceived to be congruent with the preferences and interests of actors in subgroups, the relationship between formal and informal norms will be closely coupled. The close coupling of informal norms and formal rules is what promotes high performance in organizations and economies.

In other words, if formal and informal rules are congruent, because individuals and groups within the organization or polity "buy in" to the institutions which underlie formal rules, then greater productivity, harmony, or success is the result. An example might be the relationship between informal and formal work rules, where tension often appears between the formal, corporate rules for task assignment, work speed, and use of equipment (e.g., personal use of office telephones and electronic mail) and the informal norms established by convention or tacit agreement. Coupling or congruence applies to entire institutions as well as to individual rules, as in the incongruence between formal, corporate governance and goals (directors maximizing profit for shareholders) and the actual system of incentives in place for managers and workers in their multiple contexts of laborers, coworkers, and members of households. Assessment of institutional congruence (between formal and informal institutions, as well as public and social, or local and non-local) is a part of the institutional assessment to which we turn at the end of this chapter.

Domain. The domain of an institution refers to whether it is public or social (or non-public) in nature. "Public" institutions are formulated out of a "conscious" or deliberative discourse and are politically implemented. Public institutions are the "official" articulations of normative values, as opposed to social institutions which are largely "unofficial" in the sense that they derive from unquestioned belief or taken for granted rules or norms. The meaning of "public" institutions can often be conflated with the physical "brick and mortar" facilities, and associated organizational bureaucracies, rather than their philosophy of intervention within the economy. The reason public (intangible) institutions are often conflated with public organizations is that their interventions are easily attributed to the physical organizational form. The operationalization of public institutions can be clearly identified in terms of their intervention: in the development process, in providing services to small businesses, or by way of providing capital to business ventures which are rejected in private capital markets, etc. Behind each of these interventions, however, and defining their actual meaning, is a socially-negotiated mandate, or "philosophy" such as attitudes about the definition and reasonable mix and rate of economic growth, beliefs surrounding the value of small business enterprises, and ideas about the need to direct resources to particular activities. It should be kept in mind that such policies are implemented or administered according to formal and informal institutions.

Social, or "non-public", institutions refer to "cultural norms" toward accepted behavior and conduct, beliefs, values, ideals, and aesthetic appreciation, which are both the product and wellspring of tradition, customs, and cultural-historical distinction. Social institutions are not articulated by a conscious, deliberative discourse, but by incremental adaptation and reinterpretations of tradition, myth, socially-constructed cultural memory[2]. At the level of individual rules, we can have "social" rules as distinct from "public" rules. In the example of the job interview above, social rules suggest the boundaries of the social aspects of the interaction – attitudes about proper behaviors regarding gender, appearance, language, and so on – while public rules are represented in large part by legal sanctions and allowances defining the bounds of the interaction according to public ideals (institutions) regarding equal opportunity, equality, and fairness, etc. It is easy to conflate formality with domain, or "public" with "formal" and "social" with "informal", because public institutions are most easily observed in terms of formal codification – laws, ordinances, and other "official" statements. In fact, social institutions may well be formal (having the weight of being codified, yet not derived via a deliberative public discourse), such as clauses written into contracts derived out of practices inherited from the past or out of agreements reflecting the distribution of power and continuity of social relations. Public institutions may

---

2 Greif (1994:915) defined "cultural beliefs" as "ideas and thoughts common to several individuals that govern interaction, and differ from knowledge in that they are not empirically discovered or analytically proved."

also be informal, grounded in attempts to create or persuade certain social behaviors; the state cannot effectively impose attitudes on society, but can only attempt to shape them through a mix of formal and informal practices. For example, formal laws dealing with fair labor practices might be mixed with public discourses around international competitiveness which result in particular blends of formal and informal actions shaping the concrete context of local production in a particular industry. While there may be specific formal rules of conduct, their adherence is regulated largely informally in specific contexts in line with "official" discourses which may or may not "couple" with those rules. In addition to the particular blend of formal and informal public institutions, formal and informal social institutions also shape particular production or labor contexts.

The significance of the domain dimension is the utility of separating the deliberative institutions (especially those springing from public discourse) from the "taken for granted" institutions, as this may be a point of intervention once particular institutional contexts are interpreted. An institutional analysis which focuses on public institutions only (because they can be more easily observed), uninformed by ethnographic information about social institutions, is likely to be incomplete and lead to misguided recommendations for action.

Capacity. Some institutions rely on a sense of the social collective that is difficult to sustain beyond a limited number of actors or a common social bond. At issue is the degree to which behavior can be informally monitored and sanctioned. Game-theoretic approaches such as developed by Axelrod (1984) show that

> wealth-maximizing individuals will usually find it worthwhile to cooperate with other players when the play is repeated, when they possess complete information about the other players' past performances, and when there is a small number of players" (North, 1990:12).

The relationship between this game-based finding and the social-network sources of "trust" and reciprocation is obvious. Not only are relatively informal institutions more feasible within smaller, more insulated networks, they are more cost-effective than formal, publicly monitored, institutional arrangements that often increase transactions costs. In fact, economic history contains many examples of institutional changes that increased transactions costs (by establishing the need for intermediaries, contracts, and legal assistance) but that allowed an increase in economic and spatial scale, with resultant unit-cost savings that outweigh the additional transactions costs (North 1990:65; Greif 1994). One potential difficulty in regional economic development is the decreased effectiveness of informal institutional arrangements with increases in population, social heterogeneity, economic activity, or extraregional linkages.

Spatial scale. Many institutions are localized – values, mores, beliefs, and philosophies behind the way "things are done" in a given place – while others reach across boundaries. For most people, the everyday lived world is local, and distinctly local institutions develop and evolve in the unfolding of everyday life.

Local institutions – formal and informal, public and social – articulate local culture and distinguish it from other places. We thus see institutional differences between rural, urban, and suburban places, and among subnational regions. These differences may be punctuated by language, dialect, nationalism, regionalism, and "sense of place".

In addition to the localization of everyday life and its institutions, supra-local institutions develop, often under unifying or centralizing public institutions negotiated in the political sphere. Thus we have supra-local, or "national" norms, which often bound the variability of local institutions. As a very simplistic but illustrative example, "American" institutions can be distinguished from "French", "Iranian", and "Japanese" institutions, and further distinctions can be drawn among "Yankee", "Southern", and "Midwestern" institutions, and so on. As with the recursive change between public and social institutions, change is also recursive across scales, with the local feeding change in the supra-local, and vice versa. In some literatures, this is often conceptualized as "bottom-up" or "top-down" institutional change. As the world economy globalizes, more effective global vectors of institutional change emerge. The concept of "glocalization" (Swyngedouw, 1992) refers in part to the negotiation of divergent local institutions, partly mediated through supra-local norms, in particular contexts, and the resulting institutional adaptations and changes. The "local-global" tension is largely located in institutional renegotiation and change. Returning to the example above, a global economic linkage bringing together, say, "American-southern" and "French-Occitan" institutions will lead to renegotiation and adaptation before the interaction can be "successful", thus certain linkages may be favored over others, depending on what is meant by success.

The institutional context of economic change is very much defined by the spatial scales of the institutions implicated in the economic process(es) under question. Keeping in mind that institutions distribute power, institutional negotiation or "glocalization" is a working out of new rules which distribute power to organizations and individuals.

Temporality of institutions. Some institutions are long-lasting, and deeply embedded in culture, while others are transitory. With regard to labor processes, a long-standing institution would be that of wage labor – or more specifically, the system by which one enters into contract with an employer to sell his or her labor power, which involves a widespread acceptance of beliefs about the proper work ethic, distribution of wealth and income, property rights, and so on. This long-standing institution of wage labor, however, is supported by a changing "under-structure" of beliefs about the fair terms of labor contracts – including expectations about workplace conditions, wages, status, autonomy, etc., which are more temporal and a point of adaptation to changing technology and regimes of accumulation. The distinction is that we have long-standing institutions which articulate the division of labor, integrated with other "non-labor" institutions regarding other social relationships, while shorter-term institutions incrementally adapt and adjust, allowing for changing "terms" by which the overall division of labor persists. In this example, it is important to note that expectations about the terms of labor contracts must become congruent with the fabric of the whole. For example, even if the unfolding of technological change pushes towards a recasting of the "typical" individual labor contract toward widespread home working, other institutions implicated by that change (the gendering of household responsibilities, formal rules governing home production, even neighborhood zoning) must first allow for it or later adapt to it. Identifying the temporalities within the set of economic institutions implicated by the economic process in question is important, as these together signify the degree of stability or durability of that process.

Much of the regional literature has emphasized the role of regional institutional differences. One way to capture the dimensions of spatial scale and temporality is by the concept of "regional memory". Regional memory refers to the local social construction of social and public institutions, formal and informal, which involves collectively shared interpretations of place-bound experience. Regional memory is cumulative, as it is grounded in the shared interpretation of local history, while at the same time it is "selective", in the sense that the interpretations of more powerful organizations *and* institutions are dominant.

3.2.3 What Makes a Local Labor Market Local?

The concept of the local labor market is central to regional economic analysis. Unfortunately, the definition of local labor markets is subject to the difficulty of all delineations based on potential interaction, though practicalities of data availability generally drive the implementation of the concept (Schubert et al., 1987). Regardless of the delineation method, the scale of local labor markets is totally dependent upon the individual's temporal and transport relationship between home and work, which can be generalized to bear relationships with gender, income, social networks, and occupation.

Most importantly, the extent and operation of local labor markets are socially regulated by a range of institutions. Significantly for regional economic development, these institutions are not mutually self-regulating toward the end of matching labor reproduction and labor demand in ways that allow continued regional economic growth and distribution (Peck, 1994). The following section presents four types of labor processes that operate through identifiable institutions. The processes are localized by the ways in which institutional development *and interaction* are unique to individual regions, even though many of the institutions exist at a spatial scale larger than the region (Saxenian 1994; Peck 1996, Ch.4). The role of institutionalist analysis of local labor processes, then, is to identify the components and overall nature of these institutional ensembles, to understand their congruence or incongruence, to relate a localized ensemble to perceived problems in regional labor processes, and to identify potential institutional interventions. The remainder of this chapter provides an introductory overview of these analytic and prescriptive elements.

3.3 Labor Processes

For our purposes, and following other overviews of local or regional labor analyses,[3] an institutional approach to labor markets entails four large processes: production and reproduction of labor and labor characteristics from a regional population, the generation of employment demand, allocation of potential workers to specific employment situations, and the control of labor under employment or other contract.

3.3.1 Reproduction: The "Creation" of a Regional Labor Supply

The characteristics of region-specific labor supply relies on the interaction of a number of processes. This section relates the straightforward economic processes to the localized institutional ensembles that shape the processes and their outcomes, proceeding through three linked dimensions of regional labor supply: qualities, quantity, and organization. Exploring the institutional bases for these processes facilitates both interregional comparison and attempts to intervene in these processes.

Qualities. Labor "quality" is a highly contentious and ultimately political term. There are at least two very different meanings: work-relevant skills and autonomy, and worker practices related to work (see Standing, 1992). The specific skills that are useful in a particular employment relationship cross many dimensions: craft expertise, technical knowledge, written expression, interpersonal management, abstract analysis. Regional and industrial-location analyses generally simplify these dimensions into "skilled" (in possession of any of these identifiable skills) and "unskilled" (in possession of none of these skills insofar as they are relevant to a particular set of tasks)[4] – or at times, make use of occupational designations (managerial, professional, technical, clerical, operative, and so on)[5]. Merely determining the availability and trend of workers and potential workers in these categories is difficult, and individual employers as well as regional analyses often resort to indicators or proxies such as high-school diploma, university education, prior work experience, and the self-identified occupational affiliations of population surveys. The wage premia attached to these indicators presumably reflect their

---

3 See Peck (1994, 1996, Ch. 2) and Villa (1986) for two characterizations of the processes that regulate labor markets; we have adopted a characterization somewhat closer to Villa's.

4 Storper and Walker (1983) provided examples of this problematic treatment in industrial location studies, and suggested the bias and shortcomings this introduces in industrial location models, practice, and policy.

5 Standing (1992:258) distinguished occupation ("involving a career of learning and the mastery, or possession, of the mysteries of a craft or profession") from skill, which reflects both the ability to make productive decisions autonomously and a set of characteristics bound up with social status.

relative (and changing) importance in the production process, should affect the willingness of individuals to gain specific skills or proxies.[6]

Institutional factors determine the premia paid (in the form of monetary compensation and work conditions) for certain skills or characteristics, and affect the individuals' access and incentives for skill or proxy acquisition. Formal procedures to admit limited numbers of specially-prepared candidates to medical or legal professions, supported by public support or recognition only of formally-certified professionals, forms a largely ceremonial, institutional determinant of the wage premium and access to certification – a determinant that is increasingly global. Informal, social conventions, operating through the family, peer group, educational system, and mass culture, affect skills acquisition by gender, race, and class – these conventions are potentially more localized.

Skills acquisition is a major determinant of regional economic development. Regions differ in the institutional arrangements for labor training, generally provided by governments, employers (typically through long tenure), or cooperative arrangements – or not provided at all (Peck, 1994). Scott (1988:143) described publicly provided, specialized training as "one of the sure signs of a mature local labor market and one of the conditions of its smooth internal reproduction." On-the-job training has been shown to be the source of most capabilities that most workers use regularly. This makes economic sense because: it is more efficient for people to be somewhat productive while they learn; workers' awareness of what needs to be learned is greater on the job; and the immediate application of skills reinforces learning. The ability of individuals to make autonomous decisions on the job is a function of institutional reproduction of *both* individual characteristics (decision-making, technical ability, and assertiveness) *and* the design of work, which is a function of convention and of relative power of employer/manager and employee. Therefore, the supply of skilled labor is in fact related to the demand for and use of skilled labor (Thurow, 1983). Institutional analysis allows interregional and intertemporal comparison of these institutional arrangements, as well as suggesting points of intervention to increase the production of particular labor qualities in a place or the recognition of performance-relevant skills over indirect indicators.

The second meaning of "quality" refers to the package of general attitudes and capabilities brought to and used in employment. Workers' relative attachment to waged work versus household or informal-sector tasks affects their flexibility with respect to work hours and locations, their concern for employment ladders, their ability to get consistently to a particular place at a particular time. These attachments reflect social institutions of family relationships and duties and public institutional support for transportation availability and pricing. Workers' social and cultural backgrounds influence (often, in ways distinct by gender) their communication and interaction styles and expectations regarding work conditions.

---

[6] With widespread knowledge of these premia, some individuals will obtain more schooling than they might need to perform the tasks for which they are best suited or most interested (see Arrow (1973), Stiglitz (1985).

Quantity. The neoclassical economic explanation of labor supply relies on the relationship between the prevailing wage level and the reservation wage to elicit labor-force participation within a region, and the relationship between real (or cost-adjusted) wage levels across regions to elicit migration among regions. Institutional analysis benefits our understanding the role of waged work versus other forms of livelihood, and the consequent determination of the reservation wage.

The individual's choice between (a) waged work and (b) non-waged work in a household or other family setting is mightily influenced by formal and informal social institutions' sanction of particular divisions of labor within households or extended families, typically attaching and supporting particular roles according to sex and age (Thurow, 1983; Peck, 1994). Relevant social institutions include religion, cultural tradition, and the arrangements and expectations that differ by socio-economic class. In each case, what we term "institutions" are broader than a single convention about the role of women, children, or grandparents, but are interlocking arrangements for childcare, domestic duties, labor in a family business, restaurant meals, etc. Individual conventions, capabilities, and beliefs are the components of these institutions, but the institution is their interaction. (Indeed, an important motivation for change in social institutions is incongruence among the institutional components). Public institutions play a large role in the basic supply of waged labor, through explicit labor regulation (of wages, child labor, employment practices), through the provision of alternatives to waged labor (full-time education, compulsory education, support for mothers of young children), and through support for waged labor (tax or direct support of child care, transportation, tax policy). These public and private, formal and informal institutional arrangements vary across local and especially national regions. Analysis of these institutions is messier than a statistical determination of the reservation wage and its relationship to locally prevailing wages, but the messier analysis provides a better basis for (a) intervention in the labor-supply process and (b) comprehension of the effects of a particular, unplanned change in technology, labor demand, or cost of living (Nee, 1998).

Analogously, interregional (or international) migration responds in part to real-wage differentials. However, the degree of responsiveness depends on social institutions (especially in the origin locations), social connections, and public institutions that affect the ability and attractiveness of migration. Gordon (1995) operationalized a comparative analysis of interregional migration flows by distinguishing observable labor-market processes: migration after new employment was contracted; migration in hopes of employment at destination; and migration because of the employment-related move of the household head. The prevalence of these causes depends on occupation- and labor-market-specific practices regarding employment search, and on social conventions of household relations.

Organization and expectations. Labor organization into guilds, professions, or collective-bargaining units is at once a key institutional component of labor supply and a characteristic of region- and sector-specific labor. These institutions (some clearly identified with single organizations, some more complex sets of conventions and rules) can determine labor supply, influence the characterization of labor quality, and influence the determination of wage levels and conditions of work. The presence and nature of labor organization (determined at international, national, or local scales) are critical institutional arrangements affecting localized employment and work practices. Despite the centrality of labor and its organization in regional economic change, traditional regional analysis faces a paucity of concepts for labor organization.[7] If labor organization were simply a matter of unionization or not, the empirical decline in unionization might explain the silence on labor issues. However, the expectations, relative power, and internal divisions of workers are more complex than one dichotomous variable (whether that variable be "skilled/unskilled" or "unionized/non-union"). A beginning point for regional, institutional analysis of labor organization is the public allowance for and recognition of specific forms of organization, formal institutions for labor and professional representation and quantity control, and the social practices (e.g., of exclusion) of the organizing institutions.

These organizational issues (and the process of labor reproduction generally) are an important component of "regional memory." The employment experiences of workers, particularly the social and spatial divisions of labor within firms (or government agencies, for that matter), powerfully reinforce or disturb existing institutional arrangements and divisions by class, gender, and ethnicity (Massey, 1995). Geographic agglomeration increases the strength of these institutional arrangements, at the same time that it encourages certain social and spatial divisions of labor within firms (Scott, 1988).

3.3.2 Labor Demand

The level and trend of labor demand are critical components of regional economic analysis and policy. The typical approach is to assume a causal relationship from external demand for regional production to derived demand for regional labor. The precise nature of the relationship is expressed via a production function that describes technology and labor-capital elasticities (or, in the case of a simple input-output model, via fixed production coefficients). This relationship prescribes policy interventions of increasing external demand and reduction of effective wages, via wage subsidies or wage regulation. Incentives for capital investment are also common components of regional policy, on the assumptions that newer or larger capital stocks will increase external demand for regional production and that capital-labor complementarity will outweigh the substitution of capital for labor.

---

7 In a discussion focused on the study of economic globalization, particularly in manufacturing sectors, Herod (1995:346) wrote "the overwhelming emphasis in the literature has been on capital as the producer of industrial landscapes in the global economy.... Workers have been rendered invisible and relegated to the status of little more than an afterthought in conceptualizing the making of economic landscapes."

The very different employment-growth trajectories of the U.S. and western Europe suggest the importance of institutional differences in translating economic trends into job growth or decline. Analysis and resolution of the European employment "problem" depends on the institutional arrangements that are investigated and emphasized. Most policy-oriented treatments of these differences emphasize public institutional differences in labor supply and demand: on the supply side, provisions for financial support of people outside the labor force maintain a higher real reservation wage in Europe than in the U.S.; on the demand side, regulation of corporate labor shedding and requirements for employee benefits increase the long-term cost of similarly waged labor in Europe over the U.S. In addition, however, individuals in some regions of Europe may have greater support through non-waged work in the home or farmstead than has become the norm in the U.S. (a difference that does not beg for intervention into either setting). Institutional barriers to skills or occupational upgrading include the structure of and restrictions to formal education, and informal social expectations and prejudices based on class and sex. Thus does the nature of and the very need for intervention depend on the analysis made of slower employment growth.

The nature of technological change, and the translation of innovation adoption into employment change, are both institutionally determined. The widely noted tendency toward labor-displacing technological change within a production process reflects both a general capitalist tendency toward capital investment and intensification, and a response to variable institutional characteristics of a place, a set of occupations, or a time. The analysis of technological change changes tremendously when reliance on economic signals (wage and interest levels and trends) is supplemented with an explicit recognition of (a) the power of technological change to disrupt institutional arrangements (gender divisions of labor, occupational structures, career ladders, labor organization) and (b) the relative power of institutional actors to force, withstand, or benefit from change (Scott, 1988).

"Employment" itself is, of course, an institutional arrangement among worker, organization, and the state, involving regulations and conventions of control and exit. The institution of employment is becoming more variable via increased options for payment (such as tying compensation to individual or organizational performance), contracting (indirect employment through personnel agencies, contracts with independent sole proprietors), and work location (in the home or in clients' facilities). The employment impact of increased production is mediated by employers' ability to determine the terms of employment via changing the employment institution. This ability hinges on the relative power of employers in institutional regulation, and on the ways in which labor can be deployed in production (Salais, 1992). The flexibility of these arrangements has increased with improvements and cost reductions in computing and communications technology, with implications for the incomes, public-service needs, and support-service needs in suburban, exurban, and rural locations, Beyers and Lindahl 1997a,b; Nelson 1997a,b) as well as for urban residents who may not benefit from decentralized, flexible contracting arrangements. The technology, however, is not the driver but the enabler of work relationships: institutional arrangements, cost structures, and social goals of firms, workers, and governments determine the ultimate trends in particular places (Castells, 1996:220-221).

3.3.3 Allocation: Employment Searching

The economic conception of labor markets requires only information about jobs and applicants, and the absence of government or other institutional floors or ceilings for wages, to determine at what wages and quantities markets will clear. Regional scientists have explored the effects of regional differences in labor supply, demand, and information availability. Formal and simulation approaches have modeled the relationship between local job search and interregional migration (Rogerson and MacKinnon, 1981), interregional migration and stylized information flows about labor-market conditions (Amrhein 1985) or wage levels (Maier, 1987), and searching and hiring in labor markets permeably segmented by occupation (Amrhein and MacKinnon, 1985). These approaches allow us to model the effects of labor-market segmentation and employment institutions on vacancy and unemployment rates, wage levels, and regional disparities. It remains for other research approaches to theorize and measure labor-market segmentation and employment institutions.

Labor markets are obviously segmented into submarkets, distinguished by occupation, skill, organization, and norms for pay, tenure, and control (Doeringer and Piore, 1971). It is a fairly straightforward exercise to model segmented labor markets by allowing limited movement of workers between segments (for example, through training) and limited substitution of workers from different segments. However, the *creation* of labor-market segments and the allocation of individuals to segments entail each of the labor processes presented in this chapter: social and public institutions for labor reproduction, the development and channeling of labor demand, the allocative process, and the experience of firms, regions, and individuals with different forms of labor control. For example, labor-market segmentation is enforced by differentials in the search, hiring, tenure, and control rules and norms by occupation, industry, and organizational size. The differentials are not based on strict technological requirements, but on norms established on the basis of relative power of employers and employees, and among competing employers (Scott, 1988; Gordon, 1995).

Institutional approaches entail conceptualization and measurement of the social, legal, and technological bases of information flow and institutional arrangements for job search, hire, compensation, and tenure. Indicators such as university education provide crude distinctions among people, and very imperfectly predicts the abilities and performance of individuals in a given work context. In settings where individuals' capabilities have tremendous impact on the success of firms, the crudeness of this widely used signal increases the incidence of two observed outcomes: hiring (or even internal promotion[8]) on the basis of prior observation or

---

8 Pfeffer (1977) and Rosenbaum (1981) noted empirical relationships between (a) socio-economic background and proxy measures for the closeness of social interaction with top

information from professional or personal networks (Granovetter, 1988, 1995); and entrepreneurial activity by those who feel undervalued by the employment market. On the other hand, in settings where the capabilities of importance are timing and attitude rather than training or background knowledge, and where the stakes of a poorly judged hire are fairly low, employers make use of stereotyped signals such as gender and ethnicity (Gordon, 1995; Daponte, 1996), particularly when these signals have been widely used in the particular industry and region, becoming part of the local expectations.

Observed patterns of wages and nonwage compensation can only be understood as a reflection of institutional norms and incentives. For example, Thurow (1983) proposed that the importance of informal relationships among co-workers to on-the-job training is an incentive to maintain seniority-based rules for wage and job security. More senior workers are a source of valuable, tacit knowledge about production, markets, and firm-specific procedures. They are more likely to pass this knowledge to incoming workers if their own positions and compensation are not threatened by lower wages offered to incoming workers.

Extended social networks have been shown to be a dominant source of information about job openings and job seekers (Granovetter, 1995; U.S. Department of Labor, 1975; Corcoran et al., 1980). The type of network (purely social versus work related) that is most relevant, and its importance relative to impersonal information routes vary with occupation and employment experience. The geographic scale of the job search and/or personal network of the searcher, and the scale of the recruitment mechanisms (formal, work-network, or personal network) of the employer varies tremendously by gender, race, and occupation (Hanson and Pratt, 1991, 1992).

3.3.4 Labor Control

For our purposes, "labor control" is the process by which the labor time offered by individuals becomes productive labor output. This involves the incentives to perform, monitoring of performance, control of the pace of work, and design of work tasks. Perhaps the most politically contentious part of the production process, the components of labor control take shape from formal and informal institutions of employment, pay, and divisions of labor. The outcome of these interactions affects regional economic development in at least three ways:

1. influencing labor productivity in the region's economic sectors, thereby affecting wage levels, wage increases, and regional attractiveness for capital investment;
2. becoming key components in the quality of work life for the employed population of a region; and

management and (b) rate of promotion within hierarchical organizations. Pfeffer suggested that this effect was greater in organizational settings where quantitative measures of employee productivity or skill was more difficult to observe (e.g., positions without direct responsibility for profitability and positions in service sectors).

3. shaping the reproduction of labor practices and labor expectations, with implications for skills acquisition, adaptability, and entrepreneurship.

From the microeconomic perspective, labor control represents transaction costs in the hiring and use of labor, which costs reduce the surplus from labor, manifested in foregone profits or wages. Institutional arrangements of control can be judged by the costs expended to measure and provide incentives for worker output, when in fact, the productivity of each worker is dependent on the actions of others (North, 1990). Options for labor control abound. Instititionalist analysis has yielded the concept of internal labor markets for "core" workers (Doeringer and Piore, 1971), through which employee performance is monitored within an organization, determining the rate and nature of pay raises and promotions. Gordon (1995) distinguished between such "intensive" forms of worker selection and labor control and more "extensive" forms such as hiring with little effort and using the threat of firing as a blatant sanction. While most industrial sociology has based its concepts and study in manufacturing, Coombs and Green (1989) and Kuhn (1989) provided organization-specific observations of work processes in health and financial services, respectively. These very different cases draw explicit attention to the roles of institutional context and change (labor markets, competitive strategy, modes of client payment) in determining how new technologies affect the occupational division of labor. More broadly, the nature of the skills brought to the work task (specific to the organization versus brought to the job by workers), public regulation of employee terminations and employee representation, informal conventions regarding wage reductions, and formal agreements between employer and employee organizations interact to determine the forms of labor control used in particular sectors, occupations, and nations or regions.

The design of work tasks, and the distribution of power over that design, has received much attention in industrial sociology, but little in analysis of regional productivity or regional development. Yet these issues are critical for the embodiment and development of skills and knowledge in a region's workforce. Traditional labor process theory (Braverman, 1974) postulates that capitalist development entails continually finer divisions of labor and embodiment of technology in machinery, for the purpose of reducing reliance on worker-embodied skills. Employers are thus able to change to less-skilled and less-experienced workers, by replacing workers *in situ* or by geographic relocation, paying for only minimal worker skills. Labor-process theory holds this goal to be a major motivation of technological change. The outcome of this process, especially insofar as that outcome is systematically supported by institutions, affects the reproduction of labor skill and autonomy in the region.

The result, dependent on institutional arrangements, is the redundancy of workers, the reskilling of those same workers, or the replacement of clerical workers with others (perhaps with accompanying shifts in location, tenure, gender, and work conditions). Is it possible to modify institutional arrangements so that the microeconomic outcome of technological change corresponds to local employment needs?

3.3.5 Local Labor Regimes

It is obvious that these several labor processes are interrelated, and co-determine each other's outcomes. In addition, the public and social, formal and informal institutions that shape these processes interact. Some of these institutions are national or supra-national – some increasingly so, such as firms' placing local labor forces in competition with one another through flexible sourcing arrangements. Other institutions are local, and movements toward national-government programmatic devolution provide some capacity for increasing localization, within the context of powerful national and global forces (Staeheli et al., 1997). The combination can be understood as the institutional ensemble that creates the localness of labor processes and outcomes. Recognizing the inter-dependencies that develop among institutional arrangements in a local labor market, Jonas (1996:328) introduced the concept of a "local labor-control regime, ...the gamut of practices, norms, behaviors, cultures and institutions within a locality through which labor is integrated into production." His use of the term "regime" reflects an attempt to link local labor practices to national and international regulation of production. The phrase "regional memory," introduced above, reflects the intertemporal dimension of these local/global relationships.

3.4 Overall Implications

3.4.1 Problems of Operationalization

Assessing local labor regimes. The task of an institutional regional analysis is to generate guidelines for assessing the key institutions and interactions underlying each of the labor processes. Accomplishing this in an exhaustive fashion would be difficult or impossible. However, analysis can begin with an assessment of the *outcomes* of these labor processes: sex- and age-group-specific labor force participation rates; distribution and unemployment rates by occupation; prevailing wage levels by sex and occupation, relative to national averages; proportion of employment under terms negotiated with labor organizations; cyclicality of employment by sector and occupation; numbers and numerical trends of self-employed or independent-contractor residents, by sex and occupation. Institutions that are formalized into or organized by organizations (religious doctrine, patterns of formal schooling and training, regulation of and by labor unions) and the state (tax and social-assistance policies; official rules for hiring and employment relationships) are readily identifiable, and provide the broad outlines of the overall institutional context for regional labor processes. Amin and Thrift (1994) developed the concept of "institutional thickness" as a norm for assessing the presence, number, interaction, coalitions, and agenda-sharing of labor, production, and marketing organizations within a region.

Identifying institutional incongruence. Key to such a regional analysis is a non-functionalist view of institutional arrangements. Institutions relevant to labor processes may be incongruent, reflecting: contradictions among formal and informal, or public and private institutions; contradictions in the dominant institutions of labor reproduction and labor demand; contradictions between regional memory of labor processes and current local and inter-local conditions. To the extent feasible, analysis of regional institutional ensembles should identify points of incongruence, and develop interventions that reduce the incongruence. Key to an institutional analysis of local labor regimes are the incongruences between labor reproduction and labor demand. The prevailing social and labor-control norms may not provide for the entry, migration, training, or tenure of appropriate workers for the current or changing production structure of the region.

Comparative analysis provides one way to identify institutional incongruence. Greif (1998) defined "historical and comparative regional analysis" as inductive analysis of formal and informal institutions in particular places and times, their interaction among themselves and with exogenous features to yield observably distinct outcomes in the two contexts. Using primary historical sources, Guinnane (1994) compared the experience of turn-of-the-century rural credit cooperatives in Ireland with that of the late-nineteenth-century German cooperatives after which they were explicitly modeled. He was thus able to identify important institutional differences that help explain the relative failure of the Irish cooperatives. In similar fashion, Greif (1994) compared the development of informal social networks versus formal organizations among different groups of Mediterranean traders in the eleventh through thirteenth centuries: once trade expanded beyond a certain point, the transaction cost of formal organizations was outweighed by the greater capacity of these institutional arrangements. Using published studies, DiGiovanna (1996) illustrated a comparative approach to the labor reproduction and allocation processes among the Emilia-Romagna, Baden-Württemburg, and Silicon Valley industrial regions.

3.4.2 Opportunities for Intervention

Public recognition of inappropriate or undesired labor outcomes (with respect to unemployment, wages, participation, household relationships, skills, productivity) yields calls for action. Routes of intervention depend on (a) the assessment of labor outcomes and (b) the understanding of the institutions influencing each of the key processes. The purpose of an intervention strategy is to induce institutional changes to bring the outcomes closer to desired goals. Again, simple measures may be attempted without a full assessment of outcomes or any attention to institutional relationships. However, measures such as changing the minimum wage, increasing training subsidies, or changing workers' rights to group negotiation can and do have unforeseen (or insufficient) effects when (changing) local institutional contexts are ignored.

What routes can such action take? Interventionist government policy has fallen out of favor, except insofar as local and national governments have been restricting the power of labor organizations and providing subsidy to local capital investment. However, national and subnational governments worldwide have begun to pay increased attention to "active intervention through schemes such as support for small firm networks, upgrading of work conditions and industrial relations, and industrial innovation" (Amin and Thrift, 1995:46). Public organizations (government and NGO, at local, regional, and national scales) comprise a key point at which to engage and modify local institutional practices, but must recognize the salience of the institutions themselves, which are the province of public and social organizations and "non-organizations".

Labor issues are politically charged, as they affect the power of individuals within households, of individuals and households vis-á-vis potential employers, and of organizations that play a role in labor processes. Power is augmented by a party's explicit recognition that labor markets themselves are structured by institutional interactions, and that institutions can be modified. Following that step, organizations can assess their individual power to enforce institutional change, or can enter into alliance relationships to accomplish change. In either case, institutional roles must be perceived clearly. Because institutional change inevitably favors some interests over others (Hudson, 1994), identification of goals and effects may lead to agreements for political or economic compensation to the interests that are harmed.

A fundamental characteristic of institutionalist analysis and, thus, institution-based intervention, is its context dependence. The local labor regime defines the uniqueness of local labor processes, and any intervention acts within the unique regime. Further, the interventions will inevitably be indirect and negotiated (Boekema and Nagelkerke, 1990). For example, the state can only act through its organizations and its work with other entities, none of which are the actual institutions (Amin and Thrift, 1995). The private, not-for-profit (or "third") sector must also be engaged for its non-public, quasi-capitalist role (Morgan 1997), but it too is composed of organizations that reflect and potentially affect the institutions relevant to labor processes and regional development.

3.4.3 Evolution of Regional Institutions and Labor Processes

Just as institutional components interact to yield region-specific labor processes in the midst of supra-regional regulation and movement of capital, labor, and technology, *change* in these local, regional, or larger practices or regulation affect the local processes – subject to the conservative influence of established institutions, organizations, and regional "memory". Change in local labor-related processes is jointly determined by local and supra-local conditions, and by formal, informal, public, and private institutions. Joint determination does raise the clear possibility for change to originate at the local scale, despite powerful supra-regional forces. Labor relationships and their change are ultimately local and historical (Cooke and Morgan, 1994; Massey, 1995), through the institutional arrangements presented in this chapter. Therefore, generalization and prediction are impossible: that is why this chapter has adopted a regional-case-study approach to analyses of labor processes.

The institutionally focused analysis proposed in this chapter provide many points of possible intervention by private, public, and quasi-public organizations. It is even likely that some of the interventions, undertaken by different entities for different purposes, will have countervailing influences. Of necessity, assessments of regional labor processes and attempts at intervention aim at moving targets, and interact with trends and forces outside the region. All of this makes the task more difficult, but it may also make the outcomes more forgiving: there cannot be a one-to-one correspondence between actions taken and ultimate results.

Acknowledgements

The authors thank Ian Gordon and other participants in the Uddevalla workshop for their helpful comments and suggestions on the paper on which this chapter is based.

References

Amin, A. and N. Thrift, 1994, "Living in the Global", in A. Amin and N. Thrift (eds.), *Globalization, Institutions, and Regional Development in Europe*. Oxford University Press, Oxford.

Amin, A. and N. Thrift, 1995, "Institutional Issues for the European Regions: From Markets and Plans to Socioeconomics and Powers of Association", *Economy and Society,* Vol. 24 1:41-66.

Amrhein, C.G., 1985, "Interregional Labour Migration and Information Flows. *Environment and Planning A,* 17:1111-1126.

Amrhein, C.G. and R.D. MacKinnon, 1985, "An Elementary Simulation Model of the Job Matching Process within an Interregional Setting", *Regional Studies,* 19:193-202.

Arrow, K.J., 1973, "Higher Education as a Filter", *Journal of Public Economics,* 2:193-216.

Axelrod, R.M., 1984, *The Evolution of Cooperation*, Basic Books, New York.

Beyers, W.B. and D.P. Lindahl, 1997a, "Endogenous Use of Occupations and Exogenous Reliance on Sectoral Skills in the Producer Services", Paper prepared for the RESER conference, Roskilde, Denmark, September.

Beyers, W.B. and D.P. Lindahl, 1997b, "Strategic Behavior and Development Sequences in Producer Service Businesses", *Environment and Planning A,* 29:887-912.

Boekema, F. and A. Nagelkerke, 1990, "Labor Relations and Regional Development in the Netherlands: A Network Approach", in F. Dietz, et al. (eds.), *Location and Labor Considerations for Regional Development*, Avebury, Aldershot.

Braverman, H., 1974, *Labor and Monopoly Capital*, Monthly Review Press, London.

Bush, P.D., 1987, "The Theory of Institutional Change." *Journal of Economic Issues,* Vol. 21, 3:1075-1116.

Castells, M., 1996, *The Rise of the Network Society,* Blackwell, London.

Cooke, P. and K. Morgan, 1994, "Growth Regions under Duress: Renewal Strategies in Baden Württemberg and Emilia-Romagna", in A. Amin and N. Thrift (eds.), *Globalization, Institutions and Regional Development in Europe,* Oxford University, Press, Oxford.

Coombs, R. and K. Green, 1989, "Work Organization and Product Change in the Service Sector: The Case of the UK National Health Service", in S. Wood, (ed.), *The Transformation of Work?* Routledge, London.

Corcoran, M., L. Datcher and G. Duncan, 1980, "Most Workers Find Jobs Through Word of Mouth", *Monthly Labor Review,* August, 33-35.

Daponte, B.O., 1996, "Race and Ethnicity During an Economic Transition: The withdrawal of Puerto Rican Women from New York City's Labour Force, 1960-1980", *Regional Studies,* Vol. 30, 2:151-166.

DiGiovanna, S. 1996, "Industrial Districts and Regional Economic Development: A Regulation Approach", *Regional Studies,* Vol. 30, 4:373-386.

Doeringer, P.B. and M.J. Piore, 1971, Internal Labor Markets and Manpower Analysis, D.C. Heath, Lexington.

Gordon, I. 1995, "Migration in a Segmented Labour Market", *Transactions of the Institute of British Geographers NS,* 20:139-55.

Granovetter, M., 1985, "Economic Action and Social Structure: The Problems of Embeddedness", *American Journal of Sociology,* 91:481-510.

Granovetter, M., 1988, "The Sociological and Economic Approaches to Labor Market Analysis: A Social Structural View", in G. Farkas and P. England (eds.), *Industries, Firms, and Jobs: Sociological and Economic Approaches,* Plenum Press, New York.

Granovetter, M., 1995, *Getting a Job: A Study of Contacts and Careers*, 2nd edition University of Chicago Press, Chicago.

Greif, A., 1994, "Cultural Beliefs and the Organization of Society: A Historical and Theoretical Reflection on Collectivist and Individualist Societies", *Journal of Political Economy,* Vol. 102, 5:912-950.

Greif, A., 1998, "Historical and Comparative Institutional Analysis", *American Economic Association Papers and Proceedings,* Vol. 88, 2: 80-84.

Guinnane, T.W., 1994, "A Failed Institutional Transplant: Raiffeisen's Credit Cooperatives in Ireland, 1894-1914", *Explorations in Economic History,* 31:38-61.

Hanson, S. and G. Pratt, 1991, "Job Search and the Occupational Segregation of Women", *Annals of the Association of American Geographers,* Vol. 81, 2:229-253.

Hanson, S. and G. Pratt, 1992, "Dynamic Dependencies: A Geographic Investigation of Local Labor Markets", *Economic Geography,* Vol. 68, 4:373-405.

Herod, A., 1995, "The Practice of International Labor Solidarity and the Geography of the Global Economy, *Economic Geography,* Vol. 71, 4:341-363.

Hodgson, G.M. and E. Screpanti (eds.), 1991, *Rethinking Economics: Markets, Technology, and Economic Evolution*, Edward Elgar.

Hudson, R. 1994, "Institutional Change, Cultural Transformation, and Economic Regeneration: Myths and Realities from Europe's Old Industrial Areas", in A. Amin and N. Thrift (eds.), *Globalization, Institutions, and Regional Development in Europe,* Oxford University Press, Oxford.

Jonas, A.E.G., 1996, "Local Labour Control Regimes: Uneven Development and the Social Regulation of Production", *Regional Studies,* Vol. 30, 4:323-338.

Kuhn, S., 1989, "The Limits to Industrialization: Computer Software Development in a Large Commercial Bank, in S. Wood, (ed.) *The Transformation of Work?* Routledge, London.

Lucas, R., 1988, "On the Mechanics of Economic Development", *Journal of Monetary Economics,* 22:3-42.

Maier, G., 1987, "Job Search and Migration", in M. Fischer and P. Nijkamp (eds.), *Regional Labour Markets,* North-Holland, Amsterdam.

Massey, D., 1995, *Spatial Divisions of Labour: Social Structures and the Geography of Production*, 2nd edition, Basingstoke, Macmillan, Hampshire.

Mayhew, A., 1987, "The Beginnings of Institutionalism", *Journal of Economic Issues* Vol. 21, 3:971-998.

Morgan, K., 1997, "The Learning Region: Institutions, Innovation and Regional Renewal", *Regional Studies,* Vol. 31, 5:491-503.

Nee, V., 1998, "Norms and Networks in Economic and Organizational Performance", *American Economic Review,* Vol. 88, 2:85-89.

Nelson, P.B., 1997a, "Migration, Sources of Income, and Community Change in the Nonmetropolitan Northwest", *The Professional Geographer*, Vol. 49, 4:418-431.

Nelson, P.B., 1997b, "A Multivariate Analysis of Income Structure in the Nonmetropolitan West", Paper prepared for the Nothern. American meetings of the Regional Science Association International, Buffalo NY, November.

North, D., 1990, *Institutions, Institutional Change, and Economic Performance,* Cambridge University Press, Cambridge.

Peck, J., 1994, "Regulating Labour: The Social Regulation and Reproduction of Local Labour-Markets", in A. Amin and N. Thrift (eds.), *Globalization, Institutions, and Regional Development in Europe,* Oxford University Press, Oxford.

Peck, J., 1996, *Work-Place: The Social Regulation of Labor Markets,* Guilford Press, New York.

Pfeffer, J., 1977, "Toward and Examination of Stratification in Organizations", *Administrative Science Quarterly,* 22:553-567.

Polanyi, K., C.M. Arensberg and H.W. Pearson, 1957, Trade and Market in the Early Empires, The Free Press.

Rogerson, P.A. and R.D. MacKinnon, 1981, "A Geographical Model of Job Search, Migration and Unemployment", *Papers of the Regional Science Association,* 48:89-102.

Romer, P.M., 1986, "Increasing Returns and Long-Run Growth", *Journal. of Political Economy,* Vol. 94, 5:1002-1037.

Rosenbaum, J., 1981, "Careers in a Corporate Hierarchy", in D. Treiman and R. Robinson (eds.), *Research in Social Stratification and Mobility,* Vol. 1, JAI Press, Greenwich.

Sabel, C.F., 1992, "Studied Trust: Building New Forms of Co-operation in a Volatile Economy", in F. Pyke and W. Sengenberger, (eds.), *Industrial Districts and Local Economic Regeneration,* International Institute for Labor Studies (International Labor Organization).

Salais, R., 1992, "Labor Conventions, Economic Fluctuations, and Flexibility", in M. Storper and A.J. Scott (eds.), *Pathways to Industrialization and Regional Development*, Routledge, London.

Saxenian, A., 1994, *Regional Advantage: Culture and Computing in Silicon Valley and Route 128,* Harvard University Press, Cambridge Mass.

Schubert, U., S. Gerking, I. Isserman and C. Taylor, 1987, "Regional Labor Market Modeling: A State of the Art Review", in M. Fischer and P. Nijkamp (eds.), *Regional Labour Markets,* North-Holland, Amsterdam.

Scott, A.J., 1988, *Metropolis: From the Division of Labor to Urban Form,* University of California Press, Berkeley.

Staeheli, L.A., J.E. Kodras and C. Flint, (eds.), 1997, *State Devolution in America: Implications for a Diverse Society,* Sage Publications.

Standing, G., 1992, "Alternative Routes to Labor Flexibility", in M. Storper and A.J. Scott (eds.), *Pathways to Industrialization and Regional Development,* Routledge, London.

Stanfield, J.R., 1998, "The Scope, Method, and Significance of Original Institutional Economics." Unpublished paper, Department of Economics, Colorado State University, Ft. Collins, Colorado.

Stiglitz, J.E., 1985, "Information and Economic Analysis: A Perspective", *Economic Journal,* 95:21-41.

Storper, M., 1995, "The Resurgence of Regional Economies, Ten Years Later: The Region as a Nexus of Untraded Interdependencies", *European Urban and Regional Studies,* Vol. 2, 3:191-221.

Storper, M., 1997, *The Regional World: Territorial Development in a Global Economy,* Guilford, New York.

Storper, M. and R. Walker, 1983, "The Theory of Labour and the Theory of Location", *International Journal of Urban and Regional Research,* 7:1-41.

Swyngedouw, E.A., 1992, "'Glocalization,' Interspatial Competition and the Monetary Order", in M. Dunford and G. Kafkalas (eds.), *Cities and Regions in the New Europe,* Belhaven, London.

Teece, D.J., 1986, "Profiting from Technological Innovation", *Research Policy,* 15:285-305.

Thurow, L.C., 1983, *Dangerous Currents: The State of Economics,* Random House, New York.

U.S. Department of Labor, 1975, "Jobseeking Methods Used by American Workers". Bureau of Labor Statistics Bulletin 1886.

Veblen, T.B., 1899, *The Theory of The Leisure Class: An Economic Study Of Institutions*, Macmillan.

Villa, P., 1986, *The Structuring of Labour Markets: A Comparative Analysis of the Steel and Construction Industries in Italy*, Clarendon Press, Oxford.

4 Regional Dynamics with Endogenous Knowledge in a Multi-Sector Model

Wei-Bin Zhang
Dept. of Economics, National University of Singapore
10 Kent Ridge Crescent, Singapore 119260

4.1 Introduction

The dynamic relationship between the interregional division of labor, division of consumption and the determination of price structures in the global economy is a main concern of contemporary economics. In fact, economists have produced different trade theories, which deal with various aspects of international and interregional trade. Although Adam Smith held that a country could gain from free trade, he failed to create a convincing economic theory of international trade. It is generally agreed that David Ricardo is the creator of the classical theory of trade. The theories of comparative advantage and the gains from trade are usually connected with Ricardo. In this theory the crucial variable used to explain international trade patterns is technology. The theory holds that a difference in comparative costs of production is the necessary condition for the existence of international trade. But this difference reflects a difference in the techniques of production. According to this theory, technological differences between countries determine the international division of labor and consumption and trade patterns. It concludes that trade is beneficial to all participating countries. This conclusion is in contradiction to the viewpoint about trade held by the doctrine of mercantilism. In mercantilism regulation and planning of economic activity are held to be efficient means of fostering the goals of a nation. This theory assumes that production costs are independent from factor prices and the composition of output. Since it lacks consideration of the demand side, the theory cannot determine how and at what value the terms of trade are or should be. The Ricardian theory failed to determine the terms of trade, even though it can be used to determine the limits in which the terms of trade must lie. It was recognized long ago that in order to determine the terms of trade, it is necessary to build trade theory that not only takes account of the productive side but also the demand side.

The Ricardian model and Heckscher-Ohlin model are two basic models of trade and production. They provide the pillars upon which much of pure theory of international trade rests. The so-called Heckscher-Ohlin model has been one of the dominant models of comparative advantage in modern economics. The Heckscher-Ohlin theory emphasizes the differences between the factor endowments of different countries and differences between commodities in the intensities with which they use these factors. The basic model deals with a long-term general equilibrium in which the two factors are both mobile between sectors and the cause of trade is different between countries with different relative factor endowments. This theory deals with the impact of trade on factor use and factor rewards. The theory is different from the Ricardian model that isolates differences in technology between countries as the basis for trade. In the Heckscher-Ohlin theory costs of production are endogenous in the sense that they are different in the trade and autarky situations, even when all countries have access to the same technology for producing each good. This model has been a main stream of international trade theory. The Heckscher-Ohlin theory provides simple, intuitive insights into the relationships between commodity prices and factor prices, factor supplies and factor rewards, and factor endowments and the pattern of production and trade.

It is well known that one topic that was almost entirely absent from the pure theory of international trade was any consideration of the connection between economic growth and international trade in classical literature of economic theory. Almost all trade models developed before the 1960s are static in the sense that the supplies of factors of production are given and do not vary over time; the classical Ricardian theory of comparative advantage and the Heckscher-Ohlin theory are static since labor and capital stocks (or land) are assumed to be given and constant over time. Although Marshall held that it is important to study international trade in order to be clear of the causes which determine the economic progress of nations, it has only been in the last three or four decades that trade theory has made some systematic treatment of endogenous capital accumulation or technological change in the context of international economics. This paper is concerned with interdependence between interregional trade patterns and knowledge accumulation.

Classical economists such as Adam Smith, Marx, Marshall and Schumpeter, emphasized various aspects of knowledge in economic dynamics. But there were only a few formal economic models which deal with interdependence between interregional economic growth and knowledge accumulation. But recently there have been a rapidly increasing number of publications in the theoretical economic literature concerning the relationship between knowledge accumulation and regional economic development. Issues related to regional economic growth with endogenous knowledge have recently been modeled (e.g., Krugman, 1991; Rauch, 1991; Lucas, 1986; Andersson and Mantsinen, 1986; Johansson and Karlsson, 1994). But it may be argued that many interregional aspects remain to be further examined. In this study, we propose a two-region and multi-sector model of perfect competition with endogenous knowledge to examine the role of knowledge in the interregional division of labor and consumption.

In addition to explicitly modeling endogenous knowledge, we also take account of some other important aspects of economic geography. We emphasize the geographical character of services. Services are consumed simultaneously as they are produced and thus cannot be transported like commodities. Accordingly, when explicitly modeling economic geography, we have to take account of this special character of services. Many services such as those provided by schools, hospitals and restaurants have to be consumed where they are supplied. Accordingly, services have a special location property in comparison to commodity production. We also try to explore the role of regional amenity in economic geography. Amenities such as climate and historical buildings are space-fixed (at most, slowly changing) and have an important impact on residential location. Climates may affect people in multiple ways. It is well observed in many countries that productivity and preferences exhibit different between, for instance, cold and warm regions. It may be argued that the emphasis on geographical characteristics of services and amenity is important to explain the so-called capitalization, which implies that the price of land is interdependent with local amenities, economic agents' densities, transportation costs and other local variables or parameters. Although the significance of capitalization has been noticed by location theorists (e.g., Scotchmer and Thisse, 1992), it may be said that we still have no compact framework within which we can satisfactorily explain the issue. The paper on compensating regional variation in wages and rents by Roback (1982) has caused a wide interest among regional and urban economists to theoretically investigate how the value of locational attributes is capitalized into wages and services. Since the publication of Roback's model, many empirical and theoretical studies have also shown that between urban areas wages may capitalize on differences in amenity levels or living costs (e.g., Simon and Love, 1990; Bell, 1991; Voith, 1991). In this study we show how factors such as knowledge utilization efficiency and amenity may affect the land values of different regions.

The remainder of the paper is organized as follows. Section 4.2 defines a two-region and multi-sector economic model with endogenous knowledge accumulation. Section 4.3 shows that a dynamic system may have either a unique or multiple equilibria and each equilibrium may be either stable or unstable, depending upon knowledge utilization and creativity of the different sectors. Section 4.4 examines the effects of changes in the efficiency of knowledge accumulation and amenities upon the equilibrium economic geography. Section 4.5 concludes the study. The appendix offers the proof of the results presented in Section 4.3.

4.2 The Basic Model

To treat economic geography as an endogenous and contextually determined entity, we concentrate our study on an isolated state. This method was initiated by von Thünen (1826). The isolated state consists of two regions, indexed by 1 and 2, respectively. The system produces two commodities, indexed by 1 and 2, respectively. Each region is assumed to produce only one commodity. Each region has two production sectors: industry and services. Services are produced by combining knowledge, labor and land. Industrial production is regionally specified. Region j produces commodity j. It is assumed that each region's product is homogeneous. We assume a homogenous and fixed national labor force, N. We neglect any cost for migration and professional changes. We assume perfect competition in all markets.

Each region has fixed land. The land is distributed between the service sector and housing by perfect competition. We select commodity 1 to serve as numeraire, with all the other prices being measured relative to its price. For simplicity of analysis, we neglect transportation cost of commodities between and within regions. It can be shown that we may introduce transportation costs for commodities by taking Samuelson's "iceberg" form, in which transport costs are incurred in the good transported. It should be noted that by adding transportation costs some of our results on geographical pattern formation may become stronger. As our model is already too complicated, we will assume zero transportation costs at this initial stage. The assumption of zero transportation cost of commodities implies price equality for any commodities between the two regions. As amenity and land are immobile, wage rates and land rent are not necessarily identical between the two regions.

We introduce

i, s	subscript index for industry and services, respectively;
N and L_j	the fixed total population and region j's territory size, j = 1, 2;
$F_{ij}(t)$ and $F_{sj}(t)$	the output levels of region j's industrial and service sectors at time t, j = 1, 2;
$N_{ij}(t)$ and $N_{sj}(t)$	the labor force employed by region j's industrial and the service sectors;
$N_j(t)$	region j's population j;
$L_{sj}(t)$ and $L_{hj}(t)$	the land size used by the service sector and for housing in region j;
Z(t)	the knowledge stock of the national economy;
p(t) and $p_j(t)$	price of commodity 2 and price of services in region j; and
$w_j(t)$ and $R_j(t)$	region j's wage rate and land rent in region j.

4.2.1 Production Sectors

First, we describe behavior of the production sectors in the system. We assume that service production is to combine knowledge, labor and land. We specify the production functions of the service sectors as follows

$$F_{sj},= Z^{m_{sj}} L_{sj}^{\alpha} N_{sj}^{\beta}, \ \alpha, \beta > 0, \ \alpha + \beta = 1, \ m_{sj} \ge 0, \ j = 1,2. \tag{4.1}$$

Similarly to Grossman and Helpman (1991), we only use labor and knowledge as input factors. It should be noted that to introduce other factors such as capital may cause great analytical difficulties. We use variable Z(t) to measure, in an aggregated sense, stock of knowledge in the society at time t. A similar variable has been widely used in the recent literature of growth and knowledge. In this study, knowledge is treated as a public good in the sense that utilization of knowledge by any agent in the system will not affect that by any other. We assume that knowledge utilization efficiency varies spatially and professionally. The parameter, m_{sj}, describes knowledge utilization efficiency. We call m_{sj} the knowledge utilization efficiency parameter of region j's service sector. There are obviously limitations to knowledge as public goods in modeling the innovation and imitation processes. But to model the complexity of private and public characteristics of knowledge, it is necessary to further disaggregate the knowledge variable into multiple components. This will result in high-dimensional dynamic problems.

The marginal conditions of service production are given by

$$R_j = \frac{\alpha p_j F_{sj}}{L_{sj}}, \quad w_j = \frac{\beta p_j F_{sj}}{N_{sj}}, \quad j = 1, 2. \tag{4.2}$$

We specify the two regions' industrial production functions as follows:

$$F_{ij} = Z^{m_{ij}} N_{ij}, \quad m_{ij} \geq 0, \quad j = 1, 2, \tag{4.3}$$

where m_{ij} are the knowledge utilization efficiency parameter of region j's industrial sector.

The conditions of zero-profit are given by

$$w_1 = Z^{m_{i1}}, \quad w_2 = pZ^{m_{i2}}. \tag{4.4}$$

4.2.2 Consumers' Behavior

We assume that a household's utility in a given region is dependent on the amenity level, per household's consumption levels of industrial commodities and housing conditions. We measure housing conditions by lot size (e.g., Alonso, 1964). The utility function is specified as follows:

$$U_j(t) = \frac{A_j C_{sj}^{\gamma} C_{1j}^{\xi_1} C_{2j}^{\xi_2} L_{hj}^{\eta}}{N_j}, \quad j = 1, 2, \quad \gamma, \xi_1, \xi_2, \eta > 0, \quad \gamma + \xi_1 + \xi_2 + \eta = 1, \tag{4.5}$$

in which $C_{sj}(t)$, $C_{1j}(t)$ and $C_{2j}(t)$ are respectively region j's consumption levels of services, commodity 1 and commodity 2 at time t. In (4.5), A_j denotes region j's amenity level. Amenity is an aggregated measure of regional livening conditions such as infrastructure status, regional culture and climate (e.g., Diamond and Tolley, 1981; Voith, 1991). Some locational amenities such as pollution level, residential density and transportation congestion may be dependent on economic agents' activities, while other amenities such as climate, transport structure, and historical buildings, may not be strongly affected by economic agents' activities, at least, within the short-run. Accordingly, in a strict sense, it is necessary to classify amenities into endogenous and exogenous categories when modeling economic geography. Which kinds of amenities should be classified as endogenous or exogenous also depends upon the time scale of the analysis and the economic system under consideration (e.g., Kanemoto, 1980).

The assumption that utility level is identical over space at any point of time is represented by

$$U_1(t) = U_2(t). \tag{4.6}$$

The consumer problem is defined by

$$\max U_j, \quad s.t.: p_j C_{sj} + C_{1j} + pC_{2j} + R_j L_{hj} = w_j N_j. \tag{4.7}$$

We have the following optimal solutions

$$C_{sj} = \frac{\gamma w_j N_j}{p_j}, \quad C_{1j} = \xi_1 w_j N_j, \quad C_{2j} = \frac{\xi_2 w_j N_j}{p}, \quad L_{hj} = \frac{\eta w_j N_j}{R_j}, \quad j = 1, 2. \tag{4.8}$$

The balances of demand for and supply of commodities are given by

$$C_{j1} + C_{j2} = F_{ij}, \quad C_{sj} = F_{sj}, \quad j = 1, 2. \tag{4.9}$$

4.2.3 Full Employment of Labor and Land

The assumption that labor force and land are fully employed is represented by

$$N_{ij}(t) + N_{sj}(t) = N_j(t), \quad N_1(t) + N_2(t) = N, \quad L_{sj}(t) + L_{hj}(t) = L_j, \quad j = 1, 2. \tag{4.10}$$

4.2.4 Knowledge Accumulation

Similar to Zhang (1992), we model knowledge accumulation as follows:

$$\frac{dZ}{dt} = \sum_{j=1}^{2} \left(\frac{\tau_{sj} F_{sj}}{Z^{\varepsilon_{sj}}} + \frac{\tau_{ij} F_{ij}}{Z^{\varepsilon_{ij}}} \right) - \delta_z Z, \tag{4.11}$$

in which δ_z is the fixed depreciation rate of knowledge, and τ_{sj} (≥ 0), τ_{ij} (≥ 0), ε_{sj} and ε_{ij}, j = 1, 2, are parameters.

We only take account of learning by doing effects in knowledge accumulation. The term, $\tau_{i1} F_{i1} / Z^{\varepsilon_{i1}}$, for instance, implies that contribution of region 1's industrial sector to knowledge is positively related to its production scale, F_{i1}, and is dependent on the current level of knowledge stocks. The term $Z^{\varepsilon_{i1}}$ takes account of returns to scale effects in knowledge accumulation. The case of $\varepsilon_{i1} > (<) \, 0$ implies increasing (decreasing) returns to scale in knowledge accumulation. The other three terms in (4.11) can be similarly interpreted. Knowledge accumulation may be affected by many factors in different ways (e.g., Lucas, 1988, Grossman and Helpman, 1991, Becker, Murphy and Tamura, 1990). The above specification takes account of one source of knowledge accumulation. We omit possible effects of R&D activities on knowledge accumulation.

We have thus built the model. The system has 30 variables, N_{sj}, N_{ij}, L_{hj}, L_{sj}, F_{sj}, F_{ij}, C_{sj}, C_{1j}, C_{2j}, U_j, N_j, w_j, p_j, R_j (j = 1, 2), p and Z. It contains the same number of independent equations. We now examine properties of the dynamic system.

4.3 Equilibria and Stability

This section is concerned with the conditions for the existence of economic equilibria and for stability. First, we show that for any given knowledge stock, Z(t), the division of labor and economic geography are uniquely determined at any specific time. The following proposition is proved in the appendix.

Proposition 4.1
For any given level of knowledge Z(t), all the other variables in the system are uniquely determined at any point of time. The values of the variables are given as functions of Z(t) by the following procedure: p by (A.16) $\to N_j$, j = 1, 2, by (A.5) $\to N_{sj}$ and N_{ij} by (A.1) $\to w_j$ by (A.7) $\to R_j$ by (A.8) $\to L_{hj}$ and L_{sj} by (A.10) $\to p_j$ by (A.11) $\to F_{sj}$ by (4.1) $\to C_{sj} = F_{sj} \to F_{ij}$ by (4.3) $\to C_{11} = \xi_1 w_1 N_1 \to C_{12}$ by (A.6) $\to U_j$ by (4.5).

The above proposition guarantees that if we can find knowledge Z(t), then we can explicitly solve all the other variables as functions of Z(t) in the system at any point of time. Hence, to examine the dynamic properties of the system, it is sufficient to examine the dynamic properties of (4.11). By the procedure given in proposition 4.1 we can represent $N_{sj}(t)$ and $N_{ij}(t)$ as functions of Z(t) as follows

$$N_{s1} = \gamma\beta\xi_1 ANZ^m\Psi, \;\; N_{i1} = (1-\gamma\beta)\xi_1 ANZ^m\Psi, \;\; N_{s2} = \gamma\xi_2\beta N\Psi,$$

$$N_{i2} = (1-\gamma\beta)\xi_2 N\Psi, \tag{4.12}$$

in which

$$A \equiv \left\{\left(\frac{\xi_2 L_1}{\xi_1 L_2}\right)^{\eta+\alpha\gamma}\frac{A_1}{A_2}\right\}^{\frac{1}{1-\gamma\beta}}, \;\; \Psi(Z) \equiv \frac{1}{\xi_2 + \xi_1 AZ^m}, \;\; m \equiv -\frac{m_s\gamma}{1-\gamma\beta},$$

$$m_s \equiv m_{s2} - m_{s1}.$$

Substituting these into the production functions (4.1) and (4.3), we get F_{ij} and F_{sj} as functions of Z. Substituting $F_{ij}(Z)$ and $F_{sj}(Z)$ into (4.11), we determine knowledge accumulation dynamics Z(t), as follows

$$\frac{dZ}{dt} = \sum_{j=1}^{2}\{\Omega_{sj}(Z) + \Omega_{ij}(Z)\}Z - \delta_z Z, \tag{4.13}$$

where

$$\Omega_{s1} = \tau_{s1}(\beta\xi_1\xi_2 AN)^\beta L_{s1}^\alpha Z^{x_{s1}}\Psi^\beta, \;\; \Omega_{i1} = v_1 Z^{x_{i1}}\Psi,$$

$$\Omega_{s2} = \tau_{s2}(\gamma\beta\xi_2 N)^\beta L_{s2}^\alpha Z^{x_{s2}}\Psi^\beta, \;\; \Omega_{i2} = v_2 Z^{x_{s2}}\Psi, \tag{4.14}$$

in which

$$v_1 \equiv (1-\gamma\beta)\tau_{i1}\xi_1 AN > 0,\ \ v_2 \equiv (1-\gamma\beta)\tau_{i2}\xi_2 N > 0,\ \ x_{s1} \equiv m_{s1} - m - \varepsilon_{s1} - 1,$$

$$x_{i1} \equiv m_{i1} - m - \varepsilon_{i1} - 1,\ \ x_{s2} \equiv m_{s2} - \varepsilon_{s2} - 1,\ \ x_{i2} \equiv m_{i2} - \varepsilon_{i2} - 1, \tag{4.15}$$

For simplicity of discussion, we make the following assumption.

Assumption 4.1.
In the remainder of this study, we require $\tau_{s1} = \tau_{s2} = 0$.

This requirement implies $\Omega_{s1} = \Omega_{s2} = 0$. That is, the service sectors make no contribution to knowledge accumulation. From the functional forms of Ω_{sj} and Ω_{ij} in (4.13) and (4.14) we see that the omission of service sectors' creativity does not affect essential conclusions of our model. If we don't make this omission, we have to discuss more cases of different combinations of the parameters. As these discussions will provide little new insight, we limit our analysis to the special case of $\Omega_{s1} = \Omega_{s2} = 0$.

First, we notice that the dynamic properties of the system are basically determined by the combination of the parameters, x_{i1}, x_{i2} and m. It can be seen that these three parameters may be either positive or negative. As $m = -(m_{s2} - m_{s1})\gamma/(1-\gamma\beta)$, we see that if region 1's service sector utilizes knowledge more (less) effectively than region 2's service sector, i.e., $m_{s1} > m_{s2}$ ($m_{s1} < m_{s2}$), then $m > 0$ ($m < 0$); if the two regions' knowledge utilization efficiency is equal, then $m = 0$. From $x_{i2} \equiv m_{i2} - \varepsilon_{i2} - 1$, if region 2's industrial sector is effective in knowledge utilization (i.e., m_{i2} being large) and its contribution to knowledge exhibits increasing returns to scale (i.e., $m_{i2} < 0$), then x_{i2} may be positive; if region 2's industrial sector is not effective in knowledge utilization and its contribution to knowledge exhibits decreasing returns to scale, then $x_{i2} < 0$. We can similarly discuss the sign of $x_{i1} \equiv m_{i1} - m - \varepsilon_{i2} - 1$. We say that when $x_{ij} > (<)\ 0$, region j's industrial sector exhibits increasing (decreasing) returns to scale effects in knowledge accumulation.

We can thus conclude that the parameters x_{ij} and m may be either positive or negative, depending upon various combinations of knowledge utilization efficiency and creativity of different economic sectors in the two regions. The following proposition shows that the properties of our one-dimensional differential equation, (4.13), are dependent on how these knowledge parameters are combined.

Proposition 4.2

i) If $x_{i1} > 0$, $x_{i2} > 0$ and $m < 0$, the system has a unique unstable equilibrium;
ii) $x_{i1} < 0$, $x_{i2} < 0$ and $m > 0$, the system has a unique stable equilibrium; and
iii) In each of the other six combinations of $x_{i1} < 0$ or $x_{i1} > 0$, $x_{i2} < 0$ or $x_{i2} > 0$ and $m < 0$ or $m > 0$, the system has either no equilibrium or multiple ones.

The above proposition is proved in the appendix. We interpreted the meanings of x_{i1}, x_{i2} and m. From these discussions we can directly interpret the conditions in the above proposition as implying that if the two regions exhibit increasing (decreasing) scale effects in knowledge accumulation, the system has an unstable (stable) unique equilibrium; if one region exhibits increasing scale effects in knowledge accumulation but the other region decreasing ones, the system has two equilibria.

From propositions 4.1 and 4.2, we see that for given parameter values, we can explicitly determine the properties of the dynamic system. For simplicity, we were only concerned with the cases that none of the parameters, x_{i1}, x_{i2} and m, is equal to zero. It is not difficult to discuss the cases that one or two of the three parameters are equal to zero. The following corollary, summarizing the properties of the dynamic system in the case of $m = 0$, is proved in the appendix.

Corollary 4.1
Let $m = 0$. Then we have

i) if $x_{ij} > 0$, j = 1, 2, the system has a unique unstable equilibrium;
ii) if $x_{ij} < 0$, j = 1, 2, the system has a unique stable equilibrium;
iii) if $x_{i1} < 0$ and $x_{i2} > 0$ (or $x_{i1} > 0$ and $x_{i2} < 0$), the system has two equilibria – the one with higher value of knowledge is unstable and the other one is stable.

In the remainder of this study, we will examine the impact of some parameter changes on the equilibrium structure of the economic geography.

4.4 The Impact of the Parameters, τ_{ij}

This section is concerned with impact of changes in knowledge accumulation efficiency parameters, τ_{ij}, j = 1, 2, of the industrial sector in region j on the system.

First, from (A.20) we get the impact of changes in τ_{ij} on Z as follows

$$-\Phi^{*\prime}\frac{dZ}{d\tau_{ij}} = \frac{\Phi_j}{\tau_{ij}} > 0, \quad j = 1, 2, \tag{4.16}$$

where Φ' is defined in (A.19). It is easily seen that the sign of $\Phi^{*\prime}$ may be either positive or negative, depending on the parameter values of x_{ij}. In the case of $m = 0$, we directly have the following three cases: (i) if $x_{ij} > 0$, j = 1, 2, $\Phi^{*\prime} > 0$ at the unique equilibrium; (ii) if $x_{ij} < 0$, j = 1, 2, $\Phi^{*\prime} < 0$ at the unique equilibrium; (iii) if $x_{i1} < 0$ and $x_{i2} > 0$ (or $x_{i1} > 0$ and $x_{i2} < 0$), $\Phi^{*\prime} > (<)\, 0$ at the equilibrium with the high (low) value of Z. We thus conclude that the sign of Φ' is dependent on whether each industrial sector exhibits increasing or decreasing returns. In case (i), we have $dZ/d\tau_{ij} < 0$. In case (ii), we have $dZ/d\tau_{ij} > 0$. In case (iii), we have $dZ/d\tau_{ij} <(>)\, 0$ at the equilibrium with high (low) values of Z. We see that an improvement in knowledge accumulation efficiency may either increase or decrease the equilibrium value of knowledge. For instance, in case (ii) which means that the two industrial sectors exhibit decreasing returns, an improvement in knowledge accumulation efficiency of any region's industrial sector increases the equilibrium level of knowledge. In case (i), the impact in changes of knowledge accumulation efficiency is the opposite to that in the case of stability. We see that the effects of changes in the parameters on the system are significantly dependent upon the stability conditions. To explain how the new equilibrium given by (4.16) is achieved through a dynamic process, we have to examine all the relations connecting the 30 variables in the system. We omit the explanation of (4.16) in detail.

From (A.16) we directly have

$$\frac{Z}{p}\frac{dp}{d\tau_{ij}} = (m - m_i)\frac{dZ}{d\tau_{ij}}, \tag{4.17}$$

in which $m \geq 0$. In the remainder of this section, we assume: $x_{ij} < 0, j = 1, 2$. We have: $dZ / d\tau_{ij} > 0$. If $m_i < 0$, i.e., $m_{i1} > m_{i2}$, the price of commodity 2 is increased. This implies that when region 1's two sectors utilize knowledge more effectively than the two sectors in region 2, then an increase in knowledge increases the price level of commodity 2. In the case of $m_i > 0$, from (4.17) $dp / d\tau_{ij}$ may be either positive or negative.

Taking derivatives of (A.5) and (A.1) with respect to τ_{ij} yields

$$\frac{Z}{N_1}\frac{dN_1}{d\tau_{ij}} = m\xi_2\frac{dZ}{d\tau_{ij}} > 0, \quad \frac{dN_2}{d\tau_{ij}} = -\frac{dN_1}{d\tau_{ij}} < 0,$$

$$\frac{1}{N_{sk}}\frac{dN_{sk}}{d\tau_{ij}} = \frac{1}{N_{ik}}\frac{dN_{ik}}{d\tau_{ij}} = \frac{1}{N_k}\frac{dN_k}{d\tau_{ij}}, \quad k = 1, 2. \tag{4.18}$$

The direction of migration due to the improvement in productivity is only determined by the sign of $m_{s1} - m_{s2}$. As we are only concerned with the case of $m_{s1} > m_{s2}$, we see that some of region 2's population migrates to region 1. If the knowledge utilization efficiency in the two regions' service sectors is identical, then improved productivity has no impact on the labor distribution. The labor employed by the two sectors in each region will be increased or decreased if the region's total employment is increased or decreased.

By (A.7), we get the impact on the wage rates of the two regions as

$$\frac{Z}{w_1}\frac{dw_1}{d\tau_{ij}} = m_{i1}\frac{dZ}{d\tau_{ij}} > 0, \quad \frac{Z}{w_2}\frac{dw_2}{d\tau_{ij}} = (m + m_{i1})\frac{dZ}{d\tau_{ij}} > 0. \tag{4.19}$$

An improvement in knowledge accumulation efficiency increases the wage rates in the two regions. From (A.8) we get the impact on the land rent as follows

$$\frac{Z}{R_1}\frac{dR_1}{d\tau_{ij}} = (m_{i1} + m\xi_2)\frac{dZ}{d\tau_{ij}} > 0, \quad \frac{Z}{R_2}\frac{dR_2}{d\tau_{ij}} = (m + m_{i1} - m\xi_2\frac{N_1}{N_2})\frac{dZ}{d\tau_{ij}} > 0. \quad (4.20)$$

We see that region 1's land rent is increased. If $N_1 / N_2 \leq 1$, then region 2's land rent is increased. But if the population of region 2 is larger than region 1, then it is possible for region 2's land rent to decline as more people move from region 2 to region 1. We see that in the case that region 2 employs more people, the sign of $dR_2 / d\tau_{ij}$ may be either positive or negative, depending on the structure of the whole system. From (A.11), we get the effects of changes in τ_{ij} on the prices of services as follows

$$\frac{Z}{p_1}\frac{dp_1}{d\tau_{ij}} = (m_{i1} + \alpha m\xi_2 - m_{s1})\frac{dZ}{d\tau_{ij}}, \quad \frac{Z}{p_2}\frac{dp_2}{d\tau_{ij}} = \qquad (4.21)$$

$$(m + m_{i1} - m_{s2} - \alpha m\xi_2\frac{N_1}{N_2})\frac{dZ}{d\tau_{ij}} > 0.$$

We see that as knowledge accumulation efficiency is improved, prices of services may be either increasing or decreasing.

From the above discussion, it is easy to get the impact of changes in τ_{ij} upon all the 30 variables in the system. It can be seen that the effects are very sensitive to how different values of the parameters are combined.

As we have explicitly solved the equilibrium problem, it is straightforward to examine the effects of changes in any parameter in the system as we did for τ_{ij}. For instance, by taking derivatives of (A.18) with respect to A_2, we get

$$-\Phi^{*\prime}\frac{dZ}{dA_2} = \frac{\xi_1 AZ^m - \Phi_1}{(1-\gamma\beta)A_2}, \qquad (4.22)$$

where $\Phi^{*\prime} < 1$ and $\Phi_1 > 0$. An improvement in region 2's amenity may either increase or decrease the equilibrium level of knowledge. If the term, Φ_1, associated with creativity of the industrial sector in region 1 is larger than the term, $\xi_1 AZ^m$, then dZ / dA_2 is negative. This can be interpreted as that when region 1's industrial sector has a high contribution to knowledge accumulation, then an increase in region 2's living conditions may reduce the long-run equilibrium level of knowledge. An improvement in region 2's amenity causes a change in region 1's equilibrium utility level. This will result in re-allocation of labor force. As knowledge utilization and creativity are spatially different, the re-allocation of labor force will cause shifts in the equilibrium structure. It can be seen from the above equation that whether an improvement in amenity has a positive or negative impact on the equilibrium level of knowledge stocks is dependent on a complicated interdependence between many factors.

There are two important points. One is that if we take derivatives of the equilibrium condition for knowledge accumulation with respect to region 1's amenity, A_1, then we can show that the sign of dZ / dA_1 is identical to that of $\Phi_1 - \xi_1 AZ^m$. The other one is that if we are concerned with the case that the system has an unstable unique equilibrium, i.e., $x_{ij} > 0$, j = 1, 2, and $m \leq 0$, then $dZ / dA_2 < 0$, which is opposite to the case that the system has a unique stable equilibrium.

By (A.16)

$$\frac{Z}{p}\frac{dp}{dA_2} = -\frac{Z}{(1-\gamma\beta)A_2} + (m - m_i)\frac{dZ}{dA_2}, \tag{4.23}$$

is held. To see how the price of commodity 2 is changed by changes in A_2, we first look at (4.6). As region 2's amenity is changed, the balance condition (4.6) is disturbed. Therefore, migration between the two regions will be necessary for the system to achieve a new equilibrium (because we assume the existence of a unique stable equilibrium). As the employment distribution is changed, the demand conditions (4.8) will be disturbed, which will result in adjustment of the price variables. As the prices and demand conditions are changed, the production scales will be shifted toward the new equilibrium. All of these changes further shift the balance condition (4.6). The process will thus be repeated until the system achieves at the new equilibrium. The shift in the price of commodity 2 is given by (4.22). The new price level may become either higher or lower than the old one.

From (A.5) we get the impact on N_j as follows

$$\frac{\Psi}{\xi_1 A N_2 Z^m}\frac{dN_2}{dA_2} = \frac{1}{(1-\gamma\beta)A_2} - \frac{m}{Z}\frac{dZ}{dA_2}, \quad \frac{dN_1}{dA_2} = -\frac{dN_2}{dA_2}. \tag{4.24}$$

If the knowledge utilization efficiency of the two service sectors is identical, i.e., $m = 0$, then an improvement in region 2's amenities attracts more people from region 1 to region 2. But in the case that $m > 0$ and dZ / dA_2 is large, an improvement in region 2's amenities may not increase region 2's population. Hence, the direction of migration is also dependent on the characteristics of knowledge utilization in the two regions. We may easily provide the effects of changes in A_2 on the other variables in the system. As little new insight can be gained, we omit further examination.

4.5 Concluding Remarks

This study proposed a perfectly competitive regional economic dynamic model with endogenous knowledge accumulation. We showed that the dynamic system may have either a unique or multiple equilibria and each equilibrium may be either stable or unstable, depending upon the knowledge utilization efficiency and creativity of the various sectors. We also examined the effects of changes in knowledge accumulation efficiency and regional amenity on the equilibrium structure.

The model may be extended in different ways. For instance, we may enrich the spatial structure of the model by further specifying location of the industrial section, services and residents within each region. We may also enrich endogenous dynamic processes of the regional division of labor with transportation costs.

Appendix 4.1: Proving Proposition 4.1

From $C_{sj} = F_{sj}$ together with (4.2), $N_{ij} + N_{sj} = N_j$ and $C_{sj} = \gamma w_j N_j / p_j$, we have

$$N_{sj} = \gamma\beta N_j, \quad N_{ij} = (1 - \gamma\beta)N_j, \quad j = 1, 2. \tag{A.1}$$

Substituting C_{ij} in (4.8) into $C_{j1} + C_{j2} = F_{ij}$ in (4.9) together with (4.3) and (4.4) yields

$$N_1 + Z^{m_i} N_2 = \frac{N_{i1}}{\xi_1}, \quad N_1 + Z^{m_i} N_2 = \frac{pZ^{m_i} N_{i2}}{\xi_2}, \tag{A.2}$$

where $m_i \equiv m_{i2} - m_{i1}$. As the left-hand sides of the two equations in (A.2) are identical, we have

$$\frac{N_{i1}}{N_{i2}} = \frac{\xi_1 p Z^{m_i}}{\xi_2}. \tag{A.3}$$

We see that the ratio of the employment in the two industrial sectors depends upon the difference in knowledge utilization efficiency and in marginal utility of the two industrial commodities. Substituting N_{ij} in (A.1) into (A.3) yields

$$\frac{N_1}{N_2} = \frac{\xi_1 p Z^{m_i}}{\xi_2}. \tag{A.4}$$

The ratio of employment between the two regions is related to the preference parameters, level of knowledge and knowledge utilization parameters. By the above equation and $N_1 + N_2 = N$, we get

$$N_1 = \xi_1 ANZ^m \Psi(Z), \quad N_2 = \xi_2 N\Psi(Z). \tag{A.5}$$

By (A.1) and (A.5), we solved the labor distribution, N_{sj}, N_{ij} and N_j, as functions of p and Z. We directly determine F_{ij} as functions of p and Z by substituting N_{ij} in (A.1) into (4.3). Region 1's consumption of commodity 1 is given by: $C_{11} = \xi_1 w_1 N_1$. From $C_{11} + C_{12} = F_{i1}$, we get

$$C_{12} = F_{i1} - C_{11} = \xi_0 F_{i1}, \tag{A.6}$$

in which $\xi_0 \equiv 1 - \xi_1 / (1 - \gamma\beta) > 0$. By (4.4), $C_{12} = \xi_1 w_2 N_2$ and (A.6),

$$w_1 = Z^{m_{i1}}, \quad w_2 = \frac{\xi_0 A(1-\gamma\beta)Z^{m+m_{i1}}}{\xi_2}, \tag{A.7}$$

are held. From (4.2) we get $L_{sj} = \alpha w_j N_{sj} / \beta R_j$. Substituting this and $L_{hj} = \eta w_j N_j / R_j$ and (A.1) into the last equation in (4.10), we obtain:

$$R_j = \frac{(\eta + \alpha\gamma) w_j N_j}{L_j}, \quad j = 1, 2. \tag{A.8}$$

By

$$L_{hj} = \frac{\eta w_j N_j}{R_j}, \quad L_{sj} = \frac{\alpha w_j N_{sj}}{\beta R_j}, \tag{A.9}$$

(A.8) and (A.5), we get:

$$L_{hj} = \frac{\eta L_j}{\eta + \alpha\gamma}, \quad L_{sj} = \frac{\alpha\gamma L_j}{\eta + \alpha\gamma}. \tag{A.10}$$

From L_{sj} and N_{sj}, we directly determine F_{sj}. We obtain C_{sj} as functions of Z by $C_{sj} = F_{sj}$. Utilizing $C_{sj} = \gamma w_j N_j / p_j$ in (4.8), we get

$$p_j = \frac{\gamma w_j N_j}{F_{sj}}, \quad j = 1, 2. \tag{A.11}$$

We have thus shown that the 28 variables, N_{sj}, N_{ij}, L_{hj}, L_{sj}, F_{sj}, F_{ij}, C_{sj}, C_{1j}, C_{2j}, U_j, N_j, w_j, p_j, R_j ($j = 1, 2$), are determined as functions of p and Z. We now show that commodity 2's price can be determined as a function of Z.

Substituting (4.8) into (4.6) yields

$$\frac{w_1}{w_2} = \left(\frac{p_1}{p_2}\right)^{\gamma} \left(\frac{R_1}{R_2}\right)^{\eta} \frac{A_2}{A_1}. \tag{A.12}$$

The equation gives a relation between the wage rates, amenity, service prices and land rents in the two regions.

From (4.4) we have

$$\frac{w_1}{w_2} = \frac{1}{pZ^{m_i}}. \tag{A.13}$$

By (A.4), (A.13) and (A.8)

$$\frac{R_1}{R_2} = \frac{\xi_2 L_2}{\xi_1 L_1}, \tag{A.14}$$

is held. From (A.11), (4.1) and (A.1), we have

$$\frac{p_1}{p_2} = \left(\frac{w_1}{w_2}\right)^{\beta} \left(\frac{R_1}{R_2}\right)^{\alpha} Z^{m_s}, \tag{A.15}$$

where w_1 / w_2 is given in (A.13). Substituting (A.13)-(A.15) into (A.12) yields:

$$p = AZ^{m-m_i}. \tag{A.16}$$

We have thus proved that for any given knowledge, Z, the system has a unique temporary equilibrium.

Appendix 4.2: Proving Proposition 4.2

In the case of $\Omega_{s1} = \Omega_{s2} = 0$, the knowledge accumulation is given by

$$\frac{dZ}{dt} = \{\Phi_1(Z) + \Phi_2(Z)\} Z\Psi(Z) - \delta_z Z \equiv \Phi(Z), \tag{A.17}$$

where $\Phi_j = v_j Z^{-y}$, j = 1, 2. We now examine properties of (A.17). An equilibrium is given as a solution of the following equation:

$$\Phi^*(Z) \equiv \Phi_1(Z) + \Phi_2(Z) - \xi_1 AZ^m - \xi_2 \delta_z = 0. \tag{A.18}$$

Our problem is whether or not $\Phi^*(Z) = 0$ has a positive solution.

In the case of $x_{i1} > 0$, $x_{i2} > 0$ and $m < 0$, we have: $\Phi^*(0) < 0$, $\Phi^*(+\infty) > 0$, $\Phi^{*\prime}(Z) > 0$ for $0 < Z < +\infty$. Accordingly, there is a unique positive Z such that $\Phi^*(Z) = 0$. The stability is determined by the sign of Φ'(Z) at the equilibrium. If Φ' > 0, the system is unstable; if Φ" < 0, it is stable; and if Φ" = 0, it is neutral. Taking derivatives of Φ with respect to Z yields:

$$\Phi' = x_{i1}\Psi\Phi_1 + x_{i2}\Psi\Phi_2 - (\Phi_1 + \Phi_2)\xi_1 mA\Psi^2 Z^{m-1}, \tag{A.19}$$

which is evaluated at equilibrium. We thus conclude the system has a unique unstable equilibrium in this case.

In the case of $x_{i1} < 0$, $x_{i2} < 0$ and $m > 0$, we have: $\Phi^*(0) > 0$, $\Phi^*(+\infty) < 0$, $\Phi^{*\prime}(Z) < 0$ for $0 < Z < +\infty$. We see that the system has a unique stable equilibrium in this case.

There are six other possible combinations of $x_{i1} < 0$ or $x_{i1} > 0$, $x_{i2} < 0$ or $x_{i2} > 0$ and $m < 0$ or $m > 0$. It can be checked that if the system has equilibrium, it must have multiple ones. For instance, in the case of $x_{i1} < 0$, $x_{i2} > 0$ and $m < 0$, because $\Phi^*(0) < 0$ and $\Phi^*(+\infty) < 0$ are held, the equation, $\Phi^*(Z) = 0$, has either no solution or multiple ones. Similarly, we can check the other five cases. We can directly check the stability of each equilibrium by (A.21). Summarizing the above discussion, we proved proposition 4.2.

We now check corollary 4.2. As in cases i) and ii) in proposition 4.2, we can directly show that i) if $x_{ij} > 0$, j = 1, 2, then the system has a unique unstable equilibrium; ii) if $x_{ij} < 0$, j = 1, 2, then the system has a unique stable equilibrium. We now show that if $x_{i1} < 0$ and $x_{i2} > 0$ (or $x_{i1} > 0$ and $x_{i2} < 0$), the system has two equilibria - the one with higher value of knowledge is unstable and the other one is stable.

It is sufficient to check the case of $x_{i1} > 0$ and $x_{i2} < 0$ as the case $x_{i1} < 0$ and $x_{i2} > 0$ can be similarly proved. As $\Phi^*(0) > 0$ and $\Phi^*(+\infty) > 0$, the system has either no equilibrium or multiple ones. From $\Phi^{*\prime} = (x_{i1}\Phi_1 + x_{i2}\Phi_2)/Z$, we see that $\Phi^{*\prime} = 0$ has a unique positive solution, denoted by Z_0. This implies that $\Phi^*(Z) = 0$ cannot have more than two solutions. When the system has two equilibria, the stability of each equilibrium is determined by the sign of Φ'. It can be seen that if $Z < Z_0$, then $\Phi'' < 0$ and if $Z > Z_0$, then $\Phi' > 0$. We thus proved the corollary.

Acknowledgements

I would like to express my gratitude to Professor Charlie Karlsson and Professor Börje Johansson for their insightful and helpful comments and to Ms. Kerstin Ferroukhi and Ms. Ingrid Lindqvist for their kind help.

References

Alonso, W., 1964, *Location and Land Use*, Harvard University Press, Cambridge, Mass.

Andersson, Å. and J. Mantsinen, 1986, "Mobility of Resources, Accessibility of Knowledge and Economic Growth", *Behavioural Science,* 25:353-366.

Becker, G.S., K.M. Murphy and R. Tamura, 1990 "Human Capital, Fertility and Economic Growth", *Journal of Economic Economy,* 98:12-37.

Bell, C., 1991, "Regional Heterogeneity, Migration, and Shadow Prices", *Journal of Public Economics,* 46:1-27.

Diamond, D.B. and G.S. Tolley, (eds.), 1981, *The Economics of Urban Amenities*. Academic Press, New York.

Grossman, G.M. and E. Helpman, 1991, *Innovation and Growth in the Global Economy,* MIT Press, Cambridge, Mass.

Johansson, B. and C. Karlsson, (eds.), 1990, *Innovation, Industrial Knowledge and Trade - A Nordic Perspective*. CERUM at University of Umeå, Umeå.

Kanemoto, Y., 1980, *Theories of Urban Externalities,* North-Holland, Amsterdam.

Krugman, P., 1991, "Increasing Returns and Economic Geography. *Journal of Political Economy,* 99:483-499.

Lucas, R.E., 1986, "On the Mechanics of Economic Development", *Journal of Monetary Economics,* 22:3-42.

Rauch, J.E., 1991, "Comparative Advantage, Geographic Advantage and the Volume of Trade". *The Economic Journal*, 101:1230-1244.

Roback, J., 1982, "Wages, Rents and Quality of Life", *Journal of Political Economy,* 90:1257-1278.

Scotchmer, S. and J.F. Thisse, 1992, "Space and Competition – A Puzzle", *The Annals of Regional Science,* 26:269-286.

Simon, J.L. and D.O. Love, 1990, "City Size, Prices, and Efficiency for Individual Goods and Services", *The Annals of Regional Science,* 24:163-175.

von Thünen, J.H., 1826, *Der Isolierte Staat in Beziehung auf Landwirtschaft und Nationalekonomie*, Hamburg.

Voith, R., 1991, "Capitalization of Local and Regional Attributes into Wages and Rents – Differences Across Residential, Commercial and Mixed-Use Communities", *Journal of Regional Science,* 31:129-145.

Zhang, W.B., 1992, "Trade and World Economic Growth – Differences in Knowledge Utilization and Creativity", *Economic Letters,* 39:199-206.

5 Technology and Regional Development: Theory Revisited

John Rees
Department of Geography, University of North Carolina at Greensboro, Greensboro, NC 27402; USA

5.1 Introduction

The continuing onset of globalization does not mean the End of Geography, when regions cease to matter as conceptual or empirical devices. Likewise, increasing attacks from alternative approaches to knowledge do not make theory any less important. It is a fundamental tenet of this paper that both regions and theory matter a great deal.

In this review of theories of regional development, it will be seen that we have come a long way in understanding the conceptual, empirical and policy orientation of the most important explanatory variable: the role of technological change. The relevance of the original concept of the product cycle will be assessed as it relates to regional life cycles and innovative milieus. It may be needless to say that the perceptive taken in this review reflects the School of Evolutionary Economics as originated by Schumpeter (1941) and molded by scholars like Nelson and Winter (1982) in economics before being introduced in a regional context by Thomas (1975) and others.

It will be shown that concepts that intrigued Schumpeter, like the production of entrepreneurs, have only been the focus of research in a regional context for about a decade. While factors that may link entrepreneurs and leadership potential still need attention, we also need to know more about a region's social capital and the way it may influence the propensity to produce both entrepreneurs and industrial leaders. Exemplars will be discussed as to how individuals played such a critical role in the evolution of Silicon Valley as an innovative milieu, and how Governors with minimal political powers in the state of North Carolina played such a leadership role in the recent evolution of that state's economic development. Before we turn to discuss the role of creative people in creative regions, we need to reflect on the role of theory in general and to review the theories that link technological change to regional development.

5.2 The Resilience of Theory

Before we discuss the various theories of regional development, it is important to review where we stand as theory developers. Over the past decade, most of the social sciences have seen the increasing influence of Postmodernism as a deliberate attack on the epistemology of science. Postmodernists reject the fundamental notions of truth and theory as goals, and appear even to glorify anarchy as a methodology. The philosophical origins of Postmodernism can be traced to Paul Feyerabend's (1985) polemic *"Against Method"*. Rosenau (1991) discusses the enormous consequences of the Postmodern Turn on the social sciences in general, where the continuing need for theory is central. "Social scientists of every orientation find it extremely difficult to give up theory as the skeptics require. A world without theory means an absolute equality of all discourse... the entire intellectual climate of the social sciences would be transformed" (Rosenau, 1992, p.89). While I have written elsewhere on Postmodernism as a Faulty Tower for scholars of regional development (Rees, 1998), the future of theory and empirical positivism in general seems safe among regional scientists, economists and geographers. Given the dissent from positivism represented by Postmodernism, then the need to continually assess and reassess our theoretical constructs appears paramount.

5.3 Theories of Regional Development: A Classification

Progress by regional scholars from geography, economics and related disciplines means that many of us by today have our own classification of theories. This paper is primarily concerned with *technology-based theories of regional development* based on the belief originally expressed twenty years ago (Rees, 1979) that technology is a prime motor for understanding and furthering the process of regional development. As I review the regional development literature over the past twenty years, the sheer quantity alone can be seen as a symbol of our success in both the intellectual and practical spheres. But I do not see much that deters from my original view that technology as a primary determinant of productivity is also a major motor of regional economic growth and change. I do however see a need to recognize a broad approach to the study of technology and related concepts, particularly the concepts of agglomeration, milieu, learning and leadership.

The evolution of the regional development literature over the past twenty years has shown how technology is directly related to the concepts of agglomeration and learning though some confusion seems to surround linkages between concepts. And while we now know much more about the links between entrepreneurship and regional development (see Malecki, 1997), we still need to know how leadership is linked to the regional culture of innovation.

Hence, I will focus on some of our major concepts: technology, agglomeration, milieu, learning and leadership. By definition then, I choose to omit other important concepts and how they affect the process of regional development. A major reason for this is that others in this volume will address these issues. The most important topic omitted from my review is the relationship between trade and regional development. As regional scholars, the regional application of trade theory is one of our most time honored achievements, stemming from the original work of Ohlin, 1933; and the topic has enjoyed some renewed attention lately (Krugman, 1991). Paul Krugman has already been hailed as a future Nobel laureate, and one sentence from his work sums up the importance of this volume. "I have spent my whole professional life as an international economist thinking and writing about economic geography without being aware of it" (Krugman, 1991, p.1).

The rediscovery of regions by such visible scholars as Paul Krugman (1991) and Michael Porter (1990) should also remind us of the importance of our most central concept: the region. In contrast to some scholars who see emerging information technologies as leading to the decreasing importance of place and the 'End of Geography' (O'Brien, 1992), Kenichi Ohmae (1993) argues in an influential article in *Foreign Affairs* that regions (or region states) will replace nation states as the centerpiece of economic activity while Allen Scott (1996) also sees globalization leading to a more intense regional pattern of production. In this review regions will become the Sine Qua Non of our evolving global system. And this may include central Sweden as well as Emilia Romagna, Silicon Valley and North Wales.

Now, there is a need to review the theoretical developments that relate to the important concepts of technology, product and regional life cycles, innovation milieus, agglomeration, entrepreneurship and leadership.

5.4 Innovation, Regional Cycles and Milieus

More than a generation ago, Morgan Thomas (1975) reminded us that technological change was among the Terra Incognita of regional development, as it had been a decade earlier in economics. In 1994 Erickson provided us with a comprehensive review of research on the geography of technological change and how this was related to the competitiveness of America's regions, at a time when technology policy was a contentious topic of public policy debate in many states. [See the early debate about 'High Hopes for High Tech' in North Carolina in Whittington (1985), and a comprehensive review of various state development strategies in the USA in Schmandt and Wilson (1990)]. More recently Sternberg (1996) compared the American experience with regional technology policy to that of three other G8 countries: the UK, Germany and Japan. Over the last quarter century therefore, serious interest in both the theory and practice of technology-based economic development has generated a literature so large that it can only be reviewed selectively here.

Yet most analysts agree that the theoretical trigger to the ensuing debates in theory and practice was Product Cycle theory initially developed by Vernon (1966) and introduced into the regional literature by Norton and Rees (1979) and Erickson and Leinbach (1979). Since this author was involved in the early development of this theory, un unbiased commentary on the evolution of recent research will follow together with comments on debates in the policy arena.

5.4.1 Product Cycles and Regional Life Cycles Revisited

Product cycle theory is now familiar to most students of regional development (see Rees, 1979; Norton and Rees, 1979; Rees and Stafford, 1986; Erickson and Leinback, 1979; Erickson, 1994; Malecki, 1997; Sternberg, 1996; Schmandt and Wilson, 1990, among others). Since all commentators did not agree with this interpretation and its implications (Taylor, 1986; Storper, 1997), the essence of the argument will be reviewed here, using Figure 5.1 as a guide.

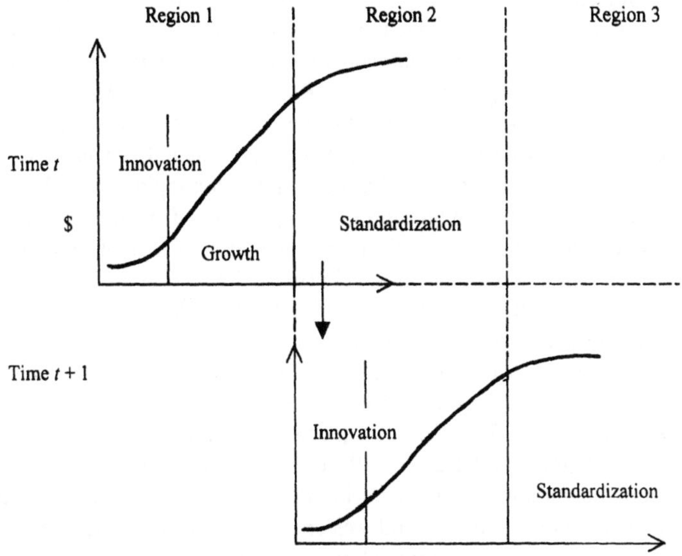

Figure 5.1 Spatial manifestation of product cycle over time

Following on from Vernon (1966), the basic premise is that both industries and companies make profits when individual products succeed in the marketplace. A typical product (but not necessarily all goods and services) evolves through three distinct stages in its life cycle: *an innovation stage*, during which a new product is both developed and manufactured in its home region where initial costs are at their highest (a stage when most new products fail); *a growth stage*, when sales increase by selling the product in an array of domestic as well as international markets, where revenues generated exceed the costs incurred in production; *a standardization phase* where production can be shifted from the original high cost home region to lower cost locations in the US or abroad. The original formulation implied the use of Fordist mass production techniques in the third stage of the product cycle, where cost could be minimized and revenues maximized. This does not imply that more flexible, post Fordist production techniques cannot be used in this stage as local conditions of capital and labor allow. For companies to survive and grow, they have to sell their products in as many new markets as possible to maximize revenues, as well as to invest profits back into R and D to ensure a succession of new product cycles.

This whole process has an important geographical dimension to it because the different stages of product development can take place in different locations (see Figure 5.1). The innovation stage in particular needs a high input of R and D and is usually carried out in the large urban industrial agglomerations of developed countries: like Boston, New York, Los Angeles and Silicon Valley in the USA and Britain's M4 corridor or Stuttgart and Munich in Germany (i.e. region 1 in Figure 5.1, also Castells and Hall, 1994). The standardized production phase of the product cycle (for both Fordist and post-Fordist production systems) can be transferred to low cost locations in the USA, Europe or abroad as well as down the urban hierarchy to rural areas (see region 2 in Figure 5.1, and Erickson and Leinbach, 1979).

A major implication of the geographical dimensions to the product cycle model relates to the way regions can change their roles over time. *"Regions can change their roles from being the recipients of innovations via branch plants to being the generators of innovation via indigenous growth"* (Rees, 1979, p.58). As production accumulates in region 2, personnel mobility, the development of local linkages and other external economies can build up there and regional demand can grow to a critical threshold where an industrial seed bed effect can develop rapidly with the spin-off of small firms or through the immigration of entrepreneurs. This then is the essence of the link between product cycle theory and the evolution of regional economies that I originally described as a regional life-cycle (Rees, 1979). Pat Norton and I showed empirical evidence for this process in the USA in 1979 where new growth industries were no longer seen as unique to the traditional Manufacturing Belt, and the seeds of innovation had spread to what has since been called 'the new industrial spaces' (Scott, 1988) of the southeastern and southwestern USA. More recent empirical evaluation in the US by Sorenson (1977, p.302) concluded that the spatial pattern he found "is consistent with Norton and Rees 1979 conclusion that innovation capacity has dispersed from the traditional Manufacturing Belt center and has led to growth in high technology sectors in the southern and western USA". Given the wisdom and simple elegance implicit in the original formulation of Regional Life Cycle Theory in 1979, it is no surprise to me that current studies lend validity to its continued relevance in the US and that others see it appropriate to extend the same theory to explain regional changes in Sweden and elsewhere (see Johansson and Karlsson, 1987, and Table 5.1 on the USA).

Table 5.1 Business start-ups per capita in sunbelt metro areas, 1997[1]

State	# Start ups per capita
CA	20
FL	12
TX	12
CO	4
TN	4
AZ	2
OK	2
NC	2
NV	2
UT	2
VA	2
GA	1
LA	1
NM	1

Calculated from IBC 1998.

[1] 67 of top 100 metro areas located in Sunbelt region (Census South and West)

One other point about the link between product cycle theory and regional life cycles shown in Figure 5.1 needs to be made more explicit because it relates to the rejuvenation of older industrial areas. After a period of industrial decentralization from region 1 to region 2, inflationary pressures can increase in region 2 (the newer industrial spaces) due to competitive pressures on land and labor. New companies spawned in region 2 will in time look for lower production costs in other regions both domestically and abroad (i.e. region 3). Since inflation in region 2 can make the relative costs of the initial region (region 1) look much more attractive, companies would reinvest in older industrial areas (particularly if they have good labor skills and other attractive production factors) instead of expanding in a new region 3. Hence, regional life cycle theory explains the potential rejuvenation of older industrial areas. It is my contention that this offers a partial explanation for the revival of the original Manufacturing Belt of the USA in recent years, the industrial Midwest (see Testa, Klier and Mattoon, 1997, and other studies by the Federal Reserve Bank of Chicago that examine the revival of the Industrial Midwest).

An useful extension of product cycle theory was provided by Ann Markusen (1985) in her articulation of Profit Cycles and oligopoly as they relate to regional development. The notion of profit cycles here extends product cycle theory by relating it to the influence of various types of industrial organization, especially the tendency towards oligopoly in the latter stages of corporate development.

Then, what of the critics of Product Cycle and Regional Cycle Theory? Storper (1997, p.60) tells us that product cycle theory is not a very good predictive tool for the development of all industries, especially if they were not a consumer durable sector during this century (a rather large group by itself). But product cycle theory was not meant to be a general theory for all products and industries as much as a way of explaining the link between product and process technology and their changing locational needs that could partly explain the growth of new industrial spaces and the revival of older industrial areas. Storper's conclusion that "the search for cycles and patterns as predictive devices in regional economics (as in Macroeconomics) had largely failed" (Storper, 1997, p.60) makes him guilty of throwing out the baby with the bath water in terms of product cycles and long cycles. His point that Product Cycle theories do not provide insights into the 'how, why and where' the *initial* agglomerated core regions of sectors build themselves is a more valid one. And this is where further research on location factors that affect high technology industries (Rees and Stafford, 1986) and recent work on 'innovative milieus' makes a contribution. It is to these research avenues that we now turn.

5.4.2 The Innovative Milieu and Related Concepts

The innovative milieu concept was formulated to explain the 'how, why and where' of new technology generation, and is a sequential follow-on to product cycle theory (se paper by Maillat in this volume).

The literature on this topic has become large and fashionable and somewhat confusing as a result (see Malmberg, 1996, 1997). The evolution of work on the milieu concept has been linked to the re-discovery of agglomeration economies by geographers and economists and how agglomeration and localization economies lead to new industrial spaces (Scott, 1988; Porter, 1990; Krugman, 1991). It appears that the

Fordism-post Fordism debate and the notion of flexible production has been unnecessarily linked to the search for agglomeration economies and industrial districts. As a result, small variations on the same theme appear in the literature as "industrial districts", regional innovations systems, growth centers, learning regions and the like. While conceptual clarity appeared in early reviews by Maillat and Lecoq (1992), Cooke (1992), Hansen (1992) and Florida (1995), this appears lost in the recent literature.

Two further points are worth noting here. Concern about the innovation milieu has made some scholars emphasize that the benefits of agglomeration are not just economic but cultural in nature, relating to the 'social capital' of a region. Fukuyama (1995) tells us that social capital is a function of trust in society and that it differs from other forms of human capital because it is transmitted through cultural mechanisms like religion and family customs. In an insightful review of rural development, Malecki (1998) tells us that we need to know more about the three concepts of local knowledge, social capital and institutional embeddedness before we can fully understand the evolution of innovative milieus or economic development as a whole. In the same way that some policy makers now advocate that investment in human capital (education and training) may be more important than investment inn physical infrastructure (roads and sewers) in order to develop an innovative milieu, our future concerns for regional development should include the nurturing of our regional social capital.

The final point about innovative milieus is that public policy is usually accepted as an appropriate vehicle to enhance such milieus in both Europe and the USA. Indeed, the concern for policies to nurture innovative environments has generated much interest around the world at both national and local levels (see Schmandt and Wilson, 1990; Nelson, 1995; Freeman, 1995). However, empirical validity of the impacts of innovative milieus is either missing or contradictory and this cannot be a help to policymakers. We are told that proximity to major universities and research centers should be an integral part of a 'scientific infrastructure' that will ensure economic development in the future. Some authors more than others see the issue of geographic proximity as crucial to the development of new technology (see Saxenian, 1994). Feldman and Florida (1994) undertook survey research that did show a positive connection between the propensity to innovate and a region's supply of university and industry R and D, but a recent study on developments in Baden-Wurttemberg was not so positive (Staber, 1996).

Policy evaluation can come up with a mixed bag of results, but it is important to report such results to policymakers. Several years ago I undertook a survey of a large number of companies to find out what experiences they had with state funded research centers in microelectronics in the USA, where some unexpected results were forthcoming (Rees, 1991; Rees and Debbage, 1992). The assumption was that access to university research was the primary reason for university-industry collaboration at that time but access to students and faculty as potential employees emerged as the major reason. Among companies who had direct experience with state-funded technology centers, most reacted positively to their experiences because collaboration was seen to stimulate related research within many companies while also leading to improvements in the quality of products and production methods. While the majority of respondents thought that states should continue to fund technology development programs, particularty at universities, *neither universities nor governments were seen as important sources of technical knowledge by many companies*. The same study however showed that small companies make greater use of university research than their larger counterparts (Link and Rees, 1990). In another evaluation of collaboration between small and medium sized enterprises (SMEs) i.e. networks, in Oregon and the northwest USA, Rosenfeld (1996) found that businesses were more open to cooperation than many skeptics contend, that small businesses are eager to learn from each other and want organizational settings in which they can interact. While a general endorsement of the Danish model of network development in Oregon, Rosenfeld's study warns against the impatience of funders for short term gains when they are involved in a process that takes a long time to bear fruit (a process that can be labeled as one of the major dilemmas of policy evaluation).

Since evaluation of innovative milieus are too sparse to generalize boldly, it still seems premature to suggest (like one reviewer does) that the interest in innovation and learning environments as a primary route to company and regional economic success has become an obsession of academics and policymakers (see Malmberg, 1997). It also seems appropriate to reflect on the words of Castells and Hall (1994, p.11) in their global review of innovative industrial milieus: "Despite all this activity... most of the world's actual high-technology production and innovation still comes from areas that are not usually heralded as innovative milieus... the great metropolitan areas of the industrialized world".

5.5 Entrepreneurship and Leadership: The Missing Variables

The previous section of this paper dealt with the conceptual building-blocks of technology-based theories of regional development as they evolved over a generation of scholarship and policy applications. But we still have missing variables and missing information about these variables that make our understanding of regional development incomplete to date. The missing variables to be discussed here are entrepreneurship and leadership. The role of the entrepreneur in the process of regional development and how that might be cultivated was a mystery to most of us only a decade ago, but a comprehensive review by Malecki (1997) and the dedication of a full journal to this topic ensure us that this is no longer the case. The way entrepreneurs become leaders and the linkages between leadership and regional development remain among our *Terra Incognita*, as suggested by Desantis and Stough (1997). Because the discussion of these concepts takes us into issues of personality and specific environmental contexts, the difficulties involved in making bold generalizations about patterns and processes as well as the plain unpredictability involved makes quantitative assessments and even theory development more challenging.

5.5.1 Entrepreneurship and Regional Development: What We Know so far

Malecki (1997, Chap. 5) reviews the salient research on entrepreneurs, networks and regional developing by focusing on the 'entrepreneurial event' and how specific locational environments may influence that event and it's more significant outcome: the formation of new firms. De facto, the concept of the entrepreneur has to be equated with risk taking, uncertainty, a willingness to not accept failure, as well as numerous other personal and environmental factors that include family background, motivations and goal orientation, educational background, access to financial capital and other experiences that dwell in the shadowy world of psychologists. Yet, "oddly absent from much of the standard research on entrepreneurship is the critical nature of the entrepreneur's *Local Context* in which he/she operates on a daily basis" (Malecki, 1997, p.164).

A number of studies have proposed critical factors that characterize an entrepreneurial environment, including: the presence of experienced entrepreneurs already in the region, the availability of venture capital and a technically skilled labor force, proximity to universities and other R and D centers, access to new markets and a variety of support services usually in large and diverse urban areas.

Table 5.2 provides a summary of factors by Dubini (1989) who distinguishes between munificent and sparse entrepreneurial environments in an Italian context. Other, more recent attempts to understand entrepreneurial environments focus on concepts introduced earlier: institutional thickness (Amin and Thrift, 1994), trust and social capital (Fukuyama, 1995). When it comes to explaining the linkage between entrepreneurs and technological innovation, Malecki (1997) emphasizes the importance of networks of small firms (drawing on the early work of Piore and Sabel (1994) and the positive experience of firm networks particularly in Denmark and Italy) as well as the necessity of various forms of risk capital. It may not be surprising to find however that the ability of government policies to create fruitful entrepreneurial environments is viewed with some skepticism among entrepreneurs as well as the academics who study them. There appears to be enough experimentation going on in Europe, Asia and North America however to suggest that the door on this topic should not be closed just yet.

Table 5.2 Characteristics of "munificent" and "sparse" environments for entrepreneurs

Characteristics of munificent environments
Strong presence of family businesses and role models
A diversified economy in terms of size of companies and industries represented
A rich infrastructure and the availability of skilled resourses
A solid financial community
Presence of government incentives to start a new business
Characteristics of sparse environments
Lack of an entrepreneurial culture and values, networks, special organizations or activities aimed at new companies
Lack of a tradition of entrepreneurship and family businesses in the area
Absence of innovative industries
Weak infrastructures, capital markets, few effective government incentives to start a new business

Source: After Dubini (1989)

Finally on this topic, anecdotes may not a theory make. But research on the success stories of Silicon Valley and Route 128 in the USA suggests that even the most rigid of economists should not assume away the notion of a regional culture that propagates entrepreneurs and innovations. Saxenian (1994) tells us that one of the major differences between the successes of the above two regions is that Route 128 has a regional culture bases on a hierarchical, authoritarian and rigid form of Puritanism whereas Silicon Valley has a culture based on the Pioneering spirit where invididuals emphasize experimentation, entrepreneurship and pure, unadulterated coolness.

5.5.2 Exemplar: The Short Story of Silicon Valley

"... the place where new ideas born in a garage can make teenagers into millionaires, while changing the ways we think"... (Castells and Hall, 19994, p.12). to the uninitiated, this is the condensed story of the *major players* initially responsible for the development of Silicon Valley as the birthplace of modern information technology.

Fred Terman, son of a Stanford professor and graduate of MIT, returned home to teach students like Charles Litton, Bill Hewlett and David Packard at Stanford. Similarly Bill Shockley left Bell Labs in Murray Hill, NJ (after inventing the transistor) to return home to Palo Alto. Shockley had eight students who founded Fairchild and then the Fairchildren, one of whom was Bob Noyce, the founder of Intel. Meanwhile, Ed Roberts first produced the PC in the "no there there" town of Albuquerque, NM, a location without an appropriate milieu that lost out to the San Francisco Bay Area as the home environment of the Home Brew Computer Club. The HBCC included Steve Wozniak and later Steve Jobs, who together invented Apple, as well as Bill Gates, the current monopoly star of software, who turned out to prefer rainy Seattle to sunny Albuquerque and the coolness of the Bay Area. If Bill Gates went back to Seattle, he could have gone back to North Wales *but* (like Shockley and Terman) his mother did not live there!

This story of people and places then begs that question: do such entrepreneurial culture regions exist elsewhere in North America or in Europe? To cite another example that's less known to many: Texas Instruments would never have existed in Dallas nor helped form the high tech complex there had it not been for Erik Johnsson, Gene McDermott and Cecil Green leaving RPI, MIT and Manchester to explore for oil in the Gulf of Mexico (Rees, 1979). But what of Europe, and European entrepreneurial culture? Since there is, de facto, no such place as Europe, the answer is simple. Then again, until there was an Italy, there were no Italians! If you believe Frank Fukuyama (1995), you will know that trust is a critical component of the social capital that creates economic development. To him, Germany along with Japan and the USA belong in a group of high trust societies while Italy, France and China each have a low trust society. Could Northern Italy be different from Southern Italy? And what about Sweden? As it happens, Fukuyama bypassed Sweden as he did many other countries. But most Americans know that Sweden, like it's American counterpart, is that region where the social capital is strong, the physical capital is good looking and all the children (or human capital) are *above average*. Enough of these allegorical tales. People matter as does their geography, but what turns people into leaders in some places more than others?

5.5.3 Leadership and Regional Development

Like the entrepreneur of a decade ago, a leader is someone who is difficult to define, yet you know one when you see one. In an exploration of the role of leadership in regional development, DeSantis and Stough (1997) define leadership as the act of motivated persons to mobilize (with or against others) economic, political, psychological and other resources to inspire, arouse, engage and satisfy the motives of followers. Like entrepreneurs, leaders cannot be understood without regard to their environmental context. Bennis and Nanus (1991) see leaders as causative, where they create institutions that empower individuals to satisfy their needs and create visions based on key values related to the social architecture that supports them. Whereas early theories of leadership were based on the individual (the 'Great Man' theories), later theory moved the focus away from the invididual and onto the environment or circumstances involved. Several consistent characteristics emerge: leaders direct by setting *goals* and defining *means*, executing a *vision* and acting as a *catalyst*.

While most regional development textbooks do not even include leadership in their indexes, a recent study of the American steel industry (Ahlbrant et al., 1996) attributes it's recent success to two factors: changes in technology (The minimill) and leadership. On the success of the newer American minimills, especially North Carolina's NUCOR and Oregon Steel Mills, the authors write: "There is an element here that stands out and seems to generalize for the most successful minimills, and this is that expressions of respect, trust and confidence in one's employees go hand in hand with the ability of firms to excel" (Ahlbrandt et al., 1996, p.71). Experimentation was rewarded and failure was accepted as a part of reality. The focus on creating a culture of continuous improvement in the steel industry reminds one of the Japanese concept of Kaizen, literally meaning a search for a better way or continuous improvement.

Apart from this intriguing account of the new American steel sector, this author does not know of any other recent sectoral study that focuses on the role of leadership. Similarly, we need to focus on the role of individual and group leaders in the growth regions and in the rejuvenation of older ones. In the context of regions, the leadership role can even come from the public sector, and an example of this will be given from North Carolina's recent experience with economic development.

5.5.4 Exemplar: North Carolina's Governors and their Megaprojects

Ever since the Reagan revolution of 1980, when the federal government cut back on its investment to individual states, both state and local governments have become much more proactive in their quest to optimize their own economic and social development (see Schmandt and Wilson, 1990). The CEO's or Governors of many of these political regions have taken the lead as catalysts of economic development, and no where is this more true than in North Carolina. A brief elaboration is made here as an example of public sector leadership in the area of economic development.

To provide some background to the issue, it should be mentioned that North Carolina is viewed as a progressive state, at least in the context of the American south (see Luebke, 1990). This implies that the people of the state prosper when business does, and North Carolina's manufacturing economy has become the largest in the Census south. Public education and public health also gets some attention as well as the sacred cow of highway construction, though unions need never apply. In an insightful interpretation of the state's political economy, Paul Luebke (1990) sees developments as a confrontation between two major groups: the Traditionalists and the Modernizers. With deep roots in rural areas and older industries like textiles and furniture, Traditionalists see continuation as preferable to change. They were opposed to the high technology development strategies of the 1980s because this meant an increase in the number of outsiders and this would also provide unwanted competition for existing low wage industries like textiles.

Modernizers, on the other hand, value individual achievement and expanding the economic base by diversifying the state's economy and have their main support base in the Piedmont Urban Crescent that runs from Raleigh through Greensboro and Winston Salem to the largest growth center of Charlotte. Modernizers have been known to support an active state government as well as taxation, at least as long as they remain in control. North Carolina has entertained all sorts of Governors over time, though since the 1950s the majority have been modernizers. Their most common characteristic is that they remain among the Governors with the least political power of any state in the nation, and many saw opportunities for legacy in the area of economic development.

As Table 5.3 shows, the success of the first megaproject launched in 1955 (though it took thirty years to really take off) ensured that future Governors would become preoccupied with their own megaprojects. It was Greensboro businessman, Romeo Guest, and a few friends who convinced Governor Luther Hodges that the state's economic base could be diversified in the mid 1950s by advertising the proximity of three universities to a cheap tract of land south of Durham that could be 'sold' with a label like Research Triangle Park. By today, the image of RTP has become so successful that it is often compared to two other innovative milieus: Silicon Valley and Route 128. It may be easy to forget that in employment terms, RTP is only one tenth the size of Silicon Valley, the equivalent of comparing Sweden with Germany.

Table 5.3 Leadership in economic development: North Carolina's governors and their megaprojects

1. Gov. Luther Hodges (1954-1961) & Research Triangle Park
   - Initiated by Romeo Guest 1955
   - Continued by Gov. Terry Sanford (1961-1965)
     National Institute for Environmental & Health Sciences, IBM
2. Gov. Jim Hunt continues "Megaprojects"
   - Hunt administration I (1976-1984)
     Targets microelectronics as key sector: MCNC
   - Hunt administration II (1992-2000)
     Leads on incentives: Hunt for Mercedes
       (Loosing a trophy = winning policy)
     Creates regional partnerships (with Dave Phillips)
3. Gov. Jim Martin (1985-92)
   Global Transpark with Jack Kasarda, to rejuvenate eastern NC?
   "Good idea, wrong place"?

The next Governor, Terry Sanford, continued to encourage the megaproject of his predecessor, by convincing the federal government (a reward for supporting President Kennedy to locate one of the new branches of the National Institute of Health in RTP.

Meanwhile, both General Electric and then IBM secured the park as a major location for branch plants if not for spin offs.(Product cycle enthusiasts will note that the earliest evidence of RTP giving birth to small companies did not come until the mid 1990s.) Governor Sanford also put his mark on another policy that has not yet been recognized as one of the most important megaprojects of all: the birth of the community college system in North Carolina. This vocational training system has since been heralded as one of the country's leading training centers for local industry.

After the post Watergate election of 1976, Governor Jim Hunt responded to an economic development problem that continued in the shadow of RTP. the fact that North Carolina had fallen to number 50 in industrial wage rankings. No one could even turn to the states of Mississippi nor Alabama for signs of hope! In 1981 Hunt focussed (like Hodges before him) on a high tech economic development strategy that included the vigorous recruiting of industry from other states (thus entering the Incentives game) and targeting the microelectronics industry as his own megaproject, culminating in the establishment of MCNC, the controversial state supported Microelectronics Center of North Carolina. "Hunt called microelectronics North Carolina's last chance, 'perhaps the only chance that will come along in our lifetime', for a 'dramatic breakthrough' to increase wages and per capita income" (Luebke, 1990, p.77). MCNC was criticized by academics and others who argued that it's net effect might be to force up wages in North Carolina's traditional industries, while some parts of the microelectronics industry were also seen to pay low wages. They also argued that the relatively well off Urban Crescent would benefit at the expense of rural areas. Later on, Governor Hunt was also lured into the bidding 'war between the states' when tax abatements and other incentives would be offered to attract industry. The most publicized case was a $120 million package offered to the Mercedes company, who eventually chose Alabama in response to a $300 million package. Despite losing other 'trophy plants' to neighboring states (like BMW to South Carolina and Motorola to Virginia), the Hunt Administration was viewed positively in this game because he refused 'to sell off the state' to Mercedes and established a more rigorous set of guidelines for incentives than other states. Hunt's megaproject, MCNC, remains a topic of contention in the Legislature, though a second R and D center: the North Carolina Biotechnology Center was also formed.

The last example of a megaproject belongs to Governor Jim Martin, a Republican who was in office during the Interregnum of Jim Hunt. Martin became enamored with the idea of a Global Transpark that would combine a high tech industrial park with a major airport where factories on site could respond to JIT orders where airplanes could meet demand from around the globe in record time. While this in many ways was an idea ready to meet the high tech industrial needs of the next century, a political decision to locate the GTP in a rural part of eastern North Carolina (near Kinston), far removed from the existing factories of the Piedmont Urban Crescent, has now put the project at a severe disadvantage. After six years of planning and over $30 million spent on feasibility studies by the federal and state governments, the GTP has only one client to date and has been mocked as catering to 'JIT strawberries' and not JIT manufacturing. Like the allocation of a major medical school to East Carolina University, the GTP story reflects the political clout of the rural lobby in the state as well as a genuine concern for the welfare of rural areas. To date however this megaproject is the only one started by a Governor that may end up in failure.

Over the past half century then North Carolina's governors have compensated for their lack of political power by showing leadership in the state's economic development plans. By launching such megaprojects as Research Triangle Park, the Microelectronics Center and even the Global Transpark, Governors have put a permanent stamp on the state's economic evolution and some have gained the envy of counterparts in other states. This type of leadership strategy is a powerful one and one that we have tended to overlook to date in our scholarly evaluations of regional development.

5.6 Conclusions

The main conclusions of this paper are similar to those of the study on the recent renaissance of the American steel industry, that the renewed competitive advantage of that sector is fundamentally a reflection of two factors: the industry's use of newly available technology, and the role of individual leadership in facing enormous challenges. My perspective on the wealth of regions is similar: the most creative regions reflect the impact of generating and/or adopting the latest available technology and this can only be done with the vision and leadership of creative individuals in both the private and public sectors. Those individuals in turn should always be attuned to the need for continuous improvement, as reflected in the Japanese concept: Kaizen. While this review shows that we know a considerable amount about the geography of technological change, we now know more about the role of entrepreneurs in that change, but we still have an empirical deficit when it comes to the link between leadership, social capital and regional development.

It is the link between the role of technological change and leadership that can lead to the growth of new industrial regions and to the rejuvenation of older ones. Since much of the research on technological change over the past twenty years has focused on the experience of successful and creative regions, our studies in the future should focus more on the experiences of less successful regions, particularly if we expect to perform in the policy arena to upgrade the general welfare of those less successful regions.

This review of technology-based theories of regional development argues that old concepts like the product cycle model and regional life cycles have as much validity today as they did twenty years ago. And while our concern with defining and creating innovation milieus has led to some conceptual and empirical confusion in the literature, we still need to refine our ideas to include the role of entrepreneurship, leadership, trust and the general social capital of the region. This can only lead to at a higher level of understanding and more informed policy.

And if some concepts are difficult to enumerate, we should not be obsessed with quantifying them, for this may lead to erroneous conclusions, always a danger in the policy arena. Policy research on networks and other instruments of regional technology development shows that North America may have much to learn from the experience of countries like Denmark and Italy. But it is important to remember that government involvement in the future of regions is more fundamental to the European culture than it is to the American. It is only when regional cultures are accounted for that we can fully appreciate the complex relationships between concepts like technological change, the role of leadership and regional development.

Acknowledgements

Thanks to Ed Malecki and participants in the 1998 Uddevalla Workshop for their thoughtful comments on an earlier draft of this paper.

References

Ahlbrant, R.S., R.J. Fruehan and F. Giarratani, 1996, *The Renaissances of American Steel,* Oxford University Press, New York.

Amin, A. and N. Thrift, (eds.), 1994, *Globalization, Institutions and Regional Development in Europe,* Oxford University Press, Oxford.

Bennis, W. and B. Nanus, 1991, *Leaders: Strategies for Taking Charge,* Harper & Row, New York.

Castells, M. and P. Hall, 1994, *Technopoles of the World,* Routledge, London.

Cooke, P., (ed.), 1995, The Rise of the Rustbelt, UC Press, London.

Cooke, P., 1996, "The New Wave of Regional Innovation Networks: Analysis, Characteristics and Strategy", *Small Business Economics,* 8:159-171.

Cooke, P., 1992, "Regional Innovation Systems: Competitive Regulation in the New Europe", *Geoforum,* 23:365-387.

Desantis, M. and R. Stough, 1997, "Modeling Leadership in Regional Economic Development", Working Paper, Institute of Public Policy , George Mason University.

Dubini, P., 1989, "The Influence of Motivations and Environment on Business Start Ups: Some Hints for Public Policies", *Journal of Business Venturing,* 4:11-26.

Erickson, R.A., 1994, "Technology, Industrial Restructuring and Regional Development", *Growth and Change,* 25:353-379.

Erickson, R. and T. Leinbach, 1979, Characteristics of Branch Plants Attracted to Nonmetropolitan Areas", in R. Lonsdale and H.L. Seyler (eds.), Nonmetropolitan Industrialization, Winston, Washington D.C.

Feldman, M.P. and R. Florida, 1994, "The Geographical Sources of Innovation in the US", *Annals of the Association of American Geographers,* 84:210-229.

Feyerabend, P., 1975, *Against Method: Outline of an Anarchistic Theory of Knowledge,* Verso, London.

Feyerabend, P., 1987, *Farewell to Reason,* Verso, New York.

Florida, R., 1995, "Toward the Learning Region", *Futures,* 27:527-536.

Freeman, C., 1995, "The National System of Innovation in Historic Perspective", *Cambridge Journal of Economics,* 19:5-24.

Fukuyama, F., 1995, *Trust: The Social Virtues and the Creation of Prosperity,* Free Press, New York.

Hansen, N., 1992, "Competition Trust and Reciprocity in the Development of Innovative Regional Milieus", *Papers in Regional Science,* 71:95-105.

Johansson, B. and C. Karlsson, 1987, "Processes of Industrial Change: Scale Location and Type of Job", in M. Fischer and P. Nijkamp (eds.), *Regional Labor Markets,* North Holland, Amsterdam.

Krugman, P., 1991, *Geography and Trade,* MIT Press, Cambridge, Mass.

Link, A. and J. Rees, 1990, "Firm Size, University Based Research and the Returns to R and D", *Small Business Economics,* 2:25-31.

Luebke, P., 1990, *Tar Heel Politics: Myths and Realities,* University of North Carolina Press, London.

Maillat, D. ,1998, "From the Innovative Milieu to the Learning Region", Paper presented at the International Workshop on Regional Development, Uddevalla, Sweden.

Maillat, D. and B. Lecoq, 1992, "New Technologies and Transformation of Regional Structures in Europe: The Role of the Milieu", *Entrepreneurship and Regional Development,* 4:1-20.

Malecki, E.J., 1997, *Technology and Economic Development,* 2nd Edition, Longman, London.

Malecki E.J., 1998, "How Development Occurs: Local Knowledge Social Capital and Institutional Embeddedness", Paper, Southern Regional Science Association, Savannah.

Malmberg, A., 1996, Industrial Geography: Agglomeration and Local Milieu", *Progress in Human Geography,* 20:392-403.

Malmberg, A., 1997, "Industrial Geography: Location and Learning", *Progress in Human Geography,* 21:573-582.

Markusen, A.R., 1985, Profit Cycles Oligopoly and Regional Development, MIT Press, Cambridge, Mass.

Nelson, R.R. and S. Winter, 1982, *An Evolutionary Theory of Economic Change,* Harvard University Press, Cambridge.

Nelson, R.R., 1995, "Recent Evolutionary Theorizing about Economic Change", *Journal of Economic Literature,* 33:48-90.

Norton, R.D. and J. Rees, 1979, "The Product Cycle and the Decentralization of American Manufacturing", *Regional Studies,* 13:141-151.

O'Brien, R., 1992, *Global Financial Integration: The End of Geography*, Pinter Press, London.

Ohlin, B., 1993, *Interregional and Regional Trade,* Harvard University Press, Cambridge Mass.

Ohmae, K., 1993, "The Rise of the Region State", *Foreign Affairs,* 92:78-87.

Piore, M. J. and C. F. Sabel, 1984, *The Second Industrial Divide,* Basic Books, New York.

Porter, M.E., 1990, *The Competitive Advantage of Nations,* Free Press, New York.

Rees, J., 1979, "Technological Change and Regional Shifts in American Manufacturing", *Professional Geographer,* 31:45-54.

Rees, J., 1979, "Regional Industrial Shifts in the U. S. and the Internal Generation of Manufacturing in Growth Centers of the Southwest", in W. Wheaton (ed.), *Interregional Movements and Regional Growth,* COUPE 2, Urban Institute, Washington D.C.

Rees, J. and H.A. Stafford, 1986, "Theories of Regional Growth and Industrial Location: Their Relevance for Understanding High Technology Complexes", in J. Rees, (ed.), *Technology, Regions and Policy,* Rowman and Littlefield, Totowa.

Rees, J., 1991, "State Technology Programs and Industry Experience in the USA", *Review of Urban and Regional Development Studies*, 3:39-59.

Rees, J. and K. Debbage, 1992, "Industry Knowledge Sources and the Role of Universities", *Policy Studies Review,* 11:6-25

Rees, J., 1999, *Regional Science: From Crisis to Opportunity,* Papers in Regional Science, forthcoming.

Rosenau, P.M., 1992, Post-Modernism and the Social Sciences, Princeton University Press, Princeton.

Rosenfeld, S.A., 1996, "Does Cooperation Enhance Competitiveness? Assessing the Impacts of Inter-Firm Collaboration", *Research Policy,* 25:247-263.

Saxenian, A., 1994, *Regional Advantage,* Harvard University Press, Cambridge, Mass.

Schmandt, J. and R. Wilson, (eds,), 1990, *Growth Policy in the Age of High Technology*, Unwin Hyman, London.

Schumpeter, J., 1942, *Capitalism, Socialism and Democracy,* Harper and Row, New York.

Scott, A.J., 1988, *New Industrial Spaces: Flexible Production Organization and Regional Development in North America and Western Europe,* Pion, London.

Scott, A.J., 1996, "Regional Motors of the Global Economy", *Futures*, 28:391-411.

Shapira, P. and J.D. Roessner, 1996, "Evaluating Industrial Modernization: Introduction", *Research Policy,* 25:181:183.

Sorenson, D. J., 1997, "An Empirical Evaluation of Profit Cycle Theory", *Journal of Regional Science,* 37:275-305.

Staber, U., 1996, "Accounting for Variations in the Performance of Industrial Districts: The Case of Baden-Wurttenberg", *International Journal of Urban and Regional Research,* 20:299-316.

Sternberg, R., 1996, "Reasons for the Genesis of High Tech Regions – Theoretical Explanation and Empirical Evidence", *Geoform,* 27:205-223.

Storper, M., 1997, *The Regional World,* Guilford Press, New York.

Taylor, M. J., 1986, "The Product Cycle Model: A Critique", *Environment and Planning A*, 18:751-761.

Testa, W.A., T.H.Klier and H. Mattoon, 1997, *Assessing the Midwest Economy: Looking Back for the Future,* Federal Reserve Bank of Chicago, Chicago.

Thomas, M.D., 1975, "Growth Pole Theory, Technological Change and Regional Economic Growth", *Papers of the Regional Science Association,* 34:3-25.

Vernon, R., 1996, "International Investment and International Trade in the Product Cycle", *Quarterly Journal of Economics,* 80:190-207.

Whittington, D., (ed.), 1985, High Hopes for High Tech: Microelectronics Policy in North Carolina, UNC Press, London.

6 History, Spatial Structure, and Regional Growth: Lessons for Policy Making

Gunther Maier
Institute for Urban and Regional Studies, Vienna University of Economics and Business Administration, Roßauer Lände 23, A-1090 Vienna, Austria

6.1 Introduction

In recent years, we have seen some fundamental developments in economic theory concerning the understanding of economic growth. The traditional growth theory, which is based on the famous Solow-model (Solow, 1956), was replaced by a set of models and arguments that are commonly known as "new growth theory". These developments are fundamental in the sense that they demonstrate that some of the basic assumptions of the traditional growth theory – and most of neo-classical economic theory in general – are inconsistent with basic phenomena of the modern economy, and that they attempt to overcome these assumptions.

In regional economics these advances in economic theory received a mixed reception (Isserman, 1996). On the one hand, they were welcomed because they are bound to reintroduce a spatial dimension into economic theory and thus may move regional economics out of its marginal position within the realm of economic sub-disciplines. On the other hand, many of the arguments that are brought forward and of the conclusions that are derived by the new growth theory have been known and discussed in regional economics for decades so that they appear to be old wine in new bottles to many regional economists.

Despite our long tradition in regional economics of discussing the seemingly new arguments of new growth theory, it seems that we have still missed some of their fundamental implications. This is probably due to the lack of a consistent framework. The recent advances in growth theory provide such a framework and they allow us to investigate more thoroughly the consequences and policy implications of our traditional regional economic argument.

This paper intends to take a step in this direction. It will look at the implications for our understanding of regional growth and of regional growth policy when we allow for scale effects and externalities. The discussion will show that this seemingly minor change in the set of assumptions has major consequences on policy and also brings into play factors like the spatial structure or the history of a region that are of no importance at all in traditional neo-classical (regional) growth theory. A very simple model is used to make and illustrate the argument.

In the next section, the basic structure of the new growth theory and its relationship to traditional neo-classical growth theory and to the polarization argument of regional economics is discussed. Section 6.3 will discuss externalities and their relationship to growth processes and to spatial structure. Section 6.4 presents the model and uses it to illustrate the main arguments that were made in the previous sections. The paper closes with a concluding section.

6.2 New Growth Theory, Traditional Growth Theory, and Polarization Theory

Since the polarization model (Perroux, 1950; Myrdal, 1957; Hirschman, 1958) of regional economics as well as new growth theory have been developed in response to the traditional neo-classical growth model, the discussion begins with a brief review of the latter model. In order to avoid unnecessary complications, a very simple version of the model is used. This version, however, represents all the crucial elements of traditional neo-classical growth theory.

A Cobb-Douglas-production- function

$$Y = K^{\alpha} L^{1-\alpha} \tag{6.1}$$

where Y is the total level of production, K is the capital stock and L is labor input is used as the starting point of the argument. This production function is linearly homogeneous and has positive but decreasing marginal products of labor and capital. Let us assume a constant input of labor ($L' = 0$). Capital increases through investment and decreases through depreciation. We assume further a constant share s of Y to be saved and invested, and a constant proportion δ of the capital stock to be depreciated. Therefore, the temporal change in capital can be written as

$$K' = sY - \delta K. \tag{6.2}$$

From this set of equations the production per unit of labor input changes according to the following time path:

$$\frac{Y}{L} = [Ae^{(\alpha-1)\delta t} + \frac{s}{\delta}]^{\alpha/(1-\alpha)} \tag{6.3}$$

where A is a constant representing the initial conditions. Since α is between zero and one, production per unit of labor converges toward a long run equilibrium that is characterized by

$$[\frac{s}{\delta}]^{\alpha/(1-\alpha)}. \tag{6.4}$$

Note that this equilibrium is determined only by exogenously given parameters. Therefore, it can be said that the traditional neo-classical growth theory is a theory that "explains" long term growth only through external factors.

Figure 6.1 shows the basic relationships of the model. Capital intensity, i.e. the amount of capital relative to constant labor input, is represented by the horizontal axis, and the vertical axis represents monetary units. The curve marked Y is the production function for different levels of capital and a constant level of labor, the other two curves represent depreciation (δK) and investment (sY). The equilibrium is characterized by the intersection of the latter two curves.

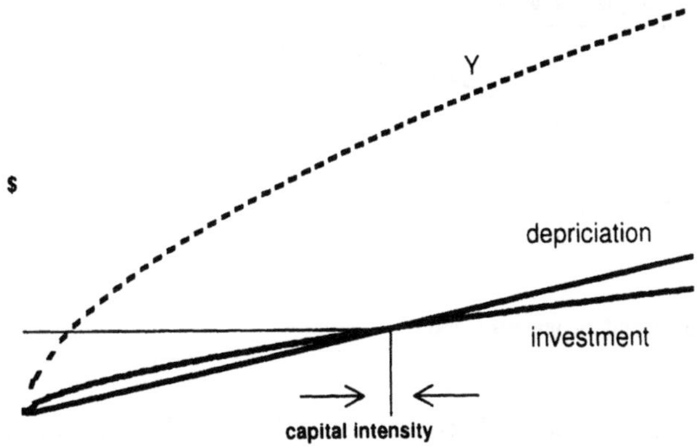

Figure 6.1 Growth and net-investment

It is easy to see from Figure 6.1 that the system will converge to the equilibrium point. To the left of the equilibrium investment exceeds depreciation, therefore the capital intensity increases. To the right of the equilibrium, depreciation is higher than investment, and the capital stock declines. This can also be seen when the corresponding time path is plotted. Starting from a point above equilibrium, production per labor input declines; starting below the equilibrium, it increases toward the long term equilibrium value.

It is easy to see that even in this simple form the model predicts convergent regional growth. Given two regions with identical parameters α, δ, and s, both starting from below the equilibrium value, but one of the regions being more advanced than the other, the less advanced region will grow more rapidly than the advanced one so that the gap in production per unit of labor diminishes over time and finally disappears. Note that this is the case even when there is no interaction at all between the regions. This convergence results solely from capital accumulation.

This convergence process can also be seen in another way. Suppose that a region is pushed off of its growth path by some historical event (a war, for example). Because of the convergence process, the impact of this event is only temporary. It is "washed away" over time by the growth process. The time path after the shock asymptotically approaches the original path.

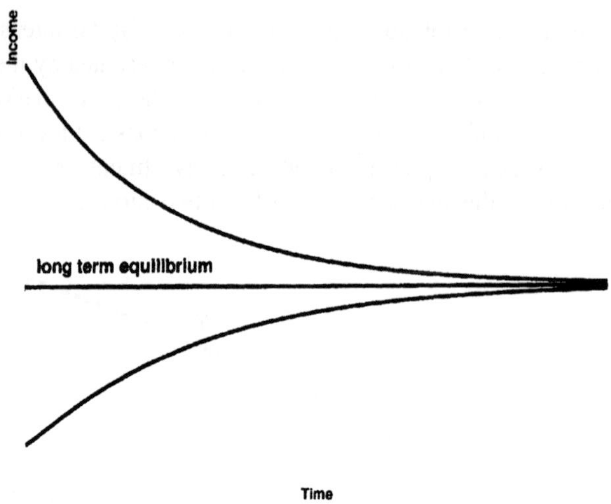

Figure 6.2 Growth and income per worker

In the version of the model discussed so far the growth process comes to a halt once the equilibrium level of capital intensity is reached. Therefore, the production function is usually augmented by a term that represents technical progress. The production function then becomes

$$Y = K^{\alpha} L^{1-\alpha} e^{\tau t} \tag{6.5}$$

where t represents time and τ the rate of growth of technical knowledge. With this alternative formulation the production function and the investment function are continuously shifted up and the equilibrium point therefore moves further and further to the right. Therefore, the long run steady state growth of the system is determined solely by the growth rate of technical knowledge – another parameter that is external to the model.

Although our simple version already allowed us to derive regional implications, the regional economic version of the neo-classical growth model (Borts and Stein, 1964; Richardson, 1969) often puts particular emphasis on factor mobility. Because of the basic assumptions of neo-classical economics, capital and labor are paid their marginal product. Therefore, when capital is relatively scarce in region I as compared to region II the rate of interest will be higher in region I, the wage rate higher in region II. When the production factors are mobile, capital will flow from region II to region I and labor in the opposite direction until the marginal products are equated between the regions. This mechanism supports the convergence that was described above.

Summarizing this discussion of traditional neo-classical growth theory, it can be concluded that according to the model the price mechanism and the process of capital accumulation lead to convergence and thus eliminate interregional differences over time. Neither spatial structure nor historical events have any implication on the long term growth path of a region. The latter is determined only by exogenous parameters.

It is not surprising that regional economists felt quite uneasy about this conclusion. Therefore, many counter-arguments to traditional neo-classical growth theory have been formulated in the discipline. This set of arguments is often referred to as "polarization theory", although it is by no means a consistent theory. Early representatives are Perroux (1950), Myrdal (1957) and Hirschman (1958), more recent versions can be found in the literature on industrial districs and on business clusters. Since Ed Feser (1998) has provided an excellent review of this literature, the discussion of polarization theory can be brief.

Contrary to traditional neo-classical growth theory the advocates of polarization theory argue that production factors are non-homogeneous, that markets are imperfect, and that the price mechanism is disturbed by externalities and economies of scale. Therefore, it is argued, deviations from equilibrium are not corrected by counter effects, but may set off a circular, cumulative process of growth or decline. A complex set of positive and negative feedback loops accumulates to a growth process whose direction is fundamentally undetermined. In a spatial context these feedback processes generate spread and backwash effects that transfer impulses from one region to another.

It is quite clear that spatial structure is an important element in such a growth process. It generates leading and lagging regions that depend upon one another in many ways. Polarization theory often argues not only economically, but brings forward also social, political, cultural, etc. arguments when it explains why some regions are more prosperous than others. Because of this rich and interdisciplinary set of arguments, polarization theory has never been able to express its arguments in a consistent economic model and has, therefore, largely been ignored by mainstream economics.

New growth theory, on the other hand, originated in the heartland of economic theory. Economists like Romer, Barro, Helpman, and Grossman were disappointed with the fact that the long term source of growth remains exogenous in traditional neo-classical growth theory. Therefore, they tried to explain technical progress endogenously in their models. As it turned out soon, there is no incentive for economically rational agents to invest resources into the production of technical progress in the standard neo-classical growth model. Production of technical progress can only be explained when one deviates from the basic neo-classical assumptions. Therefore, the models of new growth theory allow for either agglomeration effects (economies of scale and externalities) or for market imperfections. They can now explain economic growth – therefore they are also referred to as "endogenous growth models" – but the implications of this change in the basic set of assumptions is quite dramatic. New growth theory leads to outcomes that are quite similar to those that polarization theory has claimed decades before:

In a model with agglomeration effects or imperfect markets the price mechanism does not necessarily generate an optimal outcome. Depending on the structure of the model, there are either too few (e.g. Barro, 1990; Rebelo, 1991; Romer, 1986, 1990) or too many (e.g., Aghion and Howitt, 1990; Segerstrom et al., 1990;

Grossman and Helpman, 1992) resources invested in the production of technical progress in the free-market solution. The market mechanism by itself does not lead to an efficient allocation of resources.

The process of capital accumulation and free trade do not necessarily lead to convergence between regions. With positive agglomeration effects the concentration of economic activities in one region will be self enforcing because this region will become more attractive for new investments. This generates a circular, cumulative process as in polarization theory. While the traditional neo-classical growth theory can demonstrate that under its set of assumptions capital accumulation will lead to convergence, the models of the new growth theory allow for both, convergence and divergence. The type of situation that is prevalent becomes an empirical question. Therefore, there have been numerous studies analyzing the convergence or divergence of growth processes in the last years (e.g., Barro, 1991; Barro and Sala-i-Martin, 1991).

The most important implication in our context is that in an economy with agglomeration effects or imperfect markets spatial structure and historical events become important. While they are washed away quickly in the traditional theory, in the new growth theory they may trigger long term growth differentials. In the case of positive agglomeration effects, for example, a randomly occurring concentration of economic activity may attract outside investment that will make this region even more attractive (Arthur, 1994; David, 1985). Because of the positive effects between the various economic agents, the process will yield an unequal distribution of economic activity between regions that is stable in the long run. To repeat an argument from above, this outcome may not be optimal or even desirable in any way.

6.3 Agglomeration Effects, Growth, and Regional Structure

Polarization theory as well as new growth theory see economies of scale and externalities as major sources of cumulative effects in the economy. Therefore, in this section these factors are investigated more closely. Since economies of scale and externalities both result from the concentration of economic activities, they are subsumed under the term "agglomeration effects".

While economies of scale occur within an economic unit, externalities happen between economic units. In regional economics it is quite common (Hoover, 1937; Carlino, 1978) to subdivide externalities into localization effects and urbanization effects. Localization effects occur between firms of the same sector, urbanization effects between firms from different sectors or even different types of economic actors like firms and households. Obviously, this distinction is not very precise, because it depends on the definition of sectors and economic units. In a macroeconomic context, for example, when all the firms of a region are aggregated into one producing unit, agglomeration effects can only show up in terms of economies of scale.

The justification for agglomeration effects differs substantially between polarization theory and new growth theory. While polarization theory states them as empirical phenomenon, new growth theory shows that a perfect-market model without agglomeration effects is inconsistent with the production of technical progress. With this theoretical justification, agglomeration effects have to be taken into account when we are talking about dynamic aspects of the economy (Bröcker, 1994).

However, agglomeration effects are directly related to spatial structure. As will be argued below, spatial structure and agglomeration effects are actually two sides of the same coin. On the one hand, agglomeration effects lead to a spatially differentiated structure of the economy, while on the other hand spatial structure produces agglomeration effects.

Because it does not allow for agglomeration effects, the traditional neo-classical model implies a peculiar spatial structure. Mills (1972, p.113) summarizes it in the following way:

> Consider a general equilibrium model in which an arbitrary number of goods is produced either as inputs or for final consumption. The only nonproduced goods are land and labor, each of which is assumed to be homogeneous. Assume that each production function has constant returns to scale and that all input and output markets are competitive. Utility functions have the usual properties and have as arguments amounts of inputs supplied and products consumed. Under these circumstances, consumers would spread themselves over the land at a uniform density to avoid bidding up the price of land above that of land available elsewhere. Adjacent to each consumer would be all the industries necessary – directly or indirectly – to satisfy the demands of that consumer. Constant returns assures us that production could take place at an arbitrarily small scale without loss of efficiency. In this way, all transportation costs could be avoided without any need to agglomerate economic activity.

One of the key assumptions is that of production functions with constant returns to scale, i.e. the assumption that there are no agglomeration factors. If only one industry of this economy had a production function with positive agglomeration factors, the spatial structure of the economy would differ markedly. Because of positive returns to scale this industry could produce in a more efficient way when it concentrates production in one or a few locations. When it does so, however, not "all the industries necessary" would be adjacent to each consumer any longer. Some consumers or sectors would need to buy the products of the agglomerated sector or to sell their products to this sector and would therefore have to overcome the spatial distance to this sector. Those who are located closer to the location of the agglomerated sector would have an advantage over their colleagues further away. They could either save transportation costs or produce larger quantities.

In the traditional model as described by Mills, each consumer provides labor inputs to all the industries at her location. But, when one of the sectors is agglomerated, it produces at a larger scale and therefore needs more labor input than is available at this location. It has to attract additional workers who either will have to commute or to migrate to this location. In the first case the agglomerated industry will have to pay higher wages in order to compensate the workers for the costs of commuting, in the second case the sector will have to pay higher wages because with increasing density the costs for land will increase. As a consequence, interregional differences in wages and land prices, and in the density of economic activity occur. The location of the agglomerated industry turns into a market center.

Because of the increasing wages and land prices, those industries that cannot afford these would move away from the locations where the industry with economies of scale is concentrated. The result would be a specialized pattern of land use. Some of the industries that are forced out would try to locate in the vicinity of the market center because of the need to sell their products there. However, as has been shown already by von Thünen (1826), they will not locate randomly but according to their bid-rent functions. Those industries whose products are more sensitive to transportation will locate closer to the market than those whose products can be transported more cheaply.

Note that the argument has been made within the framework of the traditional neo-classical model and has allowed only one industry to have a production function with increasing returns to scale. The result, however, is a structure with spatial differentiation in land use, prices, and densities and with the need to transport products and production factors from one location to another. When other industries are allowed to show increasing returns to scale as well, or add location or urbanization effects, the results that were described above will be strengthened further, but qualitatively they will not change. Agglomeration effects in one industry are sufficient for producing spatial structure and spatial differentiation.

Starrett (1978) analyzes the relationship between spatial structure and agglomeration effects in a more rigorous way. He finds that in a system with spatial structure "the usual types of competitive equilibria will never exist" (Starrett, 1978, p.21). Whenever there are transportation costs in the system and all agents are price takers in competitive markets, all possible allocations are instable. "That is, for any set of prices and location allocation, some agents will want to move to the other location" (p.25). The reason for this result is quite simple. Because of transportation costs, and in the absence of any counter forces every economic actor sees an incentive to move closer to his suppliers and/or his markets. In a system of perfect competition this tendency pulls related actors closer together until the system collapses to the structure described by Mills, where all transportation costs are eliminated.

Market imperfections like agglomeration effects or monopoly power counteract this tendency and can stabilize a system with transportation costs. A company with economies of scale, for example, would trade off these economies of scale against potential transportation cost savings and find a location and size of production that balances these counteracting forces.

That these forces are closely related to transportation costs also shows up in Starrett's work. In addition to the above mentioned non-existence result he demonstrates the following: "The monetary incentive to move is on the order of magnitude of transport costs. The degree of market imperfection which is required in order that a location allocation be stable is also related to transport costs" (Starrett, 1978, p.25).

These arguments show that a stable spatial structure, i.e. an allocation of economic activities where the type and amount of economic activity differs between locations and where goods and production factors are exchanged between locations, and agglomeration factors are two sides of the same coin. One requires and implies the other.

To some extent our discussion in this section parallels the arguments of the new growth theory discussed above. While new growth theory shows that the production of technical innovations is incompatible with the standard assumptions of neoclassical theory, the above arguments show that spatial structure is incompatible with these assumptions as well. Allowing for agglomeration factors can overcome both of these problems. However, when agglomeration factors are allowed for, the model will generate much more complex dynamic processes and a diversified spatial structure. Agglomeration factors constitute a theoretical link between spatial structure of an economy and its growth dynamics.

In order to illustrate the relationship between agglomeration effects, spatial structure and growth dynamics, let us use a simple model by Arthur et al. (1987). Suppose we have two regions and a process that generates one new company per time period. In each period this new company is assigned to one of the two regions at random. Once a company is assigned to a region it stays there. There is no interregional mobility of companies.

Two assignment processes are identified:

1) The probability that a new company is assigned to a region is exogenously given. For simplicity it is assumed that both regions have probability 0.5.
2) The probability that a new company is assigned to a region is proportional to the region's share of companies. For obvious reasons the assignment begins with an initial endowment of one company per region.

The two assignment processes differ as far as agglomeration factors are concerned. While in the first version each assignment is independent of earlier assignments and the existing concentration of companies, in the second assignment process there is a positive feedback between a region's share of companies and its chance for the next new company. This is clearly a positive agglomeration factor.

Despite this seemingly small difference, the long term outcomes of the models as far as spatial structure and growth dynamics is concerned differ dramatically. Figures 6.3 and 6.4 show simulations of the development of the shares of companies in the two regions. Figure 6.3 illustrates the case without agglomeration effects, Figure 6.4 the case with agglomeration effects.

The long term behavior of the version without agglomeration effects (Figure 6.3) is quite clear. Since the random assignments in each period are independent from one another the law of large numbers applies and in each of the four simulation runs a region's share of companies has to converge toward the constant assignment probability for this region (0.5 in our example). The four time paths plotted in Figure 6.3 all show some major fluctuations in the early phase of the process and then converge toward the long term share of 0.5. The fluctuations result from the fact that adding one company has a larger impact on the share with a small total number of companies than with a larger one. However, these fluctuation die out over time. Any advantages in the share that a region gains because of early success in the random assignment process is eliminated quickly by later random assignments. This corresponds to the typical growth process of the traditional neo-classical theory.

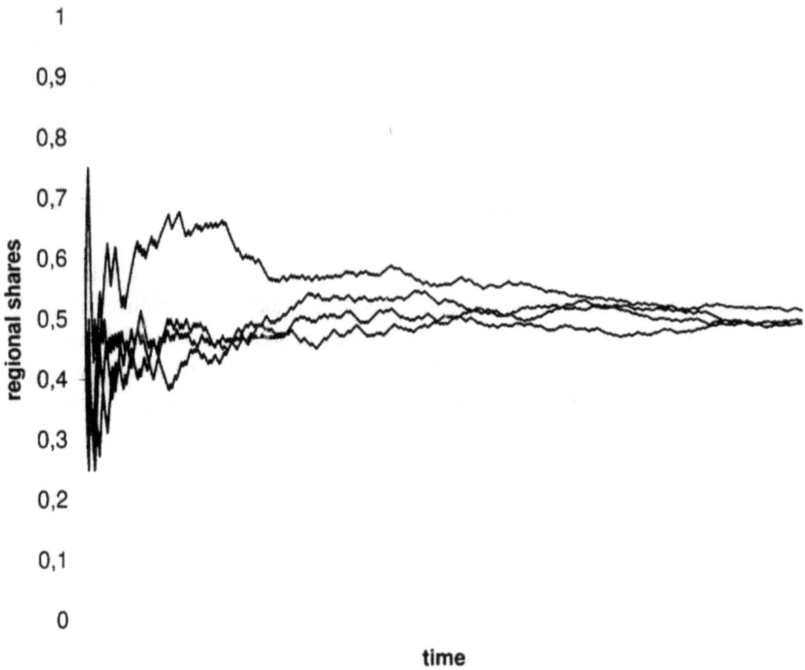

Figure 6.3 Growth paths in a system without agglomeration factors (4 simulations)

Figure 6.4 shows that the in the second case the regional shares do not converge toward a single value. Each of the four runs plotted seems to tend toward a different value in the long run. This is in fact the case. In mathematical terms the process that we have used for our second example, where the assignment probabilities at a certain point in time are equal to the shares at that time, is known as a Polya-process (Polya and Eggenberger, 1923). From Polya-theory it is known (Polya, 1931) that such a process converges to a stable set of proportions in the long run. "But although this vector of proportions settles down and becomes constant, surprisingly it settles to a constant vector that is *selected randomly* from a uniform distribution over all possible shares that sum to 1.0" (Arthur, 1994, p.102). So, although we know that the process will settle down to a certain regional distribution of companies that will then remain constant over time, each possible outcome is equally likely. Formulated differently: this process will produce a stable spatial structure, but it is unknown a-priori what this structure will be. All structures are equally likely.

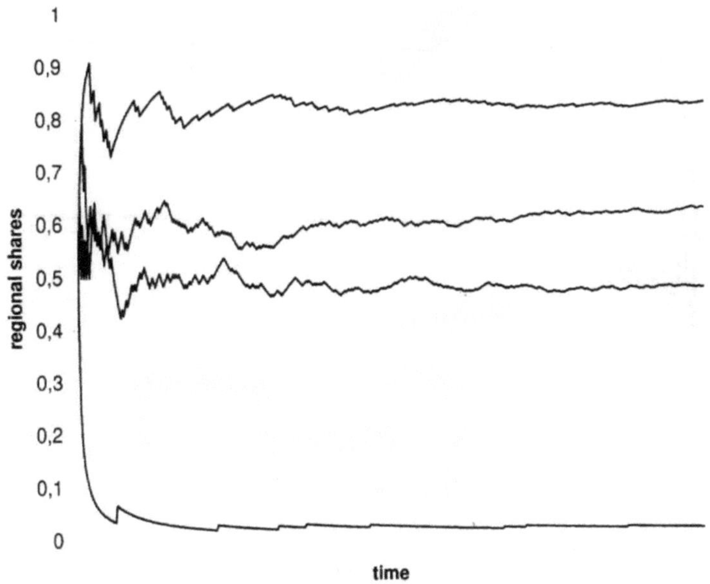

Figure 6.4 Growth paths in a system with agglomeration factors (4 simulations)

As in the case without agglomeration effects, we see strong fluctuations early on. But now these fluctuations are not just temporary phenomena, but determine the long term result of the process. A region that accumulates companies early on in the process because of good luck will end up with a high share of companies in the long run. Similarly, a region that loses early on in the growth process will also lose in the long run in the sense that it will reach only a low share of companies.

This process with agglomeration effects clearly shows path dependence. The long term fate of the process is determined early on in the process. Because of the relationship between share and assignment probability seemingly small events in the early process accumulate over time to differences in the long term outcome. The importance of early events is illustrated in Figure 6.5. The top path has been generated according to the mechanism described above by use of a series of random numbers. The first time a company is assigned it is assigned by chance to region I. After one thousand repetitions of the assignment process the share of companies of region I has settled down at 89.7%. The bottom path of Figure 6.5 has been generated from the same series of random numbers. Only for the first round the company was manually assigned to region II. As a result, the long run share of region I reaches only 57.4%. The difference between these two shares results only from whether the first company is assigned to region I or region II.

Because of the impact this assignment has on future assignment probabilities, it results in a long term difference of about 32 percentage points. Depending on whether the first company is assigned to it or not, a region will develop into a dominant location of economic activity or just a marginal one.

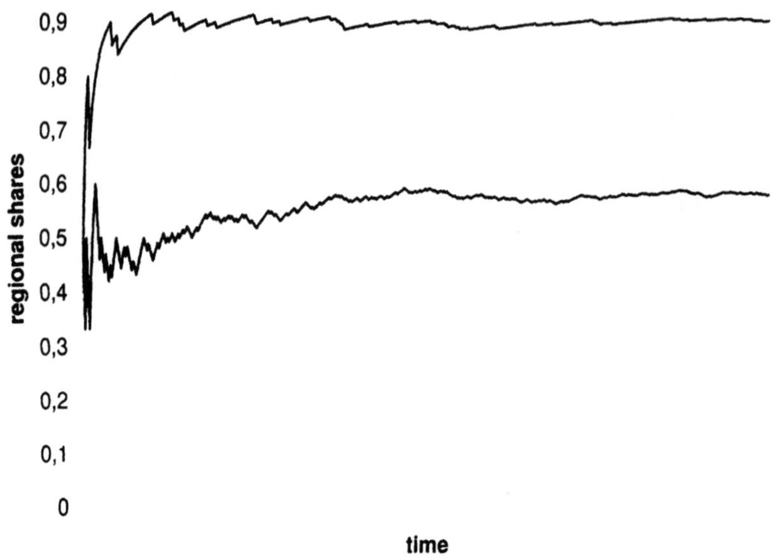

Figure 6.5 Impact of early differences in assignment

The two types of dynamic random processes discussed – a process with constant assignment probabilities and a process with assignment probabilities proportional to shares – are just two out of a number of possible variants. These variants may show different long term behavior. For example, they may possess a number of fixed points to which the process may converge in the long run. However, when there are agglomeration effects the process is path dependent and events in its early phase determine to which fixed point it will converge in the long run. It can be shown (Arthur, 1986) that if the benefits from agglomeration increase without a ceiling as companies are added to the system then one of the regions will eventually gain enough attractivity to capture all the subsequent allocations. This region will dominate the allocation of economic activity in the long run and shut out all the other regions.

Additional insights into the dynamics of the process can be gained by plotting the assignment probability of a region (vertical axis) against the share of this region (horizontal axis). Figure 6.6 gives some typical examples. The broken line marked "model 1" represents the model without agglomeration effects. Since this model had a constant assignment probability it is represented by a horizontal line (in our case at a probability of 0.5). The long term behavior of the model can be read off directly from this graph. Whatever the share of the region, the assignment probability is always 0.5. Therefore, when the share is below 0.5, it will tend to increase, whereas when it is higher than 0.5 it will tend to decrease. The fixed point of this process is at the intersection of the line with the 45°-line, the share will tend toward 0.5.

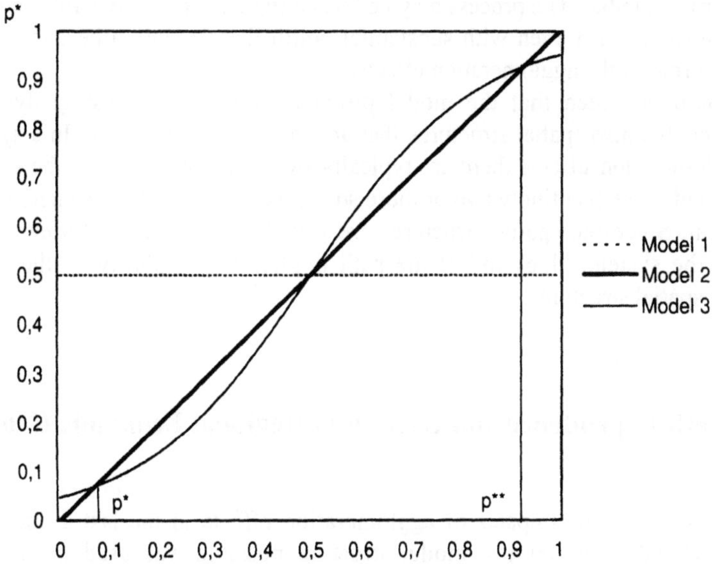

Figure 6.6 Dynamics of different models versions

The line that represents model 3 intersects the 45°-line three times. However, the fixed point at 0.5 is unstable because the slope of the line at this point is higher than 1. Therefore, when the share of the region is slightly higher than 0.5 the region's chance to get the next assigned company is higher than its current share and the share will grow. When the share is slightly below 0.5 it will tend to decrease. The other two intersections of the curve with the 45°-line represent stable fixed points. Therefore, in the long run the share will tend to a value either close to zero (p^*) or close to one (p^{**}).

For the Polya-model (model 2) the assignment probability always equals the share. Therefore, it is represented by the 45°-line. As a consequence, the Polya-model has an infinite number of stable fixed points. This implies the results discussed above.

A few points are of particular importance in this discussion of dynamic processes with agglomeration effects:

Path dependence implies that "historical events" – represented by random influence in our discussion – may play a decisive role in the development of a region. When they occur early in the process they may set the process off in a certain direction. Later on in the process, however, they may influence it only marginally.

Path dependence is paralleled by the phenomenon of "lock in". Once the process has been set off on a certain path it becomes more and more difficult to move it away from this path. Since the famous results of welfare economics about the Pareto optimal outcome of the market allocation process do not apply in the case of agglomeration effect, there is no guarantee that the path the process takes is in any form desirable. The process may be locked in to a very unfavorable path but cannot depart from it even with substantial political intervention because of the stabilizing role of the agglomeration effects.

It should be noted that the model produces not only interesting dynamic trajectories, but also spatial structures that are stable in the long run. In a system with agglomeration effects there are typically two or more paths that the system can take and correspondingly two or more spatial structures that may emerge from the system. Since the spatial structures are just the cross-sectional view of the paths of the system, "lock in" of the path implies high stability of the corresponding spatial structure.

6.4 Path Dependence and Lock-in in Regional Economic Growth

The discussion of the impacts of agglomeration effects in the last section was highly stylized. The simplest model structure possible was used in order to concentrate the discussion on the main points. Therefore, all the economic mechanisms discussed in Section 6.2 of the paper were ignored. Now, the question arises "what will be the result if these economic mechanisms are reintroduced"? Will the qualitative results of the previous section prevail?

In this section the analysis focusses on a model that combines components from Section 6.2 and Section 6.3. More specifically, the neoclassical growth model from section 2 will be augmented with an innovation process that is modeled in accordance with Arthur's model with agglomeration effects that was discussed above (see Figure 6.7).

6.4.1 Model Structure

The basic model structure is as follows:

Suppose we have two regional economies that each can be modeled by the neoclassical model. In order to avoid unnecessary disturbances and adjustment processes each regional economy is assumed to have reached its long term equilibrium. The capital stock grows according to savings and depreciation. Capital is assumed to be perfectly mobile between the two regions, labor on the other hand is assumed to be regionally immobile.

In each time period one unit of innovation is added to the system. This additional unit is allocated to one of the regions at random. The probability for a region to receive this additional unit of innovation is assumed to be equal to its share of production in this period.

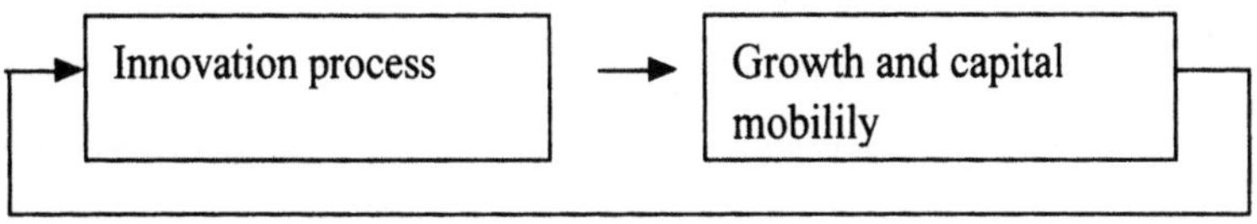

Figure 6.7 Basic structure of the model

Note that this is basically a neo-classical model. It differs from the traditional neo-classical model only in the way innovation is allocated. While in the traditional neo-classical model one unit of innovation per time period is added to every region, in our model the additional unit of innovation is allocated at random. The generation of innovation is still exogenous to the model as it is in the traditional neo-classical model.

6.4.2 Dynamic Behavior of the Model

In our model we combine two components that display quite different dynamic behavior. While the neo-classical growth model tends toward equilibrium and interregional convergence, the innovation model produces path dependence and may tend to different fixed points (see Fig. 6.6). The question arises as to which type of dynamic behavior will occur when the two models are combined.
Before looking for an analytic answer to this question it is useful to view some simulation results. Figure 6.8 shows the production share of one of the regions for six simulation runs of the model. Clearly, the model does not tend toward convergence. It seems that in the long run the region's share of production tends to either one or zero. The dynamic behavior of the total model seems to be different from that of both of its components.

In order to analyze the dynamic behavior of the model, let us look at it in more detail. First, note that the generation of capital can be divided into two components: the accumulation of capital for the system as a whole and the allocation of capital according to its marginal productivity in the region. Second, labor is assumed to be immobile. This is necessary because without keeping one of the production factors fixed all capital and labor would immediately move to the region that has a temporary innovative advantage. Additionally, it is assumed that labor is the same in both regions.

Therefore, we can write the production function for region i as

$$Y_i = (\mu_i K)^{\alpha} L^{1-\alpha} \exp(\tau I_i) \tag{6.6}$$

where μ_i is the region's share of capital and I_i is the number of units of innovation it has accumulated so far. K is the capital in the system as a whole, L is the constant amount of labor in the region. The other variables have the same meaning as in Section 6.2.

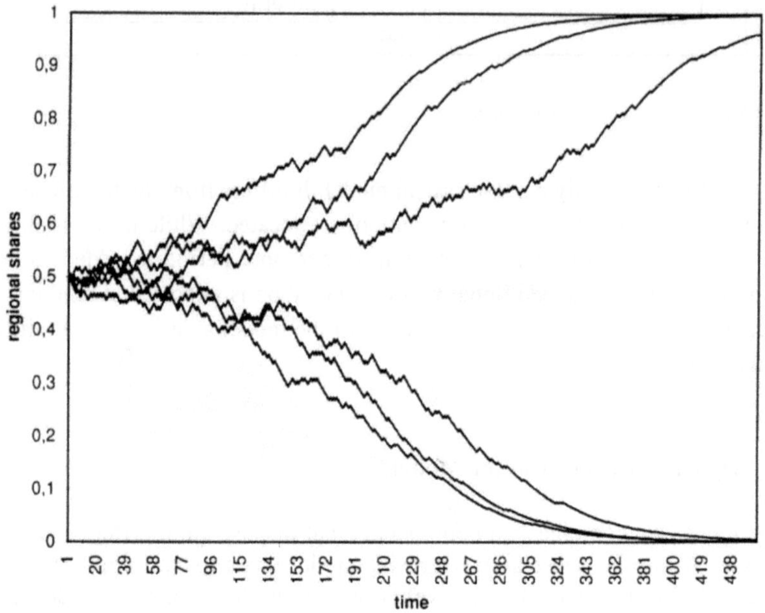

Figure 6.8 Time paths of the model (six simulation runs)

The share of capital is determined by setting the marginal productivities of capital equal in both regions. This yields the following condition for μ (we set $\mu_1 = \mu$ and $\mu_2 = 1 - \mu$):

$$[\frac{\mu}{1-\mu}]^{\alpha-1} = \exp[\tau(I_2 - I_1)]. \tag{6.7}$$

In order to learn about the dynamic behavior of the system it is necessary to find the relationship between the region's share in innovation units – $i_1 = I_1 / N$ with $N = I_1 + I_2$ – and the probability that it will receive the next unit of innovation. This probability for region 1 is P_1 and is defined as:

$$P_1 = \frac{Y_1}{Y_1 + Y_2} \tag{6.8}$$

Substituting the equation of the production function and the condition for μ, after some simplification, yields the following result:

$$P_1 = \frac{1}{1+\exp[\tau N(1-2i_1)/(1-\alpha)]}. \tag{6.9}$$

Note first that the assignment probability takes the form of a logit-model. Second, the assignment probability depends not only on the share and externally given parameters, but also on N, the number of units of innovation in the system. Since N changes over time we must expect the fixed points to change over time as well.

It is easy to see that the function increases monotonically in i_1. Moreover, it has a fixed point at $i_1 = 0.5$ irrespective of the values of τ, N and α. Whether this fixed point is stable or not can be determined by examining the slope of the function at this point. When the slope is equal or less than 1 the fixed point is stable. Conducting the respective calculations it is found that the fixed point at 0.5 is stable only for

$$N \le \frac{2(1-\alpha)}{\tau}. \tag{6.10}$$

This shows that for given parameters the fixed point at 0.5 is stable only up to a certain point in time represented by N. Up to this time the system will tend toward an equal distribution of innovations and production, after this period it will tend toward another fixed point.

Figure 6.9 plots the function of the assignment probability for different values of N (τ has been set to 0.01, α to 0.6). The dotted line represents the 45°-line, the thick broken line represents the line for $N = 80$, in this example the value when the fixed point at 0.5 becomes instable.

So, for the first 80 periods the system will tend toward an even distribution of innovation and consequently also of production. But, starting with period 80 small deviations from this distribution will imply that the assignment probability for this region will change in the same direction by more than the deviation in the share. Therefore, the share will tend to either increase or decrease depending on the direction of the deviation. After period 80 the system has two stable fixed points, one at a value above 0.5 and the other the same distance below 0.5. But, as the share tends toward one of these new fixed points with every period the fixed points move further away from 0.5. Since the fixed points tend toward zero and one as N increases, also the share of innovation (and consequently the share of production) will eventually end up in this extreme situation.

In the neo-classical model capital mobility is an important equilibrating factor. In our model, however, it amplifies the random fluctuations in innovation and therefore adds to instability. In order to see this, assume that we fix μ exogenously at 0.5. In this case it drops out of the equation for the assignment probability and this one simplifies to

$$P_1 = \frac{1}{1+\exp[\tau N(1-2i_1)]}. \tag{6.11}$$

The condition for a stable fixed point at 0.5 becomes $N \leq \frac{2}{\tau}$.

But, this threshold is obviously higher than for the model with capital mobility above. Therefore, when allowing for capital mobility the model is stable (in the sense that it has a stable fixed point at 0.5) for a shorter period than when capital is immobile.

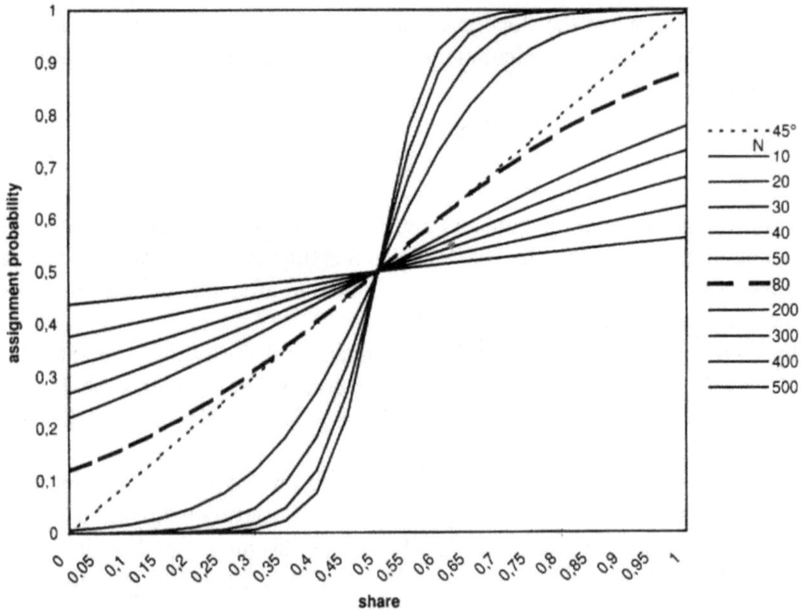

Figure 6.9 Assignment probability function for different levels of N

6.4.3 Policy Considerations

Initially the two regions distinguished in our model are identical. They start off with the same amount of capital and labor, identical production functions and the same probability for getting assigned the first unit of innovation. Therefore, a development path that keeps the level of economic activity balanced between the two regions seems like a reasonable goal for regional policy. However, in the previous discussion of the dynamic properties of the model it was shown that after an initial period where balanced growth is a likely pattern the model tends to concentrate economic activity and thus also economic growth in one of the regions. In the long run there will always be one region that eventually reaches a sufficient concentration of economic activities that it attracts practically all future innovation and therefore all growth. The other region will stagnate and because of the growth in the system as a whole constantly lose share of economic activity. Taking into account that labor was assumed to be interregionally immobile it is implied that half of the population of the system is confined to a stagnating economy and shut out from future economic gains. Obviously, because of the social and political tensions this must generate, such a situation is not sustainable.

The question arises whether regional policy can save the system from this fate. Can regional policy keep the economic activity balanced between the two regions?

When we look at Figure 6.9 we see immediately that the chances for this are slim. With growing levels of N the logit function that describes the assignment probability tends more and more toward a step function with assignment probabilities being zero for shares below one half, 0.5 when the share is exactly one half, and assignment probabilities being one for shares higher than one half. Therefore, the forces that pull the system away from a balanced distribution of economic activity become stronger and stronger over time. In this range, whenever the system departs from a balanced distribution because of some random influence it will be sucked into a state with almost all economic activity concentrated in one region. Policy's only chances to avoid this are either to eliminate the influence of innovation on the economic system or to perfectly assign innovations to the regions.

Neither of these alternatives is very practical. The first one would eliminate all growth from the system and both regions would fall into stagnation. The second alternative would require excessive authority and probably a centrally planned economy.

A particular problem for regional policy lies in the initial period of "stability". In this period the shares fluctuate around the desired value of 0.5. When the behavior of the system during this period is observed, one finds no indications that after a few more time periods it will reach a state of instability. Once the system has reached this state, it does not switch immediately into the final state of instability. Since the fixed points move away from the value 0.5 gradually over time, the final fate of the system does not become apparent immediately. Therefore, it will probably take some time before regional policy even identifies the problem. During this time the system will most likely have reached a state where it is already locked into a path toward its final fate.

The path dependency of the system suggests that the timing of a policy is important. However, it is necessary to distinguish the timing of a policy from its relative weight. Reassigning one unit of innovation, for example is a much more important policy measure in the fifth period (with $N = 5$) than in period 50. Because of the simplicity of the model only two types of policy can be analyzed, namely exogenous assignment of units of innovation either as the assignment of additional units or as reassignment from one region to the other, and transfer of capital from one region to the other.

Figure 6.10 shows the effect of the first type of policy at different points in the growth process. The graphs show the reference growth path and then growth paths for exogenous assignment of 1/50 of the units of innovation to the region at periods 50, 100, 150, and 200. Identical random numbers were used for these simulations. None of the policies can keep the system on a balanced path and none can turn the displayed region from the losing to the winning one in the long term distribution of economic activities.

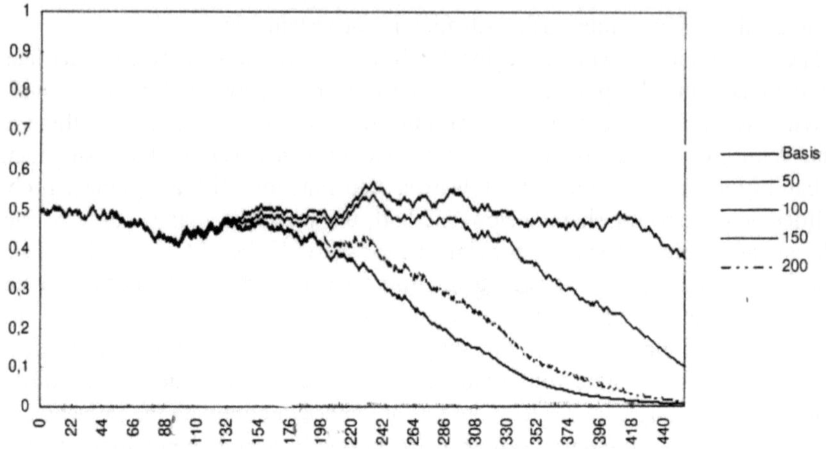

Figure 6.10 Policy implication – exogenously assigning one fifth of innovations

Figure 6.10 shows quite clearly the importance of timing of the policy measure. When the policy is applied at period 200 it has the least effect. It can raise the growth path over the baseline, but toward the end of the observation period the effect has almost vanished. Obviously, at period 200 the growth path has already moved too far down from a balanced distribution for the policy to have a major impact. Although we assign only half the number of units of innovation at that time, when the policy is applied at period 100 it has the biggest impact. The growth path remains balanced for a much longer period of time and even exceeds the 50% mark for quite a number of periods. However, at the end of the observation period the growth path starts to turn down.

When applying the equivalent policy fifty periods earlier, its impact is less pronounced. Interestingly, the same long term outcome from this timing occurs as when the policy is applied at period 150. The respective curves coincide for periods beyond 150. The reason for this seems to lie in the fact that the policy is applied already in the stable period of the system and that its effect is partly washed away before instability sets in. This indicates that in such a dynamic system a policy may not only be applied too late but also too early for its full impact.

Finally, turning toward interregional transfer of capital as a possible policy for keeping the distribution of economic activity balanced. It is assumed that policy has the authority to transfer capital from one of the regions to the other. Whenever the share of production in the regions deviates by more than a certain threshold from the ideal value of 0.5, the policy maker looks at the distribution of capital between the two regions and implements a policy that in the next period shifts a certain percentage of the capital difference from the less capital intensive region to the more capital intensive one. So, the policy measure has a time lag of one period and takes into account the situation in only one time period. Parameters of this policy are the threshold when the policy will become effective, and the percentage of the difference in capital that is transferred.

Figure 6.11 shows a typical simulation run for this type of policy for different threshold values. The second parameter was set to 0.5 which means that the policy maker attempts to balance the distribution of capital. If capital remained constant in the two regions, by transferring 50% of the difference the policy maker would balance the capital stock in the regions. As illustrated in Figure 6.11, the policy produces a lot of turbulence but no fundamental change in the long term result. When the system exceeds the threshold for the first time, the policy is implemented, capital transfer moves the production share back down under the threshold so that in the next period the policy is discontinued. This creates the fluctuations that are diplayed in Figure 6.11. After some periods, however, capital transfer cannot push the production share below the threshold any longer, the policy remains in place, the fluctuations stop, but the system continues to drift away from an unbalanced distribution – despite the policy.

As can be seen from Figure 6.11, the level of the threshold does not make much difference. When policy makers are more sensitive the tendency toward the extreme outcome of the system is generally delayed, but even very sensitive policy interventions cannot save the system from its final fate. This is the case despite the fact that the policy is assumed to transfer a substantial amount of capital. Even more powerful policies (higher values of the second parameter) yield qualitatively the same result. They only generate more severe fluctuations.

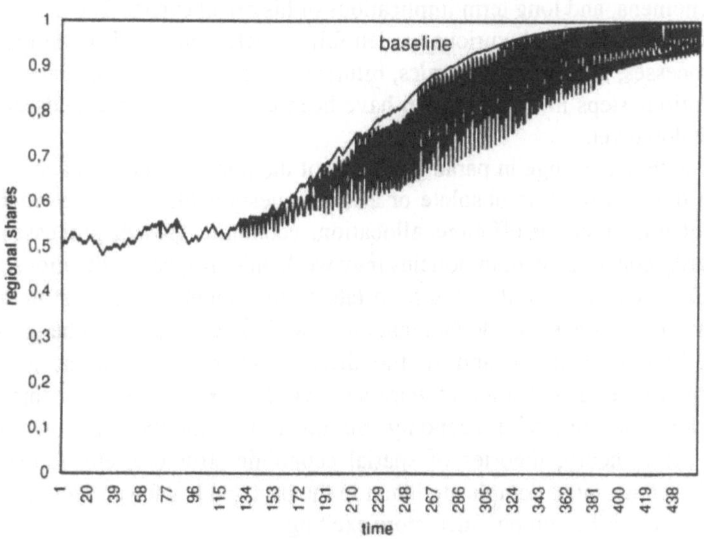

Figure 6.11 Effects of interregional transfer of capital

6.5 Concluding Remarks

In this paper different concepts of regional economic growth have been discussed. The traditional neo-classical growth theory has been reviewed and its critique from the perspectives of polarization theory and new growth theory. As it has turned out, agglomeration effects (economies of scale and externalities) play an important role in the newer concepts of economic growth. Therefore, in Section 6.3 of the paper agglomeration effects, their relation to spatial structure, and the implications they have for the long term dynamics of a process were discussed. It was concluded that spatial structure and agglomeration effects are closely related. They imply and require each other. In Section 6.4 of the paper a simple model implementing some of these concepts was discussed. The model combined a traditional neo-classical growth model with a stochastic model of innovation that implies agglomeration effects. The results of the model are quite striking. Instead of a tendency toward equilibrium and balanced growth the model produces economic disaster in the long run. It converges toward a distribution where one region has almost all production and all future growth and the other region stagnates. No meaningful policy that was able to avoid this extreme outcome was found.

The discussion in this paper illustrates a fundamental shift of paradigm that is taking place in economic theory. The work of new growth theory has shown that agglomeration effects are an essential element of a modern economy and that it is not possible to understand the functioning of an economy without allowing for agglomeration effects.

However, by accepting this argument other factors that have been outside the consideration of traditional economic theory move to the center. Agglomeration effects bring about spatial structure, path dependence of growth processes, "lock-in"-phenomena, and long term implications of historical events. The new paradigm opens up the gates to a luxurious garden full of inefficiencies, disequilibria, divergent processes, non-linear dynamics, bifurcation points, etc. In this paper only the first cautious steps into this garden have been taken. Its diverse landscape is far from explored yet.

But, with this change in paradigm many of the policy guidelines that have been used in the past become obsolete or at least questionable. The price mechanism does not guarantee an efficient allocation, economic growth processes do not necessarily converge, certain policies may work only in specific situations, good or bad luck may determine the long term fate of an economy, etc. Most importantly for regional economists is the fact that the spatial dimension cannot be ignored any longer. Spatial structure and spatial differentiation influence the amount of agglomeration effects that are at work and may therefore have major implications for the long term fate of an economy. But this brings into the play constructs like spatial price theory, theories of spatial economic structure, etc. Economics is becoming much more complicated than in the past and has fewer decisive answers to give. But, it is becoming much more exciting.

References

Aghion P., P. Howitt, 1990, *A Model of Growth through Creative Destruction.* NBER Working Paper No. 3223, Cambridge/Mass.

Arthur, W.B., 1986, *Industry Location Patterns and the Importance of History*, Center for Economic Policy Research Paper 84, Stanford University.

Arthur, W.B., 1994, *Increasing Returns and Path Dependence in the Economy*, Ann Arbor: The University of Michigan Press.

Arthur, W.B., Y.M. Ermoliev and Y.M. Kaniovski, 1987, Path-Dependent Processes and the Emergence of Macrostructure, *European Journal of Operational Research*, Vol. 30, pp.294-303.

Barro, R.J., 1990, Government Spending in a Simple Model of Endogenous Growth, *Journal of Political Economy*, Vol. 98, pp. S103-S125.

Barro, R.J., 1991, Economic Growth in a Cross Section of Countries, *The Quarterly Journal of Economics*, Vol. 106, pp. 407-443.

Barro R.J. and X. Sala-i-Martin, 1991, *Convergence across States and Regions*, Economic Growth Center, Yale University, Discussion Paper No. 629, New Haven/Conn.

Borts, G.H.and J.L. Stein, 1964. *Economic Growth in a Free Market*, New York: Columbia University Press.

Bröcker, J., 1994, Die Lehren der neuen Wachstumstheorie für die Raumentwicklung und die Regionalpolitik, in U. Blien, H. Herrmann and M. Koller (eds.), *Regionalentwicklung und regionale Arbeitsmarktpolitik, Konzepte zur Lösung regionaler Arbeitsmarktprobleme?*, Beiträge zur Arbeitsmarkt- und Berufsforschung Nr. 184, Nürnberg.

Carlino, G.A., 1978, *Economies of Scale in Manufacturing Location*, Boston: Martinus Nijhoff.

David, P., 1985, Clio and the Economics of QWERTY, *American Economic Review Proceedings*, Vol. 75, pp. 332-337.

Feser, E.J., 1998, Enterprises, External Economies, and Economic Development, *Journal of Planning Literature*, Vol. 12, pp. 283-302.

Grossman, G.M. and E. Helpman, 1992, *Innovation and Growth in the Global Economy*, Cambridge/Mass.: MIT-Press.

Hirschman, A.O., 1958, *The Strategy of Economic Development*, New Haven, Conn: Yale University Press.

Hoover, E.M., 1937, *Location Theory and the Shoe and Leather Industries,* Cambridge, MA: Harvard University Press.

Isserman, Andrew M., 1996, "It's Obvious, It's Wrong, and Anyway They Said it Years Ago"? Paul Krugman on Large Cities, *International Regional Science Review*, Vol. 19, pp. 37-48.

Mills, E.S., 1972, An Aggregative Model of Resource Allocation in a Metropolitan Area. Edel, M., J. Rothenburg, *Readings in Urban Economics*, New York.

Myrdal, G., 1957, *Economic Theory and Underdeveloped Regions*, London: Duckworth.

Perroux, F., 1950, Economic Space: Theory and Application, *Journal of Economics*, Vol. 64, pp. 90-97.
Polya, G., 1931, *Sur quelques Points de la Théorie des Probabilités*, Ann. Inst.H. Poincaré, Vol. 1, pp. 117-161.
Polya, G. and F. Eggenberger, 1923, Über die Statistik verketteter Vorgänge, *Zeitschrift für angewandte Mathematische Mechanik*, Vol. 3, pp. 279-289.
Rebelo, S., 1991, Long Run Policy Analysis and Long Run Growth, *Journal of Political Economy*, Vol. 99, pp. 500-521.
Richardson, H.W., 1969, *Regional Economics: Location Theory, Urban Structure and Regional Change*, London: Weidenfeld and Nicolson.
Romer, P.M., 1986, Increasing Returns and Long Run Growth, *Journal of Political Economy*, Vol. 94, pp. 1002-1037.
Romer, P.M., 1990, Endogenous Technological Change, *Journal of Political Economy*, Vol. 98, pp. S71-S102.
Segerstrom, P.S., T.C.A. Anant and E. Dinopoulos, 1990, A Schumpeterian Model of the Product Life Cycle, *American Economic Review*, Vol. 80, pp. 1077-1092.
Solow, R.M., 1956, A Contribution to the Theory of Economic Growth, *Quarterly Journal of Economics*, Vol. 70, pp. 65-94.
Starrett, D., 1978, Market Allocations of Location Choice in a Model with Free Mobility, *Journal of Economic Theory*, Vol. 17, pp. 21-37.
von Thünen, K.H., 1826, *Der isolierte Staat in Beziehung auf Landwirtschaft und Nationalökonomie*, Hamburg.

Part III

Interregional Processes, Scale Economies and Agglomeration

7 Locational Advantage and Lessons for Territorial Competition in Europe

Ian Gordon[a] and Paul Cheshire[b]

[a] Geography Dept., The University of Reading, Whiteknights, Reading, RG6 6AB England

[b] Geography Dept., London School of Economics, Houghton Street, London, WC2A 2AE England

7.1 Introduction

Two significant consequences of the pervasive economic changes of the past 25 years or so, associated with internationalisation if not actually globalisation[1], involve substantial increases in the intensity of *competition*, and in the economic importance attributed to *place* (i.e. to specific spatial externalities). In both cases, it is also argued that qualitative factors (involving the presence or absence of multiple *qualities*), rather than simply availability and price, have come to play an increasingly important role. Thus Porter (1990)[2] argues both that traditional concepts of comparative advantage are inappropriate to a world in which quality competition prevails, and that qualitative attributes of a firm's national and city-regional environment play an important role in enabling it to develop competitive advantage. A natural corollary is to expect the growth of forms of inter-place, or territorial competition, involving attempts to boost local economic performance through collective efforts to enhance qualitatively significant attributes of particular places – which would be relatively new to European cities and regions, if not in the United States. Even Krugman (1996) who has argued strongly that competitiveness is an attribute of firms not of collectivities, recognises that the role of agglomeration economies makes spatial outcomes potentially dependent on chance or governmental influences operating at an urban-regional (or national) scale. Since there is in this situation no unique equilibrium outcome, 'an intellectually respectable case' can then be made in support of selective interventions as a means of boosting local real incomes – though Krugman is suspicious that such cases will usually turn out to be unwarranted pieces of special pleading on behalf of more specific interests.

---

1 Globalisation is a misleading concept, both because of the extreme unevenness of the interrelations involved, most of which are actually concentrated within transnational 'regional' blocs, such as Europe, and because nation-states continue to be very actively involved in the way in which these processes evolve.

2 Though he is more concerned with arguing the need for new ways of understanding current processes of competition than with assessing how far this represents a change over those prevailing in earlier eras.

Indications of a shift to more flexible, market-mediated forms of economic co-ordination, as well as to the smaller units characteristic of post-industrial activities, imply an enhanced role for such agglomeration economies, and thereby strengthen the potential case for interventions in pursuit of territorial competition. Emphasis on the role of networks in urban economies, supported by relations of trust (representing a form of social capital, in Putnam's (1993) sense) – rather than the haphazard connections enabled by pure agglomeration – suggest an alternative perspective, in which the territorial economy may operate as a club promoting its own economic development policies (Gordon and Jayet, 1991). Its objectives would include the recruitment of additional members contributing actively to the range of potential partners, information sources, trainers and sub-contractors within the club, as distinct from 'visitors' who may free-ride on the club's facilities without embedding themselves in the area. In contrast to the agglomeration case, where high rent levels in a successful centre select for those businesses most likely to enhance interaction potential (and thus externalities), the pre-existence of a strong network appears a necessary rather than sufficient condition for policies to reinforce local competitive advantage.

Two more specific factors promoting territorial competition in Europe recently are: the crises initiated by industrial job losses in many urban areas[3], and the prospective impact of trade liberalisation on high order service activities. In the first case, the need of many ex-industrial cities for substantial additional employment has served both to increase the bargaining power of footloose businesses in a situation of excess demand from potential site areas striving to gain the external benefits of additional jobs, and to increase pressure on local politicians to be seen to be pursuing such opportunities. In the second case, completion of the European Market, in particular, is starting to open up to international competition high order service functions in which the leading national centre within each country had previously enjoyed a monopoly. This should bring the competitive situation of those cities much more in line with those of structurally comparable cities in the US, and ultimately lead to the emergence of an integrated European urban system, with particular implications for centres in border areas as well as for these national centres. Its immediate effects, however, are to increase the competitive significance of infrastructural provision and other economic externalities in those cities, creating new incentives for the pursuit of territorially competitive policies by both the cities *and* national governments concerned.

In the context of these changes, it is important to try to integrate the political aspects of territorial competition within analyses of regional development processes, rather than treating these policies either as exogenous influences, or predictable extensions of commercial competition. With that objective in mind, the next sections of this paper consider in turn:

- the potential influence of territorial agencies over currently important location factors;

---

3 also an issue in North America.

- the relationship between inter-area competition for inward investment and for market share; and
- structural influences on the capacities of areas to engage effectively in territorial competition.

7.2 Locational Advantage

As a first approximation, territorial competition may be conceived of as involving attempts by agencies representing particular areas to enhance their locational advantage by manipulating some of the attributes which contribute to their own area's value as a location for various activities. This locational advantage would vary between sectors, economic functions and types of organisation, reflecting the differing salience of various attributes for these activities, but might be thought to have similar importance for both established firms and potential inward investors – unless the agency offered targeted incentives. Relevant locational characteristics might be classified in various ways, but a useful functional distinction involves four categories:

(i) ***Production factors*** – including costs and availability of premises and labour of various types, together with locally produced intermediate inputs, and local markets;
(ii) ***Residential environment*** – affecting the ability of businesses to attract and retain some particularly mobile types of labour (thus acting through some of the production factors);
(iii) ***Business milieux*** – external, unmarketed influences on the productivity, innovativeness and dynamism of local firms; and
(iv) ***Accessibilities*** – involving the relative ability to access valued production factors, environments and milieux in other areas.

As between these categories, the second and third are believed to be of increasing importance in flexible, post-industrial economies. These shifts (together with internationalisation) are also changing the relative importance among production factors and accessibilities (e.g. raising the importance of specialised producer services, air services, and telecommunications). Each of these categories involves elements where significant agglomeration economies are expected – together with some diseconomies (e.g. the price of inelastic factors, overcrowding, regulation, and traffic congestion). The agglomeration economies are believed to be of growing importance, as the proportion of personal interactions *between* organisations increases – though the significance of diseconomies may also be growing.

In terms of the scope for non-market intervention, to enhance particular attributes (or stop depreciating them), the opportunities seem most limited in the first case, given the centrality of market processes here. However, general planning and educational policies can clearly be significant, while the public sector can also play a catalytic role in reducing the uncertainties associated with major redevelopment schemes. More 'flexible' labour markets, in which it is less clear who will bear responsibility for skill development, provide another context in which appropriate public intervention can add to the value of local productive factors.

Each of the other sets of attributes (i.e. environments, milieux and accessibilities) more obviously involves public goods. These are most straightforward in the case of accessibilities, where public agencies represented at the urban/regional scale have established capacities to analyse and respond to requirements for transport infrastructure (if not for telecommunications), and/or to manage existing capacity more effectively. For environments also, there will be a range of urban agencies responsible for factors affecting the quality of life of population groups to be attracted and retained, in terms of security, public health, culture, open space and the built environment. However, there may be less clear understanding of what is really significant, and limited ability to manipulate these in a coherent way. In the case of milieux enhancing characteristics, this is still more true, with very little experience or knowledge within typical public authorities, particularly at this level, as to how these might be promoted in practice. This may well be the case also for private businesses and their representatives in areas where these attributes have not been historically significant.

Evidence for a shift from traditional to new location factors is actually far from clear-cut. A survey of attitudes and behaviour among 'locationally sensitive businesses'[4] in a number of leading European cities (including London, Milan, Venice, Rotterdam and Stockholm), undertaken by the TeCSEM network, highlighted the continuing importance of land, labour and accessibility, whereas non-traditional factors, such as the availability of business services, business networks, proximity to competitors, housing and cultural/recreational facilities tended to be accorded the lowest importance as location factors. This was broadly true for both manufacturing and service establishments, with the main difference lying in the type of labour whose availability was seen as important – though non-traditional factors were actually less likely to be cited by service firms (Friden and Gordon, 1996). Such evidence needs to be treated with some caution, since the meaning of broad factors such as the need for good access to the major city, proximity to customers and the availability of highly skilled labour is open to a number of different interpretations.

Where a substantially more detailed listing of locational factors was presented to prospective respondents (in the Milan survey) it is notable that many were seen as 'important' by substantial numbers of firms (31 by over 40%, 51 by 20% or more) and 'crucial' by significant minorities (11 by over 20%, 26 over 15%, 40 by 10% or more). Among the four sets of location factors distinguished above, however, it was the production factors and accessibilities which were seen as important by a majority, while environmental factors, and a series of 'proximities', providing the nearest to an indicator of milieux, were clearly of minority importance (even

---

4 distinguished in terms of prospective mobility, differential growth and involvement in international trade.

among producer services). While the former are necessary conditions for competitive success, Senn argues that:

> 'the metropolitan region's particular competitive advantages rest on an array of more specialised location factors each of which is important to substantial minorities of businesses' (1995, p.136).

Of course the distinction between more and less specialised location factors is not a clear-cut one, since references to labour and premises requirements will often conceal very specialised needs, treated as reflecting some more general characteristics of the local labour and property markets. Ordinary language[5] does not as readily recognise the availability of various specialised services, information sources and opportunities for collaboration as instances of some general features of the urban economy of importance to a majority of businesses. But it is a real and important feature of leading metropolitan regions that their highly diverse economies do both generate and require a very wide range of specialist inputs available on an unprogrammed, unplanned basis. Indeed the concept of agglomeration economies, as distinct from industrial complexes or industrial/ service districts seems to be based on a model of random coupling, and the statistical likelihood of being able to satisfy a great variety of possible requirements which are individually non-crucial to the businesses concerned (Gordon and McCann, 1998). If this implies that firms may invest less in developing regular networks of information, supply and collaboration than in some smaller and/or more specialised centres, this may have important implications for their capacity to engage effectively in territorial competition (see Section 7.4). But it also implies aspirations for competitive policies that have more to do with catalysing change, or trying to remove significant constraints on enterprise, than directly enhancing locational factors reported as important by large numbers of firms.

7.3 Competitive Advantage, and the Spatial Division of Labour

Though nominally addressed to issues of national competitiveness, Michael Porter's (1990) discussion of competitive advantage has become one of the most significant external influences on work in regional development during the 1990s. This is appropriate since the sectoral clusters of internationally competitive businesses which he identifies turn out to be concentrated within specific regions, not simply within specific nations, implying that some at least of the significant environmental contingencies operate at a regional (or city-regional) scale. One respect in which his work departs from the norms of regional analysis, however, is in associating the success of major firms (commonly with multiple plants) exclusively with the area in which their 'home base' is located. The implication is that (what are for him) the crucial qualitative influences on performance are embodied

---

5 or at least the ordinary language of questionnaire surveys.

in design, investment plans, management culture, and marketing strategies developed in these home bases, whereas the impact of labour costs, worker productivity etc. in particular plants on price is a rather secondary consideration for the firm. It does not follow, however, that it is of such marginal importance to the regions concerned, since multiplant firms are not committed to a particular regional distribution of their activities.

From this perspective, Porter's theory of competitive advantage appears a natural complement, to Massey's (1984) analysis of spatial divisions of labour, which is essentially (if not explicitly) concerned with the ways in which comparative advantage affects the functional division of labour within such firms. Porter's work provides a logic for investigating the spatial distribution of home bases of successful firms, while Massey's addresses the question of where their jobs (and those of their less successful counterparts) are, and which sorts of jobs are where. A key difference between the quality-oriented notion of competitive advantage and the price-oriented concept of comparative advantage is that the latter is self-balancing – by definition everywhere possesses comparative advantage in some activities – whereas the latter is not. Thus there will be places lacking any competitive advantage, and these will then have to rely upon their comparative advantage, for some sort of branch plant activity, or perhaps for some less extensively traded good where quality is less rigorously evaluated.

Recognising the interplay between these two kinds of advantage somewhat complicates discussion of locational advantage, and how various characteristics of areas may contribute to this. But it also raises substantial interpretative questions about the significance of survey responses evaluating the relative importance of various location factors (such as those discussed in the previous section). Typically such surveys, whether addressed to all firms or actual/potential movers, ask questions about important characteristics of an area for the location of the firm concerned, or for other similar businesses, i.e. about factors which would affect *locational decisions*. Even if these were controlled for type of establishment (e.g. to focus on home bases), it is not clear that this is also the set of factors likely to confer competitive advantage on the businesses concerned. In terms of the two-dimensional character of territorial competition, involving both competition for inward investment and competition for market share for established firms, this means that two rather different sets of locational factors could be involved. On the one hand, the emphasis is likely to be on expected productivity and cost differentials evaluated in a rather static framework, and with no presumption that the mobile firm has any unmarketed dependencies on the local economy. On the other hand, whether firms are self-conscious of these or not, competitive performance is (on Porter's analysis) liable to involve issues both of dynamic locational advantage and of unmarketed dependence.

Some evidence that such a distinction does in fact operate has been derived from analyses of the TeCSEM survey data for 6 urban areas in southern Britain: London (further divided into four rings), Reading and Swindon (also within the core Greater South Eastern region), together with Birmingham and Cardiff (which are centres of two secondary regions). The approach used to assess the effects of location factors involved relating respondents' perception of significant aspects of their environment, to objective measures of locational behaviour and competitive performance.

In the first case, area of current location (the behavioural measure) was related to the importance ascribed to each of a set of location factors, and (separately) to establishment characteristics (including sector, size, function, ownership and market area). This established two coherent dimensions of comparative locational advantage across the 8 areas distinguished, involving 11 location factors, of which 7 related to aspects of accessibility – the other factors being the availabilities of premises, manual and white collar labour, and of general business services. At this level at least, very traditional factors appear to dominate locational behaviour.

In the second case, an indicator of business performance – incorporating differential growth (relative to comparable establishments), the export share in output, and the expected effect of the SEM on their market share – was related to the reported effect of locational attributes on performance. This analysis yielded very different results as between establishments in the core and secondary regions. In the latter case what was significant was the general cost of labour and the availability of manual workers, while in the former case the two significant factors were availability of specialised services and proximity of competitors[6]. The implication is that in the two secondary regions performance depended largely on exploiting comparative advantage in terms of a relatively slack labour market, whereas in the core region more specific sources of competitive advantage were involved (Gordon, 1996).

The practical significance of these results lies in the implication that action taken to improve an area's competitive position in relation to inward investment will not necessarily be effective also in boosting the competitive performance of established firms. Indeed, at least in the case of the core region, it appears that very different factors may be involved. This might not matter much if competitive performance varied little between areas, so that differences in rates of growth (in particular) were principally attributable to the movement of industry. In fact, however, the reverse appears to be true, with components of change analyses finding *in situ* growth (or contraction) rates not movement to be the major contributor to differential growth. This tendency was borne out by analyses of sources of employment change in sampled businesses within 'locationally sensitive' sectors across 10 of the TeCSEM case study cities[7], showing inward relocation to be only a very minor source of employment growth (Cheshire and Gordon, 1998).

The policy lesson is that effort invested in retaining and assisting existing firms with competitive potential can yield substantially greater returns in terms of growth, than pursuit of inward investment is likely to, in areas with any substantial economic base – as well as the longer term consolidation of the urban economy. How this can be achieved in practice is by no means straightforward, requiring a much better understanding of the circumstances, concerns and operating environ-

---

6 In the latter case the statistical relationship was negative, since firms tended to report proximity of competitors as a negative factor, though on our measure it actually seemed to boost performance.

7 the 5 British cities referred to above, plus Milan, the Central Veneto, Lille, Rotterdam and Zurich.

ments of local business than do simple marketing campaigns or the offer of generalised subsidies. Areas which can develop this capacity are, however, likely to acquire substantial competitive advantage over those who have to rely on attempts to attract inward investment as a means of boosting their local economies.

7.4 The Political Economy of Territorial Competition

Across Europe, the 1990s have seen a burgeoning policy literature on the competitive prospects and requirements of major cities, alongside a number of high profile practical examples of strategic economic initiatives from particular cities, a proliferation of public-private partnerships, and the pervasive adoption by local authorities of some form of local economic development policy. But these activities differ very markedly in significance, and it should not be presumed that they represent the diffusion of a common form of territorial competition, or urban entrepreneurialism, in which all city-regions will become equally involved. Indeed, in Europe's leading city (i.e. London) a series of substantial reports on this issue since 1991[8] have yet to produce a credible competitive agency. And the high profile examples of strategic economic initiatives – such as Barcelona, Bologna, Lyon, Lille, Paris, or London in the mid-1980s – do seem to reflect very particular local conditions, as well as (in most cases) a response to perceived opportunities from European integration (from as far back as the 1960s in the case of Paris). In particular, they tend to involve either specific political motives (associated with party competition or regionalism) or central government backing for a national standard-bearer. Among more typical cities, rhetoric has generally been more conspicuous than substantial action, except where direct incentives are provided, from above in the form of national or EU regeneration programmes, or from a major commercial beneficiary. Despite the logic of changing economic circumstances and behaviour, then, the pursuit of strong territorially competitive policies cannot be presumed to be the norm.

Where territorial competition is not yet institutionalised within the bureaucracy or the core programmes of mass parties, its existence depends essentially on the organisation of collective action, whether to assume direct responsibility for the policy, or to lobby for its institutionalisation and public funding. This process is problematic at best, since the private economic benefits of participation will rarely exceed the costs entailed by active involvement (Olson, 1965). Potential beneficiaries are liable either to count on free-riding or assume that no action will materialise, unless either they are one of a very few major stakeholders, or there are other adequate reasons for participating, whether through immediate private benefit or satisfaction of a principled commitment.

---

[8] These include: the London Planning Advisory Committee's (1991) *London: World City moving into the 21st century*; the Government Office for London's (1996) *Four World Cities: London, New York, Paris, Tokyo*, and the City of London's (1995) *City Research Project: the competitiveness of* financial services; and (1997) *London and Paris*.

This could well mean that often there will be either no action, or purely symbolic action undertaken by an authority persuaded to will the ends but not the means. The likelihood of substantive action should be greater, however, where benefits are heavily concentrated, interests/commitments are widely shared, mutual trust and habits of co-operation are strong, and/or some external (typically public sector) leadership is available to marshal the interests[9]. In terms of inter-city competition, reflecting the shared interests of a city region, this is most likely to be secured where there is:

- a tier of governance representing the functional economic region, and/or a few constituent authorities, with little constraint on their scope for action;
- a strong sense of cultural or political identity, to facilitate co-operation;
- strong representation in decision-making of rent-earners, or locally dependent groups (in Cox and Mair's (1989) terms) with a stake in the area's competitive success;
- local firms which are strongly integrated, sensitive to environmental conditions and/or agglomeration economies, and have a fairly homogeneous set of interests;
- a small number of leading local firms and other key actors; and
- a potential for significant positive or negative change in the local economy (Cheshire and Gordon, 1996).

These criteria have quite strong implications for the types of area which might be expected to build a capacity for effective territorial competition. In the context of European integration, where competitive pressures are expected to be most intense amongst leading city-regions (rather than backward or declining areas), these should have the strongest incentive to develop such a capacity. But the largest, most heterogeneous, political fragmented and internally dynamic metropoles, (such as London), may have particular difficulty in securing collective action to deal with external challenges (Gordon, 1995). And within Great Britain it is notable that each of the second tier centres (Birmingham, Manchester and Glasgow) has shown significantly greater capacity to act in this way during the 1990s.

The kinds of territorial economic coalition which emerge, and consequently the types of policy which they promote, vary substantially between places, however, and it would be rare for these simply to reflect a generalised interest in the city-regional economy – despite claims to do so. Indeed the fact that collective action is problematic entails a degree of bias in those organisations and actions which do actually emerge, and the logic of collective (in)action may be used to predict the likely biases among these coalitions and competitive policies. It would be expected, for example, that these would tend to be:

---

9 The expectation that such leadership should come from the public sector is stronger in Europe than in North America, where local (monopoly) utilities may play this role, though local governments may still be expected to financially underwrite competitive activities (Wood, 1993).

- ***localist***, since the degree of commonality of interest, sense of community and trust is likely to be greatest among those in close proximity, serving fairly local markets and facing competition from identifiable neighbours – a tendency exacerbated where units of government are fragmented;
- ***conservative***, aiming to preserve or restore staple industries, or at least a familiar kind of economic base, since those directly involved in these will have the strongest base for co-operation, while this is also a clear, single goal around which others can mobilise (in contrast to multiple alternative futures);
- ***CBD focused***, since among activities serving wider markets (other than any historically dominant sectors) these are the most likely to be integrated and share a perception of common interest, embodied in the image and standing of that area; and
- ***property and development oriented***, both because landowners can be expected to be the main ultimate beneficiary of locational advantage, and because physical (re)development is the most likely form of action to involve a small group of major beneficiaries with clearly shared interests.

In combination with various supply-side influences, including the strengths and competences of established public sector actors (e.g. in physical development activities), and the greater visibility of some policy instruments (e.g. marketing), these tendencies also imply a bias toward the attraction of inward investment, using some fairly standardised forms of attractor. Where commitments to competitive policies are weak, local economic development policy is more likely also to be ad hoc, piecemeal and placatory in character – since there is a minimum effective scale for the development of more strategic approaches – and this is especially likely where interests and units of governance are fragmented (Cheshire and Gordon, 1996).

Even in relation to a single territory, heavy emphasis on competition for inward investment is liable to be a relatively ineffective means of pursuing economic growth, since (as noted in the previous section) areas' performance is usually much more sensitive to the growth rates achieved by existing businesses than to inward or outward movement. In the context of bottom-up economic development policies, especially in periods of generally high unemployment, where most areas (not just the least advantaged) are seeking additional activity, the pursuit of such inward investment is likely to be costly – whether in terms of specific locational incentives or of policies designed to improve an area's general attractiveness. Indeed, if many areas are competing in the same market for a limited pool of mobile investment, implying excess demand, it can be expected that the required expenditure on attracting activity will be bid up to levels reflecting areas' valuation of the expected economic gains. And, there is substantial evidence that mobile firms are aware of their bargaining position, exploiting this to the full in the final stages of their locational search process, once a sub-set of sites has been identified meeting their technical requirements (Wins, 1995).

At an aggregate level, the implication is that only the firms concerned can gain significantly from this competitive process (King et al., 1993) – though within any area there are likely to be significant redistributive effects, with significant net benefits to members of the pro-growth coalition. For any country, and still more for Europe (or equivalent continental groupings) this activity thus approaches a zero-sum game – which was an adequate reason for its restraint by national governments prior to internationalisation. Indeed, for much of the competition towards which small areas are drawn, it may well be less than zero-sum, with significant real resources being expended to maintain the status quo, or effect shifts yielding no net gains for either productivity or equity.

This pattern of activity is sub-optimal in three respects:

1. it ignores negative spatial externalities arising from local policies, favouring sets of policies whose benefits can be limited to those actively promoting them and denied to outsiders, at the expense of those which would be more likely to generate positive spatial externalities;
2. by focusing on relocations, it favours measures which are purely redistributive between areas, rather than adding to productive capacity, through the development of infrastructure, human capital stocks, technical progress, or inter-linkages and co-operation between businesses; and
3. convergence of areas on a similar set of targets for inward investment and the pursuit of a standard set of locational attractors maximises the bargaining power of mobile capital, whereas a policy of developing distinctive local strengths and specialisations would both enhance potential gains to areas, and promote a more efficient spatial distribution of economic activity.

Three sorts of normative conclusion follow from this analysis. The first two of these relate to types of policy which may be ranked in terms of their tendency to be (purely) wasteful, spatially redistributive and/or productivity enhancing. Individual territories would be well advised to switch away from the purely wasteful, while higher levels of government should more carefully regulate territories' use of purely redistributive types of policy (Cheshire and Gordon, 1998). Thirdly, these superior levels of governance ought to provide both assistance and incentives for the development of strong agencies able to pursue effective strategies of territorial competition over areas corresponding more closely to functional urban regions.

In addition, however, there are positive implications which may be drawn about the types of area that are likely to gain from a situation of heightened territorial competition, namely those that are able to build a strong coalition of interests in support of policies to reinforce potentially competitive elements within the regional economy and develop its distinctive strengths.

7.5 Conclusions

Continuing integration of the EU economies can be expected to increase the degree of differentiation within an emergent European urban system and reinforce the position of currently strong urban economies. One interpretation of what this might mean in practice focuses on the hierarchical nature of the urban system, and the scope for one or more dominant 'continental cities' to emerge as key centres of corporate control and servicing, along the lines of Friedman's (1986) and Sassen's (1991) 'global cities'. Once the period of integration is complete, agglomeration economies might be expected to reinforce whatever patterns of dominance have emerged. With such high stakes, territorial competition has been expected to play a significant role in deciding which of the contenders assumes these dominant roles, and in the process to boost the position of all the leading cities involved vis-à-vis secondary cities which have less prospect of gaining significant international roles.

This scenario may be challenged on several grounds. In the first case, the focus on hierarchy ignores other dimensions of the urban system which are likely to be reinforced in the process of integration, notably that of specialisation. If Adam Smith's dictum that the limit to specialisation is set by the scale of the market can be applied to cities as well as businesses, this process may well reinforce the specialisation, and competitive advantage, of a number of second and third order centres which will now have the potential to compete on a European scale – and thereby with leading national centres, including their own. The second big question relates to the presumed capacity of leading cities to develop and sustain effective policies of territorial competition. One argument of this paper is that this capacity can never be taken for granted, and that (even with the sponsorship of their national governments) leading metropoles may be among those who find particular difficulty in generating a commitment and competence to pursue policies likely to boost the competitive position of more than a few atypical sectors of activity. Correspondingly, less diversified centres with the potential to compete across Europe on a specialised basis may be among those with the greatest likelihood of achieving this.

Though other special, political factors may intervene, the main implications of incorporating processes of territorial competition within analyses of regional development processes is probably to reinforce recent arguments about the contribution of embeddedness and social capital to urban competitiveness. Effective territorial competition requires forms of political capital (including institutional structures, as well as a sense of shared interests, and practice in collaboration) which can be expected both to draw on and develop economically relevant forms of social capital. Where areas have any potential sources of competitive advantage, neither of these sets of organisational assets is likely to be advanced by the undifferentiated pursuit of inward investment or physical development.

Acknowledgement

This paper is based on work undertaken with members of the TeCSEM network (Territorial Competition in the Single European Market), including Ilaria Bramezza, Paolo Filippini, Lennart Friden, Gianluigi Gorla, Hubert Jayet, Angelo Rossi, Lanfranco Senn and Jan van der Borg, to be reported in a jointly authored forthcoming book. The support of the UK Economic and Social Research Council (under grants L113251003 and L11351027) is gratefully acknowledged.

References

Cheshire, P.C. and I.R. Gordon, 1996, "Territorial Competition and the Predictability of Collective (In)action", *International Journal of Urban and Regional Research*, 20:383-399.

Cheshire, P.C. and I.R. Gordon, 1998, "Territorial Competition: Some Lessons for Policy", *Annals of Regional Science*, 32:321-346.

Cox, K.R. and A. Mair, 1989, "Urban Growth Machines and the Politics of Local Economic Development", *International Journal of Urban and Regional Research*, 13:137-146.

Friden, L. and I.R. Gordon, 1996, "Locational Factors and Territorial Competition", Paper presented to the Regional Science Association European Congress, Zurich.

Friedmann, J., 1986, "The World City Hypothesis", *Development and Change*, 17: 9-83.

Gordon, I.R., 1995, "'London World City': Political and Organisational Constraints on Territorial Competition", in P.C. Cheshire and I.R. Gordon (eds.), *Territorial Competition in an Integrating Europe*, Avebury, Aldershot.

Gordon, I.R., 1996, "Territorial Competition and Locational Advantage in the London Region", Paper presented to the American Association of Geographers annual conference, Charlotte.

Gordon, I.R. and H. Jayet, 1994, "Territorial Policies between Co-operation and Competition", Working Paper No. 12E/94 CESURE, Lille .
University of Science and Technology.

Gordon, I.R. and P. McCann, 1998, "Industrial Clusters, Complexes, Milieux and Agglomeration", Regional Science Association, British-Irish section conference, University of York.

King, I., R.P. McAfee and L. Welling, 1993, "Industrial Blackmail – Dynamic Tax Competition and Public Investment", *Canadian Journal of Economics*, 26:590-608.

Krugman, P., 1996, *Pop Internationalism*, MIT Press, Cambridge, Mass.

Massey, D., 1984, *Spatial Divisions of Labour*, Macmillan, London.

Olson, M., 1965, *The Logic of Collective Action: Public Goods and the Theory of Groups*, Harvard University Press, Cambridge, Mass.

Porter, M.E., 1990, *The Competitive Advantage of Nations*, Free Press, New York.

Sassen, S., 1991, *Global City*, Princeton University Press, Princeton.

Senn, L., 1995, "The Role of Services in the Competitive Position of Milan", in P.C. Cheshire and I.R. Gordon (eds.), *Territorial Competition in an Integrating Europe*, Avebury, Aldershot.

Wins, P., 1995, "The Location of Firms: An Analysis of Choice Processes", in P.C. Cheshire and I.R. Gordon (eds.), *Territorial Competition in an Integrating Europe*, Avebury, Aldershot.

Wood, A., 1993, "Organising for Local Economic Development: Local Economic Development Networks and Prospecting for Industry", *Environment and Planning A*, 25:1649-1661.

8 Geographic Transaction Costs and Specialisation Opportunities of Small and Medium-Sized Regions: Scale Economies and Market Extension

Börje Johansson and Charlie Karlsson
Jönköping International Business School, Box 1026, SE-551 11 Jönköping, Sweden

8.1 Introduction

In recent decades an embryo of a new theory of specialisation and trade has been developed with an emphasis on differentiated competition, scale economies and size of the "home market" (e.g. Krugman, 1990). As regards the size of the home market several contributions focus on the country as the potential home market (e.g. Porter, 1990), which may be described as an ad hoc approach without any clear theoretical underpinning. In response to this type of deficiencies, this paper introduces a theoretical framework based on the concepts of a functional region and its product-specific market potentials. This approach is developed to improve the analysis of small and medium-sized regions, with an emphasis on their internal and external market interactions.

8.1.1 Economies of Scale and Market Potentials

There is one central idea in the new approach to analysing economic growth, trade and location. This idea is represented by a focus on increasing returns. It has become obvious that in models without increasing returns it is virtually impossible to explain the geographical concentration of firms, regional specialisation and the importance of the home market. Increasing returns in the form of internal and external economies of scale are a basic factor in the analysis of trade between regions. In particular, increasing returns play a crucial role in models with two-way trade, a phenomenon referred to as cross-hauling in classic models in regional economics.

Internal economies of scale is a technological phenomenon of an individual firm or establishment, and implies that the productivity increases (the cost decreases) as output gets larger. This phenomenon is usually related to some fixed (indivisable) production factor in a firm and can be compared with a catalyst, which must be present in the production while generating a fixed cost – often a start up cost. Examples of such indivisible resources are patents and similar knowledge resources, brand names, and networks – both material and non-material. One may also note that set up costs refer to both material resources based on investments in buildings and equipment and non-material resources based on training and organisation of networks.

It is not the absolute size of the fixed costs that matter. Instead the size of the fixed costs should be related to the potential size of demand (Stigler, 1951; Krugman, 1990a, 1991). In this paper we focus on how geographic transaction costs differ between different types of products and how this affects the internal and external market potential of individual regions. In the models that we apply, location advantages depend on a combination of scale economies and market potential of a region. According to this framework it is essential to classify products with regard to their distance sensitivity as regards interaction costs. Based on such an approach one can identify size-specific categories of products, which can develop a supply advantage in small, medium-sized and large urban regions, respectively.

The indicated analysis cannot rely solely on internal scale economies. Of equal importance is the existence of external scale-economies which characterise much of the economic life in large urban regions (in the form of urbanisation economies). In addition, these scale economies are vital for a sustainable and prosperous development in most small and medium-sized regions. External economies of scale is a systems phenomenon which occurs when several firms, producing similar products, are located in the same functional region, i.e., in the same "industrial district". The theory of external scale-economies is based on suggestions made by Marshall almost a century ago (Marshall,1920). These positive externalities appear when firms producing the same type of products (same industry) locate themselves in the same functional region. Such external scale effects are therefore sometimes called localisation economies. In addition, external scale-economies are essential in the formation of industrial clusters.

Our analysis will show that localisation economies are vital for specialisation processes in small and medium-sized regions (when they are not resource-based). Combinations of three major phenomena are assumed to cause external scale effects: (i) specialised labour markets, (ii) specialised neighbourhood firms, and (iii) information spillover. The first two factors give rise to intra-market effects, whereas spillover of information/knowledge between firms is a collective, extra-market effect. This implies that we need a reliable method to identify and delineate functional regions. In this endeavour we analyse interaction costs and define "geographic transaction costs" to include transportation costs and distance-dependent transaction costs related to the interaction between seller and buyer.

In our analysis we also stress that the formation of clusters is based on combinations of internal and external economies of scale, which may be thought of as making Marshall (1920) and Chamberlin (1933) a couple.

8.1.2 Towards a new Theory of Specialisation and Trade

The emerging new theory of specialisation and trade emphasises the role of the functional region rather than the nation. The pertinent models are based on the assumption that the economy of a region primarily develops through self-organised processes (decisions by households, companies and local policy makers), while interacting with other self-organised regional economies. This approach is valid both for large and small functional regions, although the major part of a small region's market potential is distributed among areas outside the region itself.

In the theoretical framework outlined here, "functional (urban) region" (FUR) is a prime concept. It is distinguished by its concentration of activities and of its infrastructure which facilitates a particularly high factor mobility within its interaction borders. In particular, the FUR is an integrated local labour market, sometimes called a commuting region.

Commuting and all other forms of interaction that generate nearby contacts give rise to interaction costs. The size of these determines the geographic extension of an urban region. Since our focus is on market interaction, the pertinent costs are called transaction costs. In our definition, geographic transaction costs (GTC) comprise both transportation costs and transaction costs which vary with regard to the geographic distance between seller and customer, and the properties of each spatially specific interaction link. With the two concepts of FUR and GTC as starting points, the following assumptions are employed in an analysis of regional economic specialisation, trade and growth:

(i) The border of a functional region is identified by its overall pattern of geographic transaction costs (GTC). For products with contact-intensive transactions, the GTC-level is considerably higher outside than inside the border of the region.

(ii) The overall market potential of a region is identified by the population size and associated purchasing power inside the region. In addition, a specialised region may develop a large internal market potential for very specific products.

(iii) A region's total market potential is composed by its internal and external markets. Each individual region is connected to its external markets in networks for trade and economic interaction. The interaction intensity varies across such networks, and this makes it possible, for each individual region, to identify a hierarchy of sequentially widening transaction or affinity areas, such that transaction costs rise in a step-wise sequence.

(iv) Location of activities and specialisation in a functional region is a process which is influenced by two basic conditions (i) technology and scale effects, (ii) semi-invariant (or durable) regional characteristics.

One may make a distinction between transaction and interaction costs in the following way. Generally, interaction between a supplier and a customer gives rise to transaction costs both for the supplier (seller) and the customer (buyer). The sum of these costs (including all parties involved in the transaction) can then be referred to as interaction costs. I addition, one may argue that interaction costs comprise costs for which we cannot identify any transaction. The formal models outlined in this contribution refer to supplier-customer interaction, and the associated costs are defined as transaction costs.

Figure 8.1 illustrates our theoretical assumptions. Location, specialisation and regional growth is organised as dependent on technology and scale effects together with influences from durable regional characteristics. Modelling in the traditional theory of international trade and regional growth has focused on one of the aspects in the tree of Figure 8.1. The focus has been on the supply side, i.e., on the availability of production factors and especially regionally trapped factors. These factors are labelled "durable capacities" in the figure. Models in this tradition adhere to the resource-based theory of specialisation.

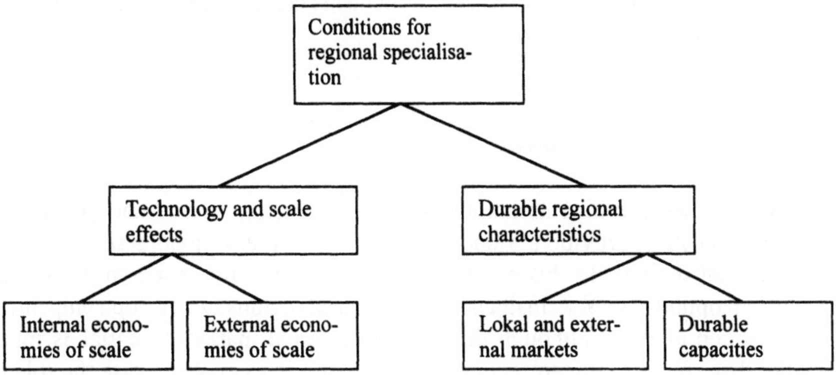

Figure 8.1 Four basic concepts of a new theory of location and trade

Up to the early 1980's comparative advantages have, with few exceptions, been derived from resource-based models. In recent decades, a partly revolutionary change has occured within international economics. The new approach claims that economic specialisation to a large extent is based on increasing returns and that differences in resources (factor intensities) can explain only certain parts of trade flows and the location of production (Dixit & Norman, 1980; Lancaster, 1980; Krugman, 1979, 1980, 1981; Ethier, 1982; Helpman, 1984). According to the new theory of trade with its scale-based models, imperfect competition and increasing returns are pervasive features of contemporary industrialised economies. With increasing returns as a basic explanation, trade develops because there exist advantages of specialisation also among economies and regions that are very similar to each other as regards resource endowments. If specialisation and trade are driven by economies of scale rather than by comparative advantage, the gains from trade arise because production costs fall as the scale of output increases.

With durable capacities as the only explanatory factor of trade patterns it is very difficult to understand why almost identical products are produced in different regions and then traded between these same regions. The limitations of the traditional theory based upon resource-based advantages imply that it cannot explain national and regional specialisation in an acceptable way. The traditional theory, for example, does not and cannot predict what sort of goods that will be exchanged by countries that have similar resource endowments. Neither does the traditional theory have anything to say about products that are not traded between nations or regions, i.e. about specialisation in so called "non-tradables". With increasing returns as a complementary basic explanation there exists a much broader assortment of comparative advantages and trade flows. Furthermore, with the new trade theory that stresses increasing returns the earlier intellectual borders between international economics and regional economics has been erased in a fundamental way.

In the sequel we show how the internal market potential of a FUR is a prime home market which together with increasing returns to scale can bring about processes of endogenous growth (and decline). In this context we also indicate how resource-based and scaled-based mechanisms combine in a dynamic, interdependent development process.

8.1.3 Outline of the Chapter

In this paper we combine assumptions about increasing returns and geographic transaction costs to provide a framework for analysing endogenous specialisation of functional regions. In this context we argue that external economies of scale provide an opportunity for small and medium-sized regions to develop competitive specialisation clusters, although their overall internal market potential is much smaller than in metropolitan regions. Our ambition is to make precise the role of internal and external scale economies in combination with product-specific geographic transaction costs in the development and survival of small and medium-sized functional regions.

Section 8.2 provides a theoretical background to internal and external scale economies, to how these phenomena are assumed to combine and to how they together generate cumulative specialisation processes of FURs. In particular, Subsection 8.2.4 provides insights to the specialisation conditions of small and medium-sized regions.

Section 8.3 contains the basic contribution of this paper. It develops the concept of geographic transaction costs and shows that this concept is essential for the understanding of the specialisation opportunities in regions (of different size) and for the analysis of scale-based specialisation. This analysis is continued in Section 8.4, which provides empirical illustrations that support the model formulations in Section 8.2 and 8.3. Section 8.4 illustrates how contact-intensive products (goods and services) can be systematically ordered with regard to their dependence on the size of the internal market potential of regions. It is also shown how the flow intensity of long-distance trade drops discontinuously at the borders of affinity-classified transaction areas, where such borders are interpreted as affinity barriers.

Section 8.5 suggests further empirical research along the lines illustrated in Section 8.4. The concluding remarks also contain some hypotheses for such future research.

8.2 Scale, Market Potential and Durable Capacities

This section tries to make precise the importance of increasing returns for regional economic development and for regional specialisation and growth. First scale economies are introduced as factors influencing specialisation processes in large and small functional regions. The analysis ends with a suggestion about a dynamic interdependence between a region's durable characteristics and the exploitation of scale economies.

8.2.1 Internal and External Economies of Scale

In this subsection we discuss the two phenomena internal and external economies of scale. We start by asking: how do internal economies of scale influence specialisation and growth in functional regions? For products with distance-sensitive transaction costs the relevant demand depends on the region's internal market potential. Hence, many small and sparsely populated regions are impossible locations for products with a thin demand, in the sense that the set of potential customers represents a small share of any region's total set of customers. For such products the only possible locations are large and dense regions with a high purchasing power.

A related phenomenon was discussed already by Adam Smith who made it clear that increased division of labour improves efficiency (and competitiveness) in a model where increasing returns are realised due to decomposition of production activities into systems of sub-activities. At the same time increased division of labour presumes increased scale which can only be realised if demand grows to a sufficient level. Hence, the degree of division of labour is limited by the size of the market (Stigler, 1951; Arrow, 1979).

How can internal economies of scale be explained? Internal economies of scale are a consequence of technological features of an individual firm or establishment, and are related to one or several fixed (indivisible) production factors in a firm's production (and cost) function. A fixed production factor can be compared with a catalyst, which must be present in the production while generating a fixed cost – often a start up cost. Examples of such resources are patents and similar knowledge resources, brand names, and networks – both material and non-material.

To shed some further light on the analysis of internal economies of scale, let us introduce geographic transaction costs. If these costs are very high, it is of course rational to split up and spread out the production in local micro firms, which can supply each local market. Kiosks and drugstores are good examples of activities (with high transaction costs) where the fixed costs are small relative to local demand. Hence, these establishments can be spread out to many locations. If on the other hand the geographic transaction costs are very low then it does not matter where production is located, although it may be important to concentrate production in large establishments, if internal scale economies are present. An example in this case is call centres.

The concept of economies of scope is related to internal scale economies. Such economies exist in their simplest form when the total costs for producing two different products within one firm is lower than the sum of the total costs of producing the two products in two different firms. Economies of scope, which arise in similar ways as internal economies of scale, are basically the result of several products using the same fixed resource within a firm. Hence, economies of scope is a basic motive for product differentiation within firms. In this way, product differentiation often gives rise to very large internal economies of scale and in practically speaking all firms there are examples of economies of scope.

Finally, one should recognise that sometimes it is essential to distinguish between fixed costs related (i) to a firm and (ii) to an individual establishment of a firm, observing that a firm may consist of many establishments.

Economies of scale imply that average costs (cost per unit output) decrease as output increases. Formula (8.1) provides the basic description of production conditions characterised by internal economies of scale.

$$c = v + F / x \tag{8.1}$$

where x denotes output, c unit cost of output, v marginal variable cost and F fixed costs. Figure 8.2 below illustrates a cost curve satisfying the conditions in (8.1). If a firm with such a cost curve has accessibility to a market demand described by D_1, then such a demand is classifed as insufficient. On the other hand, with the demand structure given by D_2 the willingness to pay exceeds the cost per unit output in the interval between x_1 and x_2. This demand is sufficient. Accessible demand is, as will be discussed later, determined by geographical transaction costs within a region and between the same region and its environment of other regions.

Formally, the demand schedule facing a region can be represented by the demand function $p^D = p(x)$. The demand schedule D_1 is characterised by $p(x) < c + F / x$ for all $x > 0$, whereas D_2 satisfies $p(x) > c + F / x$ for all x such that $x_2 > x > x_1$, and this explains why D_1 constitutes insufficient demand, whereas the second schedule offers sufficient demand for a location in the region.

In our framework internal economies of scale are considered as a generic feature of all production, although the output level x_2 in Figure 8.2 may represent a very small scale of operation. The scale effect must be evaluated in relation to the size of demand. Because of this, scale effects can be important also in micro firms whenever the pertinent products (goods and services) have a non-dense demand, which means that in the relevant population of firms or households one can find only a small number of customers at a given point in time. In such cases the demand may easily be insufficient as D1 in the Figure 8.2 illustrates.

In order to describe Adam Smith's assumption of increasing returns we can introduce two technique options ι_1 and ι_2, where the second technique is associated with decomposition of activities and a larger scale. The two techniques will then satisfy the following condition:

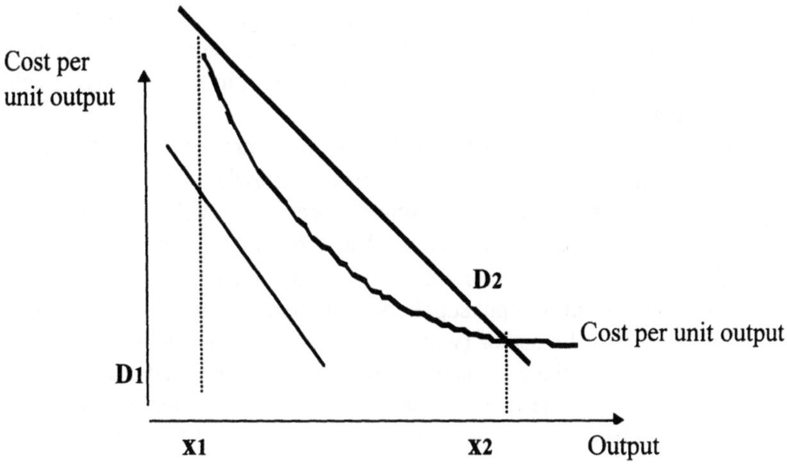

Figure 8.2 Decreasing average cost combined with insufficient and sufficient demand

$$c(\tau_1) = v(\tau_1) + F(\tau_1)/x > c(\tau_2) = v(\tau_2) + F(\tau_2)/x,$$
$$v(\tau_1) > v(\tau_2) \text{ and } F(\tau_2) \geq F(\tau_1) \tag{8.2}$$

In addition to internal scale economies, a certain production activity may also be characterised by external economies of scale, along suggestions made by Marshall almost a century ago (Marshall, 1920). External economies of scale appear when firms producing the same type of products (same industry or sector) locate themselves in the same functional region. Such external scale effects are therefore sometimes called localisation economies. As discussed earlier, combinations of three major phenomena are here assumed to cause external scale effects: (i) specialised labour markets, (ii) specialised neighbouring firms supplying inputs, and (iii) information spillover between the firms in the same sector. The firms in a sector with external scale economies form a cluster together with two types of neighbouring firms characterised as input suppliers and customers.

External economies of scale refer to advantages that a set of firms can have just by being localised close together. This form of localisation in a region has several consequences. A cluster of firms can attract to the region a rich composition of labour categories which are specialised to suit the industry in question. The concentrated demand from the pertinent industry also attracts neighbouring firms, which are first of all input suppliers (of various kinds). These input suppliers may have their own internal economies of scale, and then it is important to have accessibility to a sufficient demand, which in this case is provided by the localised firms. Neighbouring firms may in addition include specialised customers, which are attracted by the concentrated and varied supply from similar firms in the same region. But why does it matter for these specialised neighbouring firms to locate in the same region as the given industry? As we shall make explicit in the next section, the desire of specialised input suppliers to be in the same region as their customers is determined by a combination of frequent interaction with their customers and distance-sensitive transaction costs. This is of course the same motive as we find for the regional labour market, which is formed because of distance-sensitive commuting (interaction) costs. These observations help to explain the so called cluster phenomena.

This last observation about distance-sensitive interaction brings us to the third factor, i.e., information spillover between firms. Knowledge is spread without being priced in an intraregional neighbourhood, because in such an environment it becomes prohibitively costly to privatise all information. Hence, some of it will spillover. The information of importance concerns a wide area such as production technique, product attributes, input suppliers, customers and market characteristics.

If all firms in a specific sector (with external scale economies) reach roughly the same operation scale, the external scale effect can be expressed as a function of the number of firms in the same specialised industry. This is formally described for an individual firm j belonging to sector $J = 1,...,\bar{J}$ in formula (8.3) below

$$c_j = v(x_J) + F(x_J)/x_j \; ; \; x_J = \sum_{j \in J} x_j \tag{8.3}$$

where $x_j \approx x_J/\bar{J} = x^*$ denotes the output of firm j and x_J the total output from all firms in the given sector, all located in the same region. Intra-market, pecuniary externalities (as well as information spillover effects) imply that $\partial v/\partial x_J < 0$. Other externalities may reduce start-up and other fixed costs, implying that $\partial F/\partial x_J < 0$. For $\bar{J}$ sufficiently large we can assume that $v(x_J)$approaches its minimum value v_o and set $v(x_J)= v_o$. Moreover, we may approximate the fixed-cost component by setting $F(x_J)/x_j = V/x^*[x_J]^{\theta}= V_o/x_J^{\theta}$, where $\theta \leq 1$. Hence, setting $c = c_j$ yields formula (8.3'), which has a similar structure as the expression in (8.1).

$$c = v_o + V_o/x_J^{\theta} \tag{8.3'}$$

The similarity between (8.1) and (8.3') means that we can include both internal and external economies of scale in our subsequent analyses of cumulative specialisation and change in functional regions.

8.2.2 Endogenous Specialisation and Growth in Large FURs

A major concern in this paper is that different products have different distance-sensitivity. This sensitivity will be derived from more basic properties of products such as the contact intensity of the interaction between supplier and customer or, in other words, the complexity and contact requirements of the transaction process.

Certain products are exchanged under complex (and contact-intensive) transaction conditions which may involve many varieties of nearby contacts. As a consequence, the associated deliveries can be assumed to have distance-sensitive geographical transaction costs. In this subsection we show why the supply of such products have a location advantage in large functional regions. Our analysis is based on a two-region model and studies the location of the supply of a single product.

Let the internal market potential be M_1 in region 1 and M_2 in region 2. Moreover, let $\bar{p}_r$ denote the average supply price in region r = 1, 2, whereas p_r refers to the supply price of firms located in the region. The overall demand, D_r, in each region r is assumed to depend on the market potential and the average price as follows:

$$D_r = f(\bar{p}_r)M_r,\ df / d\bar{p}_r < 0 \tag{8.4}$$

Next, assume that we can specify a market share coefficient, $0 \leq m_{rs} \leq 1$, such that the delivery flow, x_{rs}, from r to s satisfies $x_{rs} = m_{rs} f(\bar{p}_s) M_s$. In equilibrium the following condition has to be satisfied:

$$x_r = \sum\nolimits_s x_{rs} \tag{8.5}$$

where x_r denotes total supply in region r. Assuming non-negative profits implies that the average selling-price, $\bar{\bar{p}}_r = \Sigma_s\, p_{rs} x_{rs} / x_r$ must satisfy

$$\bar{\bar{p}}_r \geq v + F / \Sigma_s\, m_{rs} f(\bar{p}_s) M_s + \ \Sigma_s\, c_{rs} m_{rs} f(\bar{p}_s) M_s / x_r \tag{8.6a}$$

Moreover, consider that the link price is defined as $p_{rs} = \ p_r + \ c_{rs}$ where the last term signifies the geographic transaction cost per unit delivery between region r and s. With this form of delivery prices, formula (8.6a) implies that the supply price, p_r, in region r will satisfy

$$p_r \geq v + F / \Sigma_s m_{rs} f(\bar{p}_s) M_s \tag{8.6b}$$

In addition, assume that the allocation is governed by a pure neoclassical mechanism such that

$$m_{rs} = \begin{cases} 1 \text{ if } p_{rs} < p_{ks}\,, & \text{for all } k \neq r \\ 0 \text{ if } p_{rs} > \bar{p}_s \end{cases} \tag{8.7}$$

Given the specification in (8.1) and (8.4)-(8.7), one may investigate under which conditions all supply of the commodity is located in region 1. Assume that $c_{rr} = 0$ and $c_{rs} = c_{sr}$, for $s,r = 1,2$ and $s \neq r$. Suppose that the same investor compares a location in 1 and 2 before any capital has been committed. Then all output is located in region 1 if

$$\frac{F}{x_{11}+x_{12}} < \frac{F}{x_{22}+x_{21}} \text{ and } \frac{F}{x_{22}} - \frac{F}{x_{11}+x_{12}} > c_{12} \tag{8.8}$$

where

$$x_{rr} = f(p_r)M_r,\ x_{rs} = f(p_s)M_s, \text{ and } p_r = v + F/[x_{rr}+x_{rs}]. \tag{8.9}$$

The equilibrium solution related to (8.8)-(8.9) implies that $p_1 = \bar{p}_1$ and that $M_1 > M_2$. Moreover, $p_{12} = p_1 + c_{12} = \bar{p}_2 < p_2$.

Let us now define link-specific price, p_{rs}, as viable and competitive if $p_{rs} = F/x_r + v + c_{rs}$, where x_r is total supply from region r. Consider now a start situation where $x_1 = x_{11}$, and $x_2 = x_{22}$. Assume that $M_1 > M_2$. Under what condition will this situation switch to a case with $x_{12} > 0$ and $x_{21} = 0$? Let us assume that all prices are link-viable and competitive. The firm in region 1 can select a viable price pair, (p_{11}, p_{12}), which is lower than the pair, (p_{22}, p_{21}), if

$$p_{12} = F/\left[f(p_{12}-c_{12})M_1 + f(p_{12})M_2\right] + v + c_{12} < \hat{p}_{22} \tag{8.10a}$$

$$\hat{p}_{22} = F/\left[f(\hat{p}_{22})M_2 + f(\hat{p}_{22}+c_{21})M_1\right] + v \tag{8.10b}$$

where (8.10b) represents the price that the firm in region 2 would select if it could supply both markets alone. Obviously, (8.10a) implies that the pair $(\hat{p}_{22}, \hat{p}_{21})$ is not viable. It is also clear that (8.10a) implies that $\hat{p}_{21} = \hat{p}_{22} + c_{12} > p_{11}$. Formulas (8.10a)-(8.10b) illustrate a situation where the fixed costs are much more important than the geographic transaction costs, which means that

$$F/[x_{11}+x_{12}] + c_{12} < F/[x_{22}+x_{21}] \tag{8.11}$$

This situation obtains when $(x_{11} - x_{21})$ is sufficiently larger than $(x_{22} - x_{12})$, which requires that the difference $M_1 - M_2$ is sufficiently large. We may observe that our assumptions about the market potentials ascertain that $x_{11} > x_{22}$, and $x_{12} > x_{21}$. Hence, the switch will always come about if c_{12} is reduced enough, as compared with its initial level.

In order to illustrate endogenous specialisation and growth in relation to the described case, let $M_1(0)$ signify the internal market potential of region 1 at date t = 0, before the supplier in region 1 has captured the market in region 2 and started its deliveries $x_{12}(t) > 0$. As a consequence the output in region 1 becomes larger and that may generate a growth in the market potential, $M_1(t)$ as $t > 0$. The growth process, which we shall sketch here, will share some properties with the classic export base model, but the endogenous feedback mechanism is stronger in our model example.

Let z_1 denote the output from all other sectors in region 1 so that the sum $x_1 + z_1$ denotes total production in the region. Given this, assume that the size of the market potential can be calculated from a linear equation with an intercept $\overline{M}_1$, and a term $\eta(x_1 + z_1)$such that there is an equilibrium size $M_1^* = \overline{M}_1 + \eta(x_1 + z_1)$ which obtains (after an adjustment process), if the sum $x_1 + z_1$ remains unchanged. The coefficient η is a "market-size multiplier" relating the economic activity level in a region to the size if its internal market potential. Next, let the output from all other sectors at time t be specified as $z_1(t) = \bar{z}_1 + \beta M_1(t)$, where β is a demand coefficient and $\bar{z}_1$ represents exports (of all other products) to markets outside region 1.

Assume that we start with an equilibrium at time $t = 0$. Moving to $t > 0$ implies that

$$M_1(t) > M_1(0) = \overline{M}_1 + \eta[x_{11}(0) + z_1(0)] \tag{8.12}$$

since $x_1(t) = x_{11}(t) + x_{12}(t) > x_1(0)$, as $t > 0$. The inequality in (8.12) signifies the start of a cumulative adjustment process. The realisation of these adjustments may be modelled as instantaneous, but here we shall assume that the adjustments follow a gradual process with a form that is frequently observed. The simplified specification of $z_1(t)$ is motivated by our ambition to focus on the effects of how $x_1(t)$ develops and how this process generates a growth in $M_1(t)$.

With the assumptions introduced above, one may solve for an equilibrium M_1^*, z_1^*, and $x_1^* = x_{11}^* + \bar{x}_{12}$, where we assume that that all demand in region 2 is captured immediately so that $\bar{x}_{12} = f(p_{12})M_2$, with $p_{12} = F/[x_{11}^* + \bar{x}_{12}] + v + c_{12}$. Given this, let $\overline{\overline{M}}_1 = \overline{M}_1 + \bar{z}_1 + \bar{x}_{12}$, and consider this sum as fixed. Then we have that $M_1^* = \overline{M}_1 + f(p_1)M_1^*$, which yields

$$M_1^* = \overline{\overline{M}}_1 / [1 - f(p_1) - \beta] \tag{8.13}$$

In general, the market potential cannot be assumed to adjust to its equilibrium value instantaneously. As an illustration we shall introduce an S-shaped adjustment process in which M_1, x_1 and z_1 adjust in response to each other. For this purpose we introduce three adjustment speed parameters, labelled α_M, α_x and α_z to obtain the following system of differential equations:

$$\dot{M}_1 = \alpha_M[M_1^* - \eta(x_1 + z_1)]M_1 \tag{8.14a}$$

$$\dot{x}_{11} = \alpha_x[f(p_1)M_1 - x_{11}]x_{11} \tag{8.14b}$$

$$\dot{z}_1 = \alpha_z[\beta M_1 - z_1]z_1 \tag{8.14c}$$

where we may assume that the market potential adjusts on a slower time sale so that α_M is smaller than α_x and α_z. This means that output is slaved by the growth in M_1.

The system in (8.14) contains some hidden dynamics. Observing that $x_{11}(t)= f(p_1)M_1(t)$ and specifying that $p_1(t)= F/[x_{11}+\bar{x}_{12}]+ v$, one can conclude that $dp_1/dt < 0$ as long as $x_{11}(t) < \overset{*}{x}_{11}$. This means $x_{11}(t)$will grow due to two forces: both $f(p_1)$ and $M_1(t)$ will increase. The price reduction process will be included in (8.14), if we assume that $p_1(t)$ adjusts on a very fast time scale.

Our assumption that the supplier captures the market in region 2 quickly by setting a low price gives the producer in that region an incentive to close down. This triggers the endogenous growth process. The export-delivery price may, for example, be set to satisfy $p_{12} = F/[x_{11}+x_{12}]+v+c_{12} < p_2$, which means that the supplier in region 1 predicts the subsequent development. Moreover, $z_1(t) = \bar{z}_1 + \beta M_1(t)$ describes the demand for deliveries from all other sectors, where the internal deliveries grow by βdM_1 during short time intervals dt. With price-dependent demand and scale economies in other sectors we might consider an additional growth effect due to $d\beta/dM > 0$

In other words, as activities in the region expand, the productivity is enhanced and the cost level reduced due to scale economies. To the extent that this makes prices fall, output will grow even further. At the same time the regional growth may bring about both congestion and increasing costs of floorspace, a phenomenon that retards the growth rate. At the same time, discontinuing the activity x_{22} in region 2 implies that that the purchasing power of this region may diminish, which may provide incentives for migration of resources from region 2 to region 1.
That points in the direction of cumulative interregional change, such that the larger region expands at the expense of the smaller region.

Figure 8.3 provides a general picture of the cumulative process in (8.13). The concept "economies of scale" in the figure is meant to signify both internal and external scale phenomena. In the discussion related to the figure, internal economies of scale are a major vehicle in the change process. As such the figure rather refers to metropolitan and other large urban regions than to medium-sized and small functional regions. At the same time, by reversing the direction of the arrows in the figure, it will describe a spiral of cumulative decline.

8.2.3 Endogenous Specialisation and Growth in Small and Medium-Sized FURs

The specialisation and location mechanisms outlined in the previous subsection are relevant for regions of all sizes, but work in favour of large functional regions. In particular, for metropolitan regions scale economies imply a location advantage with regard to all products with a "thin demand". Hence, large urban regions can "specialise in diversity" and can rely on the double force of internal and external scale economies

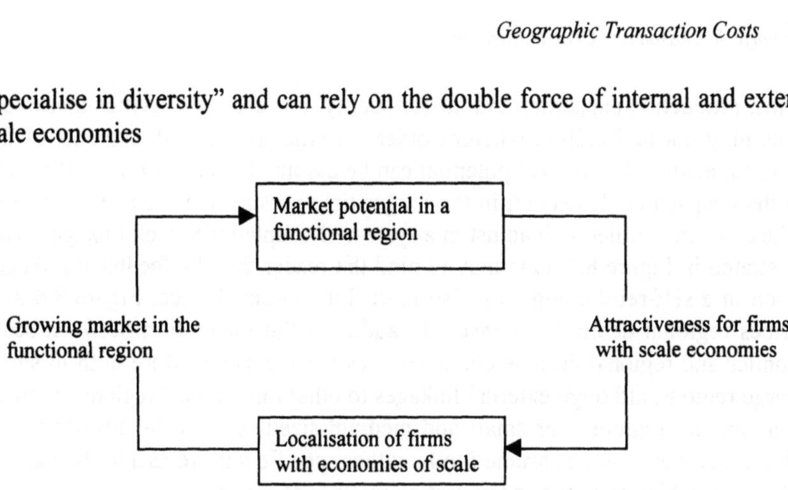

Figure 8.3 Cumulative interaction between scale economies and market potential

Scale economies are vital when we want to explain the existence of geographical concentrations and, in particular, large urban regions. Of course, high land values in regions with a high economic density is a counteracting force. Thus, firms in large cities must be more productive in such environments in order to be able to cover those extra costs for land and premises. However, scale economies constitute an equally important phenomenon for urban regions which are small or medium-sized.

For small and medium-sized regions, specialisation can have two basic forms. The first may be thought of as the classic Ricardian case, in which a small or medium-sized region hosts industries which are natural-resource based and for which internal economies of scale are important. The second form of specialisation refers to Marshall's idea about external economies of scale. In this case also a small region may develop a specialisation in a self-organised way. In such a development one can observe a narrow set of sectors (or product areas) which are characterised by external economies of scale (localisation economies). The cumulative sequence in Figure 8.3 may in this case be specified as follows:

> Firms in a core sector with localisation economies locate together in a region ⇒ Input suppliers and labour categories which are specialised with regard to the pertinent sector are attracted to the region. (Sometimes one can also observe how sector-specific customers are attracted to the region) ⇒ The environment of sector-specific input suppliers and employment categories as well as the core firms themselves form an economic milieu (a cluster) which attracts such sector-specific firms to locate and to expand in the region.

With Myrdal's (1957) terminology, the above sequence can also be referred to as cumulative causation. It represents self-reinforcing, circular dynamics which is constrained by the development of demand in the region and in its external markets. This form of positive feedback is in general constrained by the existing capacities in the form of built environment, accessibility based on transportation systems, production capacities and labour supply. For certain activities these constraints may not be binding, whereas other activities require adjustments of the durable capacities. The market potential can be assumed to adjust on a faster time scale than capacities. However, in the longer time perspective regional capacities and the economic milieu will adjust in a system of coupled feedback linkages. This is illustrated in Figure 8.4. One may remind the reader that the feedback linkages function in a self-reinforcing way also in decline phases. Hence, Figure 8.4 also describes regional decline processes. In addition, the interaction between scale economies and regional durable characteristics has the same nature both in small and large regions, although external linkages to other (and larger) regions are more vital in smaller regions. For small and medium-sized regions the adjustment of durable capacities may be assumed to be rather specific with regard to the narrow set of sectors which form the specialisation nucleus of each region.

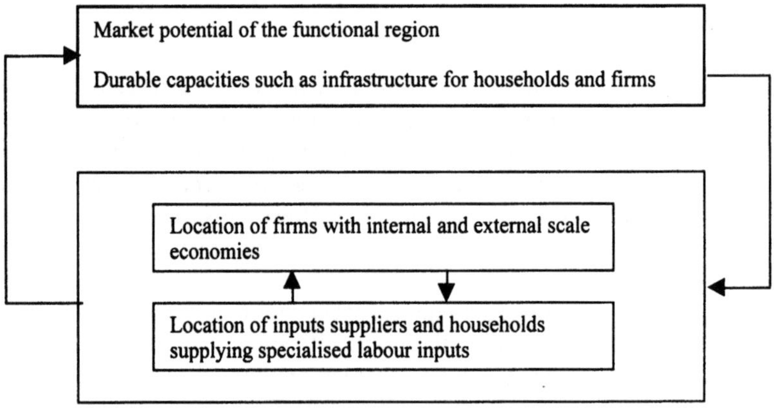

Figure 8.4 Cumulative dynamics of market potential, firm location and regional capacities

External economies of scale and associated clusters are of particular importance for small and medium-sized regions. We shall end this subsection by showing how one can formulate a model with external scale economies with such properties that our previous analysis of cumulative specialisation remains relevant also in the case of external scale economies. In this endeavour, let J be an index set denoting firms belonging to a specific sector, and apply (8.3) so that x_{rJ} denotes total output from this sector such that

$$x_{rJ} = \Sigma_{j \in J}\, x_{rj}$$

where x_{rj} is the output from a typical firm in the sector cluster, called J-cluster. Following (8.3'), the unit cost per output from any specific firm $j \in J$ is given by formula (8.15) below:

$$c_{rj} = v_0 + V_0 / x_{rJ} \tag{8.15}$$

where v_0 denotes variable costs and the ratio V_0 / x_{rJ} signifies fixed costs per unit output. Assuming that all firms in the *J*-cluster has the optimal size, the following expression for the costs per unit output from sector *J*, c_r, can be used as an approximation of (8.15):

$$c_r = v_0 + V / x_r \ ; \ \ x_r = \Sigma_s m_{rs} f(\bar{p}_s) M_s \tag{8.16}$$

where *V* is treated as a constant , and where the ratio V / x_r declines as x_r (x_{rJ} with a more precise notation) grows. We can see that formula (8.16) is the same as formula (8.1). This means that we have transformed the cost function with external scale economies to the cost specification for internal scale economies.

External scale economies are assumed to bring about lower input prices (intra-market effect) as well as spillover effects. Other externalities may reduce start-up costs and similar fixed costs. All this is captured by a fall in V / x_r.

As a final remark we should observe that with external economies of scale, the total demand for inputs from the firms in the localisation cluster provides a specialised market potential for firms supplying inputs to the cluster. Hence, these input suppliers have a market potential which is not directly dependent on the size of the region in question. Instead, small regions with a large localisation cluster can have a correspondingly large "specialised market potential", and it is partly in this way that the firms in the cluster generate their own localisation advantage.

8.2.4 Dynamic Interdependence between Durable Characteristics and Scale Economies

So far very little has been said about durable regional characteristics such as infrastructure and production factors which have a fixed or semi-fixed location in a region. In Figure 8.1 durable regional characteristics are divided into (i) local and external markets, and (ii) durable capacities. The latter consists of the supply of different labour categories, existing production capacities with specific techniques, R&D organisations, and infrastructure in the form of facilities and networks. The durable capacities generate comparative advantages in the Ricardo sense and influence the specialisation profile of a region. Although these characteristics are more or less exogenously given in the short and medium term, a major part of the durable capacities change gradually over time and are to a large extent created by investment and migration-like processes. This is illustrated in Figure 8.4.

What about the size and purchasing power of a functional region's local and external markets? In the short and medium term the properties of markets are durable phenomena which create comparative advantages in pertinent regions. However, the dynamics of the interdependence between market size and economies of scale is essential. The interaction between market potential and the location of firms with scale advantages is indeed intriguing. The process may be described as follows. Firms with internal economies of scale are attracted to locate in a region with a large market potential and a region in which firms want to locate develop a large market potential. This observation is further emphasised if we note that for many activities, internal and external economies of scale coexist.

An important message in Figure 8.4 is that if we study the process of specialisation on a slow time scale, the size and composition of a region's market potential becomes a variable and it evolves in a dynamic process. Similarly, we can observe an evolutionary process in which specialised labour supply gradually adjusts in concordance with the specialisation profile of the region.

In a sense, the insights about this type of basic cumulative endogenous courses of events are not new. They were suggested early in this century by Alfred Marshall (1920). Similar ideas were later formulated by Myrdal (1957). Another illuminating contribution can be found in Kaldor (1970), when he examines the differences in economic growth between different regions in Great Britain during the 1950's and the 1960's. This field of research is still in a stage of emergence, but it seems fair to mention contributions such as Krugman (1979), Fujita (1988), Venables (1995), Faini (1984), and Rivera-Batiz (1988). These and other related contributions show that internal scale economies and cumulative processes imply that regions that basically have the same production resources may specialise in different ways.

With this view of the world, we have indicated in this subsection how the location of a particular industry is history dependent. This property is caused by the prevalence of multiple solutions (multiple equilibria). But once a pattern of specialisation is initiated, for whatever reason, that pattern gets "locked in" by the cumulative gains from trade. There is a strong tendency toward "path dependence" in the patterns of specialisation and trade between countries and regions.

8.3 Geographic Transaction Costs

Geographic transaction costs form an important background to the analysis in Section 8.2. In the present section evidence will be presented as regards the shape of the geographic transaction costs curve (GTC curve), its steepness as well as how it may differ between specific products. We shall distinguish between products (including services) with regard to their distance sensitivity. When the distance sensitivity is low, the geographic transaction costs are more or less the same irrespective of whether the pertinent interactions take place inside the region or with actors in other regions. High distance sensitivity signifies that there is a considerable advantage of executing a transaction and associated contacts inside the region as compared with external interaction. In brief, input delivery transactions and selling costs are much lower within the region, given that suppliers and customers are located in the region.

8.3.1 Transaction Costs, Accessibility and Interaction Frequency

As already indicated, if geographic transaction costs are very high, it is of course rational to split up and spread out the production in local micro firms which can supply each local market. As mentioned before, kiosks and drugstores are good examples of such a solution. If on the other hand scale phenomena are strong and the geographical transaction costs are very low, then it does not matter where production is located as long as the production can be concentrated in one or a few locations. Call centres provide an example of such establishments.

Before continuing, observe that firm activities include both the development of new product variants and production and the sales of established products, although these activities usually are treated in separate formal models. Potentially, both development and production activities comprise a wide spectrum of interaction with other firms and other types of actors. A large part of all interactions occur between suppliers and their customers, and this motivates that we use the term transactions.

Many products are exchanged under complex (and contact-intensive) transaction conditions which may involve many transaction phenomena such as inspection, negotiations and contract discussions, legal consultation and documentation of agreements. Such products may themselves be complex and have a rich set of attributes, but the basic thing is that from a transaction point of view they are not standardised, and the interaction procedures are not routine. A special case of a contact-intensive transaction is when a product is customised and designed according to specifications by the customer in a process of supplier-customer interaction. On the other hand, for products with standardised and routine transaction procedures very little or no direct contact between buyer and seller is necessary. Moreover, when the same supplier and customer repeat the same delivery, the interaction between these two actors can be routinised, and hence the contact intensity goes down, causing transaction costs to decline.

Now, contact-intensive products and contact-intensive deliveries can be assumed to have distance-sensitive geographic transaction costs. For such products (goods and services), trade within a functional region is more advantageous than interregional trade. In the subsequent presentation we attempt to classify products according to the shape and size of the geographic transaction costs. Figure 8.5 depicts a typical curve of a product's geographic transaction costs. Where the curve starts to bend sharply upwards we recognise some sort of barrier (an interaction barrier). For contact-intensive products this barrier represents the outer ring of a functional urban region.

The GTC curve in Figure 8.5 implies that for the pertinent product there is a significant difference in transaction and hence selling costs when the product is delivered inside the region as compared with a delivery to a more distant market. In Figure 8.6, on the other hand, we recognise a product for which the distance sensitivity is small or moderate. There is a minor, perhaps insignificant barrier at the border of the region (point A). Such a product is a candidate for deliveries to distant markets.

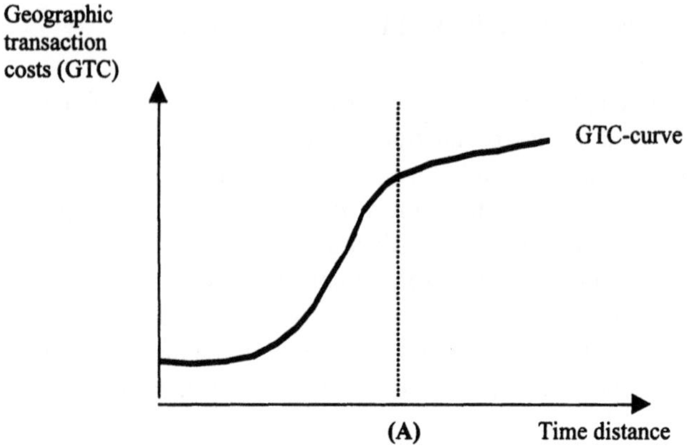

Figure 8.5 Geographic transaction costs related to distance

The distance marked by (A) in Figures 8.5 and 8.6 approximates the outer border of a functional region. The general internal market potential is determined by the size of the population inside the region and its purchasing power as well as by the size of demand from firms in the region. For a large urban region, the size of the regional income or the regional product may be a relevant proxy for the market potential. For small and medium-sized regions it becomes relevant to specify the specialised demand for inputs from localisation clusters inside the region. In the following subsection the market potential is specified with regard to a particular product.

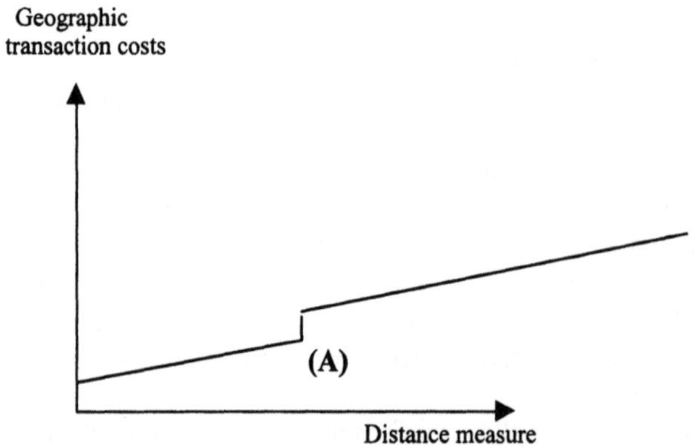

Figure 8.6 Geographic transaction costs for a product with a low distance sensitivity

8.3.2 Internal and External Market Potential

Consider the internal market potential of region s, M_s, as introduced in (8.4). Next, observe that for a given constellation of prices a certain share of this market potential, $0 \leq m_{rs} \leq 1$, may be captured by suppliers in region r. Given this, let x_{rs} be specified as follows:

$$x_{rs} = m_{rs} f(\bar{p}_s) M_s \tag{8.17}$$

Previously a pure neoclassical market-share mechanism was introduced in (8.7). Here we shall, as an alternative, consider a mechanism which adheres to the tradition of spatial-interaction models. In particular, this alternative market-share coefficient, m_{rs}, will be defined as a variable which depends on (i) transaction barriers and/or affinity conditions, (ii) the supply price, p_r, in each supply region r, (iii) the average demand price, $\bar{p}_s$, in each demand region s, and (iv) a mark-up coefficient, $\kappa_{rs} \geq 0$, for each link (r,s). Assume that the supply price, p_r is viable in the sense of (8.6b), and let in addition the delivery price, p_{rs}, be viable with regard to link (r, s) such that

$$\begin{aligned} p_{rs} &\geq v + \kappa_{rs} + c_{rs} + F / x_r \\ x_r &= \textstyle\sum_s m_{rs} f(\bar{p}_s) M_s \end{aligned} \tag{8.18}$$

which means that the link price must be at least as large as the variable costs per unit flow on link (r, s) plus the fixed cost per unit output.

Given that prices p_{rs}, for all s, are viable, the following expression shows how the market-share coefficient is specified in our version of a spatial interaction model:

$$m_{rs} = \frac{q_{rs} \exp\{-\mu_2(p_r + \kappa_{rs} + c_{rs})\}}{\sum_k q_{ks} \exp\{-\mu_2(p_k + \kappa_{ks} + c_{ks})\}} \tag{8.19}$$

where an affinity relation between r and s is represented by the coefficient $q_{rs} \geq 0$ and $\sum_{r,s} q_{rs}=1$ (cf. Snickars and Weibull, 1977).

The market-share formula in (8.19) can be derived as a partial equilibrium outcome based on standard assumptions about economic behaviour in markets with non-perfect competition (e.g. Batten and Johansson, 1985; Bröcker, 1989). In the terminology of Batten and Johansson, (8.19) describes a demand-oriented trade model according to which the delivery share m_{rs} is reduced as

$$\ln q_{rs} - \mu_2(p_r + \kappa_{rs} + c_{rs}) \tag{8.20}$$

is reduced. Using our notation, we may observe that in their model Batten and Johansson (1985) derives the expression in (8.21)

$$\ln q_{rs} - \gamma(p_r + c_{rs}) \tag{8.21}$$

which is equivalent to (8.20) if $\kappa_{rs} = 0$. In Bröcker's model version expression (8.20) corresponds to (8.22)

$$\ln Q_r - \gamma(p_r + c_{rs}) \tag{8.22}$$

where Q_r signifies the supply capapcity of region r.

As a next step we may contemplate the following specification of the geographic transaction-cost term c_{rs}:

$$c_{rs} = \delta_{rs} + \mu_1 d_{rs} \tag{8.23}$$

where c_{rs} is the sum of a barrier cost element, δ_{rs}, and a distance-dependent cost element, $\mu_1 d_{rs}$, where d_{rs} is some relevant measure of the distance on the link (r, s). Inserting this into (8.19) yields

$$\tilde{m}_{rs} = \frac{q_{rs} \exp\{-\mu_2(p_r + \kappa_{rs} + \delta_{rs} + \mu_1 d_{rs})\}}{\sum_k q_{ks} \exp\{-\mu_2(p_k + \kappa_{ks} + \delta_{ks} + \mu_1 d_{ks})\}} \tag{8.24}$$

We can see that the affinity and barrier factors play a similar role in the nominator and denominator of (8.24). Referring to (8.21), the difference $\ln q_{rs} - \delta_{rs}$ expresses the compound effect of affinity conditions and barrier effects. Hence, in an ordinary estimation procedure we cannot distinguish between these two effects. Having reached this conclusion we also observe that the delivery intensity will increase as: (i) $\Delta\mu_1 < 0$, (ii) $\Delta d_{rs} < 0$, and (iii) $\Delta q_{rs} > 0$ or $\Delta\delta_{rs} < 0$. Increased delivery intensity of this kind may be interpreted as the result of falling geographic transaction costs, which means that increased affinity is indirectly recognised as a fall in GTC.

Going back to (8.19), note that this formula is equivalent with (8.24), because of the specification in (8.23). Observe next that the latter expression yields the same solution as (8.25) below

$$\tilde{m}_{rs} = \frac{q_{rs} \exp\{-\mu_2(p_r + \kappa_{rs} + c_{rs} - \bar{p}_s\}}{\sum_k q_{ks} \exp\{-\mu_2(p_k + \kappa_{ks} + c_{ks} - \bar{p}_s)\}} \tag{8.25}$$

because $\bar{p}_s$ is a constant factor both in the nominator and denominator. This means that we can interpret the market mechanism as a response to the difference between $p_r + k_{rs} + c_{rs}$ and $\bar{p}_s$.

Using (8.23) or (8.18) one can now discuss the difference between the internal and external market potentials of region r. According to our previous assumptions the following conditions should be expected to prevail: $q_{rr} > q_{rs}$ and $c_{rr} < c_{rs}$ for all $s \neq r$. Moreover, external markets could be ranked according to the size of affinity or barrier factors. With such a ranking one could distinguish between local and global products. The latter would be recognised by having small differences between c_{rr} and c_{rs} as well as between q_{rr} and q_{rs}. Local products would instead be characterised by large differences with regard to the same pairs of parameters.

In order illuminate this discussion further define $M_{rs} = q_{rs}\exp\{-\mu_2 c_{rs}\}$ and formulate

$$m_{rs} = \frac{M_{rs}\exp\{-\mu_2(p_r + \kappa_{rs})\}}{\Sigma_k M_{ks}\exp\{-\mu_2(p_k + \kappa_{ks})\}} \tag{8.26}$$

Consider than a case where $p_k + \kappa_{ks} = p_r + \kappa_{rs}$ for all $k \neq r$, which means that all supplying regions have the same mill price. If this holds, we may identify $\tilde{M}_r$ as a "pure measure" of region r's total market potential, as specified in (8.27)

$$\begin{aligned} \tilde{M}_r &= \Sigma_s \tilde{M}_{rs} \\ \tilde{M}_{rs} &= \frac{M_{rs}}{\Sigma_k M_{ks}} M_s \end{aligned} \tag{8.27}$$

8.3.3 Market-Potential Switches in the Spatial Interaction Model

In the previous section an attempt is made to establish a platform for identifying specialisation opportunities in regions which differ with regard to the size of their internal market potential as well as with regard to their distances to large external markets. Before we elaborate further on this issue, let us investigate to which extent the market delivery model in (8.17)-(8.19) can generate market-potential switches in a similar way as the model based on the pure neocalssical market mechanism in (8.7).

With this as our objective, recall from (8.6a) that the condition about non-negative profits requires that the average selling price, $\bar{\bar{p}}_r$, is viable. We assume that the supply price satisfies that $p_r \geq v$, which means that the price for each link, $p_{rs} = v + c_{rs}$, will be link viable. With these assumptions, consider a two-region model where s = 1, 2, and inspect the market-share measure, m_{1s} as specified in (8.28)

$$m_{1s} = \frac{q_{1s}\exp\{-\mu_2(p_1 + \kappa_{1s} + c_{1s})\}}{q_{1s}\exp\{-\mu_2(p_1 + \kappa_{1s} + c_{1s})\} + \tilde{q}_{2s}\exp\{-\mu_2(p_2 + \kappa_{2s} + c_{2s})\}} \tag{8.28}$$
$$m_{rs} = 0 \text{ if } \bar{\bar{p}}_r \text{ is not viable}$$

According to (8.28) m_{rs} is zero if the average selling-price is non-viable, as defined in (8.6).

Let us now illustrate the possibility of a "viability catastrophe" in our two-region model. First, assume without loss of generality that $c_{11} = c_{22} = 0$, and that all mark-up coefficients, κ_{rs}, are zero. In addition we introduce the assumption that $q_{11} > q_{12}$, and $q_{22} > q_{12}$. As a start, we let $M_1 = M_2$. Given these preliminaries, observe that m_{rs} is a function of p_1 and p_2 in the two-region model, and that viability obtains for region 1 when

$$p_1 x_{11} + (p_1 + c_{12})x_{12} = v(x_{11} + x_{12}) + c_{12}x_{12} + F \tag{8.29a}$$

and this implies that

$$(p_1 - v)x_1 = b_1 x_1 = F \tag{8.29b}$$

where b_1 is the supply price net variable costs. Observing that $x_1 = m_{11}f(\bar{p}_1)M_1$ + $m_{12}f(\bar{p}_2)M_2$, it is evident from (8.28) that $\partial x_1 / \partial b_1 =$ $[(\partial m_{11}/\partial p_1)f(\bar{p}_1) + m_{11}f'(\bar{p}_1)]M_1 + [\partial m_{12}/\partial p_1 f(\bar{p}_2) + m_{12}f'(\bar{p}_2)]M_2 < 0$. Suppose now, as an experiment, that M_1 declines and M_2 grows. Then, from (8.29), the supplier in region 1 cannot compensate this event by increasing the price if M_1 dimishes sufficiently. Next, can a decrease in p_1 generate a compensation such that the new price is viable? To answer this, observe that for m_{11} the market share elasticity is

$$\frac{\partial m_{11}}{\partial p_1}\frac{p_1}{m_{11}} = -\mu_2(1 - m_{11}) \tag{8.30}$$

which can obviously be small in absolute value if m_{11} is close to unity. If q_{12} is small, we can disregard effects outside region 1. In such a case a reduction in p_1 will increase too little to make $b_1 x_1$ larger. Hence, a reduction of the internal market of region 1 can make it impossible to find a viable price, which then implies a viability "catastrophe" with a switch from an initial position $x_1 > 0$ to a situation with $x_1 = 0$. In the new equilibrium region 2 conquers the market in region 1, such that $x_{21} = m_{21}f(\bar{p}_1)M_1 > 0$.

Another way of generating this switch can be illustrated as follows. Assume that $M_2 > M_1$ and that F is large. Given this we may consider a gradual reduction of the geographic transaction costs as given by c_{21} until $\bar{\bar{p}}_1$ loses its viability. We may also observe that with the spatial interaction model employed in Section 8.3, it is not possible to consider ordinary "link catastrophes". This would require another model specification.

8.4 Product-Specific Analyses of Market Potential Patterns

There exists a long tradition within the field of economic geography to develop theoretical measures of a region's market potential and to empirically estimate such potentials. A pioneer in this field was Harris (1954), who developed concepts and measures to compute geographical market potentials with the argument that the attractiveness of a region as a production location could be found in its market potential. There has been many followers with similar approaches (see, e.g. Dixon & Thirlwall, 1975). In contrast to these earlier approaches we have – in the preceding presentation – been able to explain why and how the size of a region's market potential becomes an important characteristic. In the sequel a hierarchy of market potentials is introduced and related to a classification of how products differ with regard to their contact intensity and the associated geographical transaction costs.

8.4.1 Specialisation Opportunities in Three Categories of FURs

How do the specialisation conditions differ between large and small regions? In the preceding sections we have made a distinction between resource-based and scale-based specialisation. In the latter case scale economies can be utilised by regions which have the most favourable accessibility to demand, i.e., by regions with a market-potential advantage. Such advantages are often based on a large internal market potential as indicated in Figure 8.7.

What are then the specific specialisation conditions of small and medium-sized regions? A distinguishing feature of such regions is their limited internal market potential. As illustrated in Figure 8.8 these regions have two options. First, they can have a Ricardian advantage based on a pre-located natural resource in combination with high costs of transporting that resource. A second option is that the region succeeds in developing an economic milieu which can make use of external economies of scale in one or several sectors. Such a milieu is a host of specialised clusters. In this context a cluster is a core sector with localised firms together with their specialised input suppliers and specialised labour force. In addition, the regional infrastructure may also be specialised.

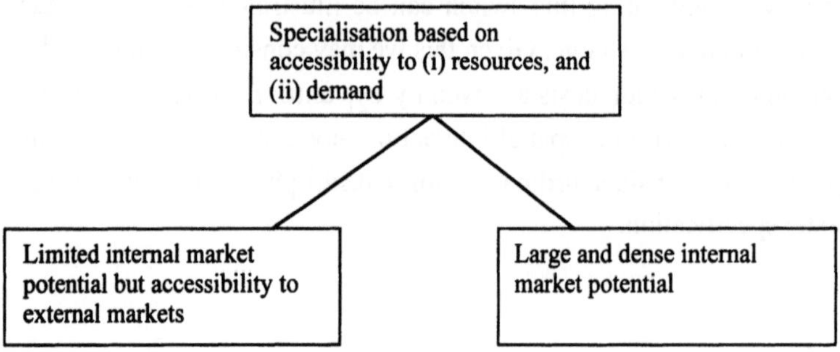

Figure 8.7 Specialisation based on demand and pre-located resources

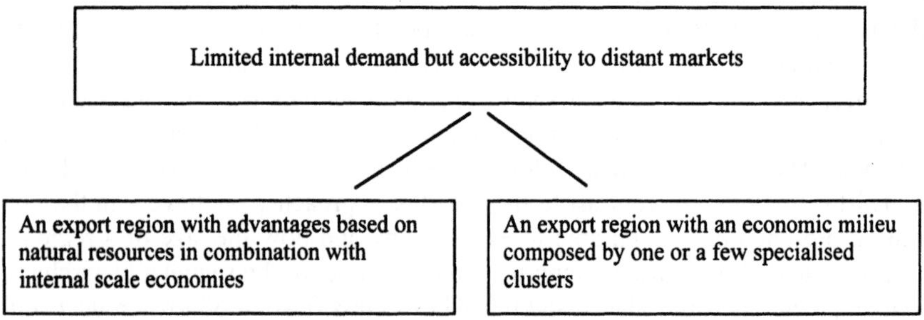

Figure 8.8 Specialisation conditions for small and medium-sized regions

8.4.2 Identifying Contact Intensive Products

In this subsection a straightforward approach is suggested. At a very fine specification of economic subsectors (around 750) three aspects are focused. First, we examine for each subsector (product group) the density of demand as a fraction of a region's population. In this way the internal market potential is identified. Second, for each subsector we relate the size of scale economies to the size of the internal market potential. Third, we refer to (8.27) when examining the importance of the internal market, which can be measured as the market-potential ratio $M_r / \widetilde{M}_r$. Moreover, we observe that for products which are very insensitive to distance the market-potential ratio will always be low.

Formally, let P_r denote the population in a functional region r, statistically referred to as a so-called LA-region (local labour-market region). For product i we identify a fraction σ_i, such that the internal market potential is given by $M_r^i = \sigma_i P_r$. Next, consider a group of distance-sensitive products for which $M_r / \widetilde{M}_r$ is close to one. For such products we can determine the minimum size, below which a region is an unfeasible location. From (8.1) we have that a feasible location must satisfy

$$p_r^i \geq v + F / f(p_r^i)\sigma_i P_r$$

where p_r^i denotes the supply price in region *r* and where $f_i(p_r^i)$ signifies the price dependency of product *i*. Thus , the population size must be large enough to satisfy the following condition:

$$P_r \geq F / f(p_r^i)\sigma_i(p_r^i - v) \qquad (8.31)$$

which shows that the density of demand, σ_i, is a critical parameter together with the price sensitivity of demand as given by $f_i(p_r^i)$.

We want to show that the production and supply of any particular product group is more likely to be located in regions with a large internal market potential than in regions with a small internal market. Therefore, let the LA-regions be ordered in a sequence 1,..., N such that $P_1 > \ldots > P_N$. In order to simplify our argument further, let all products be arranged into groups for which the distance sensitivity is (i) very high, (ii) medium and (iii) low.

First, for very distance-sensitive products we could order products according to the size of fixed costs such that all such products are supplied in the region(s) with the largest population, while the number of products that can be supplied declines as we move down the size hierarchy of regions.

Second, for products with medium distance sensitivity, $M_r / \widetilde{M}_r$ is much smaller than one and the external market potential becomes more important as a location determinant. Without any lengthy argument, let us assume that for products of category (ii) large LA-regions are large because they have large external market potentials and that small regions regions are small because their total market potential is small. To the extent that this is true, we can find a proportionality factor ϕ such that on the average we have that $\widetilde{M}_r = \phi M_r$. Given this, we should again assume that a greater variety of products are supplied from region *r* than region *r* + *k*, if $P_r > P_{r+k}$.

Third, consider now product category (iii). Many circumstances may influence the location of the associated production, e.g. located durable resources, historical "accidents" and the development of pertinent traditions of skill and knowledge. In this case we assume that the probability of such conditions to prevail is proportional to the size of the region. As a consequence we can assume that the expected number of products of the third category is proportional to the size of the LA-region.

The assumptions presented above have been tested by Johansson, Strömquist and Åberg (1998) by means of the following regression equation

$$Q_r = a + b \ln P_r \tag{8.32}$$

Q_r denotes the number of individual sectors from a specification of 747 such sectors, where each sector can be interpreted as a product group. The estimated parameters of the linear equation in (8.32) are highly significant with values a = -804 and b = 95.2. The R^2 = 0.96. Similar equations are estimated for subsets such as service and industry sectors separately, with equally good fit.

Having accepted the obvious conclusion about the importance of being a large functional region, what can be said about specialisation and economic sustainability of small and medium-sized regions? Following the previous analysis, one can suggest that there are two areas in which both small and medium-sized regions are possible locations:

- Distance-sensitive products (sectors) with small scale effects (small fixed costs)
- Groups of sectors which can form an interdependent cluster.

We observe that a cluster has one or several core sectors. Such a core sector should have low distance sensitivity, while external economies of scale dominate in importance over internal scale economies. When such firms in sector i are localised in a region r this implies that the coefficient σ_j (in formula (8.31) of an input-supplying sector j has to be replaced by a coefficient $\sigma_j^r >> \sigma_j$. In other words, when localisation economies develop this has consequences also for input suppliers which face a larger local demand than what is predicted by the population size of the region.

Finally, for medium-sized regions positive location conditions should also be expected to prevail for the following type of sectors:

- Sectors with medium distance sensitivity and with small and medium-sized scale economies.

8.4.3 Interaction Barriers, Affinity Borders and Transaction Discontinuities

In the models formulated in this paper, a lot of conclusions depend critically on the link-parameter geographic transaction costs, labelled GTC. Section 8.3.2 describes a spatial interaction model in which the GTC-level varies between different delivery links, either in the form of link barriers or link-specific affinity differentials. Such differentials are easily detected when a country's trade flows to and from other countries are compared. We shall illustrate this phenomenon by presenting results from a study by Johansson and Westin (1994).

We start by examining a model of export deliveries, x_{sr}, from Sweden to a set of European countries. The flows are measured in quantities and the model, which is based on trade data during the period 1970-1987, has the following form:

$$\ln x_{sr} = a + \theta_1 \ln Y_s + \theta_2 \ln Y_r + \theta_3 \ln p_{sr} + \theta_4 t + \theta_5 d_{sr} + \theta_6 D_L + \theta_7 D_B \qquad (8.33)$$

where Y_s and Y_r denote GDP in Sweden and an importing country, respectively. In view of the theoretical framework of this paper, Y_r represents country r's demand potential, discounted with distance and barrier effects, whereas Y_s represents Sweden's supply capacity. Moreover, p_{sr} signifies the Swedish export price, t signifies time, D_L is dummy for cultural and language similarities, and D_B is dummy for countries which have a common border with Sweden. The estimation was made separately for different groups of commodities. For all groups except one, the distance variable has a low t-value. All other parameters come out as highly significant. This implies that the "barrier variables" D_L and D_B do indeed influence trade patterns as indicated by the spatial interaction model introduced in Section 8.3.2.

As a second way of of showing the relevance of GTC-differentials on different trade links, we present another export model, estimated with the same data and for the same time period as (8.33). This second model has Y_s and Y_r as arguments together with time and the export price for each link. In addition the model includes a fix factor, a_r, a parameter which is assumed to reflect the affinity level for each destination region r.

$$\ln x_{sr} = a_r + \omega_1 \ln Y_s + \omega_2 \ln Y_r + \omega_3 \ln p_{sr} + \omega_4 t \qquad (8.34)$$

All parameters of the estimated model in (8.34) are statistically different from zero. The goodness of fit is very high. Taking an average across different commodity groups the value of the export affinity parameter a_r has the following value with regard to 9 importing regions (countries and groups of countries):

- Norway = 5.7
- Denmark = 5.4
- Finland = 5.3
- Netherlands, Belgium, Luxembourg, Ireland, Switzerland and Austria (as a group) = 4.1
- United Kingdom = 3.8
- Germany = 3.5
- France = 2.9
- Spain, Portugal, Italy and Malta (as a group) = 2.9
- Yugoslavia, Greece, Turkey and Israel (as a group) = 2.8

Clearly, the affinity parameter reflects to some degree the export distance in geographic terms. However, the above list shows that there is a stepwise decline in export intensities, where the steps also reflect other types of distances such as language, cultural similarity and size.

8.5 Conclusions: GTC and Specialisation Opportunities

In this paper we shown that regions with a large internal market potential has an absolute advantage in finding a diversified specialisation. Moreover, when a region has a large internal as well as external market-potential the competitive advantage increases even further, with increased possibilities of becoming a host of a wide range of sectors, many of which will export to other regions.

These observations point in the direction of emphasising the importance of the size of a region. Where does that leave the small and medium-sized regions? They may of course develop resource-based specialisation whenever located durable capacities generate an advantage. However, our major conclusion is that small and medium-sized regions have the opportunity to develop a narrow specialisation based on external economies of scale with specific cluster formations. This option can materialise for products where inputs to the production have large GTC-levels, whereas the output deliveries are characterised by comparatively low GTC-values. This is a form of niche-specialisation.

Based on this conclusion one may design empirical studies which identify products characterised by the described combination of high input and low output GTC-values. Given such a background the study should identify the location of such production.

References

Balassa, B. and L. Bauwens, 1988, "The Determinants of Intra-European Trade in Manufactured Goods", *European Economic Review*, 32:1421-1437.

Batten, D.F. and B. Johansson, 1985, "Price Adjustments and Multiregional Rigidities in the Analysis of World Trade", *Papers of Regional Science Association*, 56:145-166.

Batten, D.F. and D. Boyce, 1986, "Spatial Interaction and Interregional Commodity Flow Models", in P. Nijkamp (ed.), *Handbook of Regional and Urban Economics*, Vol. I, North-Holland, Amsterdam.

Bröcker, J., 1988, "Interregional Trade and Economic Integration: A Partial Equilibrium Analysis", *Regional Science and Urban Economics*, 18:261-281.

Burenstam-Linder, S., 1961, *An Essay on Trade and Transformation*, John Wiley and Sons, New York.

Chamberlin, E.H., 1933, *Theory of Monopolistic Competition*, Harvard Economic Studies, Vol. 38, Cambridge, Mass.

Davies, D. and D. Weinstein, 1997, *Economic Geography and Regional Production Structure: An Empirical Investigation,* Harvard University and NBER.

Dixit, R. and V. Norman, 1980, *Theory of International Trade, Cambridge,* Cambridge University Press, Cambridge.

Dixon, R. and A.P. Thirlwall, 1975, A Model of Regional Growth Differences along Kaldorean Lines, *Oxford Economic Papers*, 27:201-214.

Ethier, W., 1982, "National and International Returns to Scale in the Modern Theory of International Trade", *American Economic Review*, 72:389-405.

Faini, R., 1984, "Increasing Returns, Nontraded Inputs, and Regional Development", *Economic Journal*, 94:308-323.

Fujita, M., 1988, "A Monopolistic Competition Model of Spatial Agglomeration: A Differentiated Product Approach", *Regional Science and Urban Economics,* 18:87-124.

Harris, C., 1954, "The Market as a Factor in the Location of Industry in the United States", *Annals of the Association of American Geographers*, 44:315-348.

Harrod, R., 1939, "An Essay in Dynamic Theory", *Economic Journal*, 49:14-33.

Helpman, E., 1981, "International Trade in the Presence of Product Differentiation, Economies of Scale and Monopolistic Competition: A Chamberlain-Heckscher-Ohlin Approach", *Journal of International Economics,* 11:305-340.

Helpman, E., 1984, "Increasing Returns, Imperfect Markets and Trade Theory", in R. Jones and P. Kenan (eds.), *Handbook of International Economics*, Vol. 1, North Holland, Amsterdam, 325-365.

Johansson, B., C. Karlsson and L. Westin, (eds.), 1993, *Patterns of a Network Economy,* Springer-Verlag, Berlin.

Johansson, B. and L. Westin, 1991, "How Network Properties Influence Trade Patterns", Paper presented at the 31st European Regional Science Association Congress in Lisbon.

Johansson, B. and L. Westin, 1993, "Revealing Network Properties of Sweden's Trade with Europe", in B. Johansson, C. Karlsson and L. Westin (eds.), *Patterns of a Network Economy*, Springer-Verlag, Berlin, 125-141.

Johansson, B. and L. Westin, 1994, "Affinities and Frictions of Trade Networks, *Annals of Regional Science*.

Jones, R. and P. Kenan, (eds.), 1984, *Handbook of International Economics*, Vol. 1, North Holland, Amsterdam.

Kaldor, N., 1970, "The Case for Regional Policies", *Scottish Journal of Political Economy,* 17:337-347.

Krugman, P., 1979, "Increasing Returns, Monopolistic Competition, and International Trade", *Journal of International Economics*, 9:469-479.

Krugman, P., 1980, "Scale Economies, Product Differentiation and the Pattern of Trade", *American Economic Review*, 70:950-959.

Krugman, P., 1981, "Trade, Accumulation and Uneven Development", *Journal of Development Economics*, 8:149-161.

Krugman, P., 1990, *Rethinking International Trade*, MIT Press, Cambridge, Mass.

Krugman, P., 1990a, "The Hub Effect: Or, Threeness in Interregional Trade", mimeo, *Department of Economics*, MIT.

Krugman, P., 1991, *Geography and Trade*, MIT Press, Cambridge Mass.

Krugman, P., 1992, "A Dynamic Spatial Model", National Bureau of Economic Research.

Krugman, P., 1993, "First Nature, Second Nature, and Metropolitan Location", *Journal of Regional Science*, 33:129-144.

Lancaster, K., 1980, "Intra-Industry Trade under Perfect Monopolistic Competition", *Journal of International Economics*, 10:151-175.

Marshall, A., 1920, *Principals of Economics*, 8 ed., Macmillan, London.

Myrdal, G., 1957, *Economic Theory and Underdeveloped Regions*, Duckworth, London.

Ohlin, B., 1933, *Interregional and International Trade,* Harvard University Press, Cambridge, Mass.

Porter, M.E., 1990, *The Competitive Advantage of Nations*, Macmillan, London.

Rivera-Batiz, F.L., 1988, "Increasing Returns, Monopolistic Competition, and Agglomeration Economies in Consumption and Production", *Regional Science and Urban Economics*, 18:87-124.

Snickars, F. and J. Weibull, 1977, "A Minimum Information Principle: Theory and Practice", *Regional Science and Urban Economics,* 7:137-168.

SOU 1997:13, *Regionpolitik för hela Sverige*, Fritzes, Stockholm.

Stigler, G., 1951, "The Division of Labour is Limited by the Extent of the Market", *Journal of Political Economy*, 59:185-193.

Storper, M. and A.J. Scott, 1995, "The Wealth of Regions, Market Forces and Policy Imperatives in Local and Global Context", *Futures*, 27.

Sölvell, Ö., I. Zander and M.E. Porter, 1991, *Advantage Sweden*, Norstedt, Stockholm.

Venables, A.J., 1995, "Economic Integration and the Location of Firms", *American Economic Review*, 85:296-300.

9 Trade and Regional Development: International and Interregional Competitiveness in Brazil

Eduardo A. Haddad[a] and Geoffrey J.D. Hewings[b]

[a] FIPE, University of São Paulo, Brazil and Regional Economics Applications Laboratory, University of Illinois, Urbana, IL 61801-3671, USA

[b] Dept. of Geography, Regional Economics Applications Laboratory, University of Illinois, Urbana, IL 61801-3671, USA

9.1 Introduction

As the process of global integration has reached the boundaries of developing countries, there has been concern about the role to be played by these nations in the new world economic order. In many parts of the developing world, efforts are being made to transform economic activities so as to enhance their international competitiveness. Parallel to these efforts, market-oriented policies have been generally adopted based on World Bank and IMF recommendations, supported by the recognition of the distorting effects of government intervention.

Distributional effects of such policies have been neglected on the grounds that greater efficiency would lead to rapid growth which would ultimately benefit the population in the lower income groups (Baer and Maloney, 1997). At the regional level, the desire to maximize economic growth, implied by the aim of increasing international competitiveness, is very likely to negatively impact the distribution of income among regions in developing countries (Baer et al., 1998). As these countries present strong evidence of regional dualism, the more developed regions are those that concentrate the resources that can foster export-led national growth.

Brazil was late in its efforts towards integrating the country into the global economic network, as was the case for most Latin American countries.[1] In this paper, the more open policies of the 1990's and related national strategies for increasing international competitiveness are examined. An interregional computable general equilibrium (CGE) model is used to analyze the short-run and long-run regional effects of trade liberalization policies, represented by simulations of tariff cuts. The choice of this policy was made based on the relevance for the Brazilian case. It is part of a broader economic reform program that is being carried out in the country and whose *regional* effects have yet to be considered in an integrated formal framework. The general equilibrium nature of economic

---

1 Sachs and Warner (1995) provide a categorization of developing countries according to the timing of trade liberalization – defined by the absence of certain characteristic impediments to open trade – in which most Latin American countries present relative late opening, often not until the 1990s.

interdependence and the fact that the policy impacts in various regional markets differ are considered in the results presented below.

The remainder of the paper is organized as follows: Sect. 9.2 presents an overview of the effects of trade liberalization policies; in Sect. 9.3, the assumptions underlying the interregional CGE model are presented; Sect. 9.4 discusses the simulation of the effects of reducing tariff-barriers and further dicussion of the results and their implications for Brazil are presented in Sect. 9.5.

9.2 Trade Liberalization Policies

The effects of trade reforms have been extensively studied in the international trade literature. Trade liberalization processes are said to have long-run benefits derived from gains in both the production side (there is an overall increase in the foreign exchange revenue earned in export industries, or saved in import industries, per unit of labor and capital) and the consumption side (the same basket of products can be obtained at lower cost). However, the liberalization process also involves two kinds of short-run costs to the economy: distributional costs (protected sectors tend to lose); and balance of payments pressures due to the rapid increase in imports (Bruno, 1987). In a cross-country study, Sachs and Warner (1995) reach the general conclusion that economic reform, in which trade liberalization plays the major role, leads to economic growth. However, the short-term growth consequences of a trade reform will depend on the structure of the reforming economy. Evidence for trade liberalization and macroeconomic performance in developing countries is documented in Agénor and Montiel (1996). It is found that, contrary to general beliefs, there is no strong evidence suggesting that liberalization, in a broader sense, is associated with sharp reductions in employment and a short-run contraction in output. Another finding refers to the short-run improvement in the balance of payments, as the growth of exports outpaces the increase in imports, in developing countries where trade reforms were undertaken after World War II.[2]

At the regional level, the (computable) general equilibrium framework is considered to be more adequate as a theoretical tool for tariff-policy-type of analysis than partial equilibrium and traditional fixed-price general equilibrium models such as economic base and input-output models. First, and more obviously, the latter do not accommodate, by definition, shocks in relative prices, and, therefore, fixed-price models do not represent a viable alternative. Secondly, Walrasian models provide an ideal framework for appraising the effects of trade and tax policy changes on resource allocation and for assessing who gains and loses, policy impacts not well covered by empirical partial equilibrium models (Shoven and Whalley, 1984). Moreover, at the sub-national level, many aspects of regional economies, which are taken as exogenous in partial equilibrium analysis,

---

[2] In the successful cases, the adoption of more liberal commercial policies in the countries surveyed was accompanied by the depreciation of the real exchange rate.

are endogeneized; this process captures feedback effects that occur throughout a regional economy that are ignored in partial equilibrium analysis, attempting to determine, more precisely, the regional impact of such macroeconomic policies.

Experiments on trade liberalization policies adopting the CGE approach are common for national models. In general, the question posed refers to the effects of changes in trade policy, and the results show that the welfare effects are relatively small compared to the effects of other kinds of policies, such as taxes – distortions that affect a relatively small portion of total activity can be expected to have small distorting effects (Shoven and Whalley, 1984). Studies that consider the regional impacts of trade policies are less common (See, for examples, Dixon et al., 1982; Liew, 1984; Whalley and Trela, 1986; Gazel, 1994; McGregor et al., 1997.][3] In Brazil, Guilhoto (1995) explored the implications of tariff reductions on specific sectors, but, with the exception of Sousa (1987) who explored the effects of protectionism on the distribution of income between urban and rural areas, no attention has been directed to differential regional impacts. Guilhoto's model, styled on the ORANI system developed for Australia (Dixon et al., 1982) found less flexibility in the Brazilian economy than in Australia in adjusting to tariff changes.

9.3 The Brazilian Multisectoral and Regional/Interregional Analysis Model (B-MARIA)

Many modeling approaches designed to address economic impact analysis in a regional system have been developed, initially, from international trade models. They evolved from the simple economic-base framework, through input-output and the general social accounting framework, to the more sophisticated econometric input-output and computable general equilibrium (CGE) models. In a sense, these models are all related to each other in that they might either form a chain of theoretical links established in a consistent way, or they might simply play a role as a module of larger, integrated set of models (Hewings, 1985).

The CGE approach treats the economy as a system of many interrelated markets in which the equilibrium of all variables must be determined simultaneously. Any perturbation of the economic environment can be evaluated by recomputing the new set of endogenous variables in the economy. Optimizing behavior of consumers and producers is explicitly specified, as well as the institutional environment. Thus, demand and supply functions are derived consistently with prevaling consumer and production theories with both production and consumption decisions responding to changes in prices. Regional interactions can be introduced through the interregional framework, allowing for regional imbalances and feedback effects from the other regions to be captured.

3 For a survey, see Haddad (1997) and Patridge and Rickman (1998).

The **B**razilian **M**ultisectoral **A**nd **R**egional/**I**nterregional **A**nalysis Model (B-MARIA) is the first fully operational interregional CGE model for Brazil.[4] The model is based on the MONASH-MRF Model, which is the latest development in the ORANI suite of CGE models of the Australian economy. B-MARIA contains over 200,000 equations, and it is designed for forecasting and policy analysis. Agents' behavior is modeled at the regional level, accommodating variations in the structure of a three-fold regional division of the Brazilian economy: North, Northeast, and Center-South (Rest of Brazil). Results are based on a bottom-up approach – national results are obtained from the aggregation of regional results. The model identifies 40 sectors in each region producing 40 commodities, a single household sector in each region, regional governments and one federal government, and a single foreign consumer who trades with each region. Special groups of equations define government finances, accumulation relations, and regional labor markets. In the Brazilian tradition of modeling, it benefits from the work by Guilhoto (1986, 1995), that provides a computable national model of the Johansen type with the solutions given in growth rates. Besides the Moreira and Urani (1994) model for the Northeast Brazil, which is rooted in the requirement analysis framework, and, therefore, does not provide any supply-side constraint, B-MARIA is the first attempt to model the Brazilian economy in an interregional general equilibrium framework, taking into account both demand and supply constraints. [For a survey of CGE models applied for the Brazilian economy, see Guilhoto and Fonseca (1990), Moreira and Urani (1994), and Guilhoto (1995).]

9.3.1 Theoretical Structure

B-MARIA is based on the multiregional version of the MONASH Model, the MONASH Multiregional Forecasting Model – MONASH-MRF (Naqvi and Peter, 1996). The equations of the CGE core module of the model are defined following the same structure of the ORANI Model (Dixon et al., 1982), with a regional subscript added, when appropriate. It may be considered a Johansen-type model, in that the solutions are obtained by solving the *linearized* equations of the model. A typical result shows the percentage change in the set of endogenous variables, after a policy is carried out, compared to their values in the absence of such policy, in a given environment.

The schematic presentation of Johansen solutions for such models is standard in the literature. What follows is a summary of its contents in order to see how these models work. More details on the general modeling structure can be found in Dixon et al. (1982, 1992), and Dixon and Parmenter (1994) while the specific details of the Brazilian version may be found in Haddad (1997) and Haddad and Hewings (1997). A summary of the form of the B-MARIA is provided next to assist the reader interpret the results contained in succeeding sections.

Underlying the model is a set of interregional input-output tables developed in part from regional input-output tables for the North (SUDAM, 1994) and

---

[4] The complete specification of the model is available in Haddad and Hewings (1997), and Haddad (1997).

Northeast (BNB, 1992) regions, and from the national input-output tables for Brazil (FIBGE, 1995), for the year of 1985.[5] The specification of the equations of the model is presented in five different integrated blocks of equations: the CGE core module, the government finance module, the capital accumulation and investment module, the foreign debt accumulation module, and the labor market and regional migration module.

CGE core module. The basic structure of the CGE core module comprises three main blocks of equations determining demand and supply relations, and market clearing conditions. In addition, various regional and national aggregates, such as aggregate employment, aggregate price level, and balance of trade, are defined here. Nested production functions and household demand functions are employed; for production, firms are assumed to use fixed proportion combinations of intermediate inputs and primary factors are assumed in the first level while, in the second level, substitution is possible between domestically produced and imported intermediate inputs, on the one hand, and between capital, labor and land, on the other. At the third level, bundles of domestically produced inputs are formed as combinations of inputs from different regional sources. The modeling procedure adopted in B-MARIA uses a constant elasticity of substitution (CES) specification in the lower levels to combine goods from different sources.

The treatment of the household demand structure is based on a nested CES/linear expenditure system (LES) preference function. Demand equations are derived from a utility maximization problem, whose solution follows hierarchical steps. The structure of household demand follows a nesting pattern that enables different elasticities of substitution to be used. At the bottom level, substitution occurs across different domestic sources of supply. Utility derived from the consumption of domestic composite goods is maximized. In the subsequent upper-level, substitution occurs between domestic composite and imported goods.

Equations for other final demand for commodities include the specification of export demand and government demand. Exports are divided into two groups: traditional exports (agriculture, mining, coffee, and sugar), and non-traditional exports. The former faces downward sloping demand curves, indicating that traditional exports are a negative function of their prices in the world market. Non-traditional exports form a composite tradable bundle, in which commodity shares are fixed. Demand is related to the average price of this bundle.

One new feature presented in B-MARIA refers to the government demand for public goods. The nature of the input-output data enables the isolation of the consumption of *public goods* by both the federal and regional governments. However, productive activities carried out by the public sector cannot be isolated from those by the private sector. Thus, government entrepreneurial behavior is dictated by the same cost minimization assumptions adopted by the private sector. This may be a very strong assumption for the Brazilian case but the liberalization process of the 1990's offer some enhanced credibility for this assumption. Public good consumption is set to maintain a (constant) proportion with regional private

---

5 See Haddad (1997).

consumption, in the case of regional governments, and with national private consumption, in the case of the federal government.

A unique feature of B-MARIA is the explicit modeling of the transportation services and the costs of moving products based on origin-destination pairs. The model is calibrated taking into account the specific transportation structure cost of *each* commodity flow, providing spatial price differentiation, which indirectly addresses the issue related to regional transportation infrastructure efficiency. Other definitions in the CGE core module include: tax rates, basic and purchase prices of commodities, tax revenues, margins, components of real and nominal GRP/GDP, regional and national price indices, money wage settings, factor prices, and employment aggregates.

Government finance module. The government finance module (drawing on data assembled by Dinsmorr and Haddad, 1996) incorporates equations determining the gross regional product (GRP), expenditure and income side, for each region, through the decomposition and modeling of its components. The budget deficits of regional governments and the federal government are also determined here. Another important definition in this block of equations refers to the specification of the regional aggregate household consumption functions. They are defined as a function of household disposable income, which is disaggregated into its main sources of income, and the respective tax duties.

Capital accumulation and investment module. Capital stock and investment relationships are defined in this module; however, only the comparative-static version of the model produces reliable results, restricting the use of the model to short-run and long-run policy analysis. When running the model in the comparative-static mode, there is no fixed relationship between capital and investment. The user decides the required relationship on the basis of the requirements of the specific simulation.[6]

Foreign debt accumulation module. This module is based on the specification proposed in ORANI-F (Horridge et al., 1993), in which the nation's foreign debt is linearly related to accumulated balance-of-trade deficits. In summary, trade deficits are financed by increases in the external debt.

Labor market and regional migration module. In this module, regional population is defined through the interaction of demographic variables, including interregional migration. Links between regional population and regional labor supply are provided. Demographic variables are usually defined exogenously, and together with the specification of some of the labor market settings, labor supply can be determined together with either interregional wage differentials or regional unemployment rates. In summary, either labor supply and wage differentials determine unemployment rates, or labor supply and unemployment rates determine wage differentials.

---

[6] For example, it is typical in long-run comparative-static simulations to assume that the growth in capital and investment are equal (see Peter et al., 1996b).

9.3.2 Closures

B-MARIA can be configured to reflect short-run and long-run comparative-static, as well as forecasting simulations. At this stage, two basic closures for alternative time frames of analysis in single-period simulations are available. A distinction between the two closures relates to the treatment of capital stocks encountered in the standard microeconomic approach to policy adjustments. In the short-run closure, capital stocks are held fixed, while, in the long-run, policy changes are allowed to affect capital stocks.

Short-run. In addition to the assumption of interindustry and interregional immobility of capital, the short-run closure would include fixed regional population and labor supply, fixed regional wage differentials, and fixed national real wage. Regional employment is driven by the assumptions on wage rates, which indirectly determine regional unemployment rates. These assumptions describe the functioning of the regional labor markets as close as possible to the Brazilian reality. First, changes in the demand for labor are met by changes in the unemployment rate, rather than by changes in the real wage. This seems to be the case in Brazil, given the high level of disguised unemployment in most of the areas of the country; excess supply of labor has been a distinct feature of the Brazilian economy. Secondly, labor's interregional immobility in the short-run suggests that migration is not a short-term decision. Finally, nominal wage differentials in Brazil are persistent, reflecting the geographical segmentation of the workforce (Savedoff, 1990). On the demand side, investment expenditures are fixed exogenously – firms cannot reevaluate their investment decisions in the short-run. Household consumption follows household disposable income, and government consumption, at both regional and federal levels, is fixed (alternatively, the government deficit can be set exogenously, allowing government expenditures to change). Finally, since the model does not present any endogenous-growth-theory-type specification, technology variables are exogenous.

Long-run. A long-run (steady-state) equilibrium closure is also available in which capital and labor are mobile across regions and industries.

9.4 Simulation Results

Trade liberalization is an important element of the range of structural changes that the Brazilian economy has undergone in the last years, especially in the context of the creation of regional trade agreements (MERCOSUR). The rapid growth in the volume of trade among the MERCOSUR countries, recorded since the *Asuncion Treaty* was signed, benefited, to a great extent, from reductions in trade barriers. Although, in Brazil, most of the gains tend to be spatially concentrated in the South and Southeast regions, more remote regions, such as the Northeast, have also increased their exports to the MERCOSUR (Haddad, P. 1997). To explore the effects of such policies, B-MARIA is used to simulate the impacts of tariff changes in the Brazilian economy.

The model is applied to analyze the effects on the Brazilian economy of a uniform 25% decrease in all tariff rates. All exogenous variables were set equal to zero, except the changes in the power of tariffs, i.e., one plus the tariff rates, which were set such that the percentage change decrease in each tariff rate was 25%. Results of the simulation, under short-run and long-run closures, are shown as percentage deviation from the base case (which is the situation without policy changes). The analysis is concentrated on the effects on industrial activity levels, and on some general macro and regional variables.[7]

Because of the nature of the data base, it should be pointed out that the model deals with changes in the *real* tariff rates (the ratio of import tax collected over the volume of imports), as opposed to *nominal* tariff rates, which are much higher. Moreover, the model does not consider non-tariff barriers. Thus, the real tariff rate in 1985 (benchmark year) was close to 5% as compared to the average nominal rate of over 25%.

Table 9.1 provides summary information about the market share of commodities that are the most sensitive to changes in tariff rates. While the Center-South captures a significant share of these sectors, there are some important concentrations in the other two regions. A complete list of the sectors is provided in the Appendix.

Table 9.1 Market-share of selected commodities, by source, 1985 (in %)

Commodity	Source			
	North	Northeast	Center-South	Foreign
Mining	4.4	14.6	33.2	47.8
Electrical Equipment	2.8	5.2	77.0	15.0
Electronic Equipment	24.3	1.1	62.1	12.5
Trade	4.9	40.8	43.4	10.9
Transportation	2.8	7.4	79.1	10.8
Chemicals	0.6	15.2	73.6	10.6

Source: Interregional input-output table, 1985

7 The volume of information that the model produces in each simulation is overwhelming. To interpret the results, we tried to focus the analysis on a few interesting issues associated with the respective simulations, in order to rationalize particular results in terms of the model's theoretical framework and its underlying data base. This process, apart from giving insights into a particular economic phenomenon, serves to act as an informal verification of the simulations' results.

9.4.1 Short-Run

Industry activity results show that, in general, export sectors benefit most from the tariff cut, while import-competing sectors are the main losers. Explanations for specific sector results should consider structural and parametric aspects of the data base. Sectors that present higher increases in their output tend to have high share of imports in their cost structure, and greater export demand elasticity (in the case of traditional exports). On the other hand, sectors with high import substitution elasticities, high import shares in their domestic markets, and high percentage changes in their tariff rates are more likely to be harmed by the policy change.

The tariff reduction stimulates the export sectors in different ways. These sectors benefit, to a great extent, from reductions in the cost of production. Imported intermediate inputs, which have import duties levied upon them, experience a relatively sharp reduction in their cost. Since imports play a prominent role in Brazil, especially in the Center-South, the general price level decreases with the tariff cut, and through indexation mechanisms, the cost of primary factors of production will also decrease.

The mining sector in the North and in the Center-South (0.162% and 0.222%, respectively), the coffee sector in the Center-South (0.187%), and the sugar sector in the Northeast (0.282%) are examples of traditional export sectors that gain from increased competitiveness in the international markets. On the other hand, import-competing industries are adversely affected by the substitution away from regionally produced goods towards imports; examples here would be chemicals and chemical-based products that face significant competition from foreign products (see Table 9.1). Moreover, their base case tariff rates are relatively high, which also contributes to their weak performance in the simulation. When interregional feedback effects are considered, the electric-electronic industries of the North, which also face relatively high international competition, can reverse the adverse situation by increasing their sales to the markets in the other regions, especially to household consumption in the Center-South. However, substitution towards imported products in the latter region are underestimated by the model; the import coefficients in the benchmark year reflect the protectionist policies towards the Zona Franca de Manaus (North region), showing negligible consumption of imported electronic equipment by households. Moreover, these sectors face small elasticities of substitution of the Armington type.[8]

In general, the group of industries including nonmetallic minerals and the chemical complex (chemicals, petroleum refining, pharmaceuticals), because of the greater change in their exposure to international competition due to prevailing higher protection levels, reveals weak performance in the simulation. In the case of the mining sector, which faces the strongest competition, the tariff rate is close to zero and hence, direct substitution effects from changes in tariff-based market distortions are negligible. Trade and transportation also face zero tariff rates. For

---

8 The Armington assumption suggests that similar commodities produced in different regions or countries should be treated as different goods in the estimation process, thus facilitating evaluation of substitution by region/country of origin.

these two margin commodities, performance is closely related to the economy-wide performance.

Finally, the non-traditional export industries reveal results that are distinct from each other; the explanation lies in the diverse nature of each sector's demand and cost structure. For instance, the footwear industry, in each region, increases its output far above the regional average: 0.432% (North), 0.319% (Northeast), and 0.458% (Center-South). As a consumption good, it is affected by the increase in real household expenditure; it also benefits from prevaling low tariff rates and its already established international competitiveness (even facing insignificant tariff rates, the market-share of foreign footwear accounts for only 3.0% in the base year). Output in the steel industry in the Center-South and in the Northeast also increases significantly (0.360% and 0.435%, respectively), reflecting the sales orientation of the sector towards international markets. In, general, sectors with already established higher share of sales to other countries benefit from tariff reduction, as do sectors producing consumption goods.

According to the model, industry employment levels expand/contract in the same direction as activity levels. However, the expansion of these changes are more intense for employment, in both directions. The absolute value of the percentage changes in employment by industry is higher than the absolute value of percentage changes in activity by industry, for all industries in all regions. The explanation for the more intense changes in the level of employment lies in the nature of the closure adopted in the simulation. It reflects the combined effects of fixed capital stocks and the general reduction in the price of *hiring* labor, which captures movements in the nominal wage paid to workers relative to movements in the producers' product price.

Figures 9.1-3 show the percentage changes in employment, for each sector in each region. Given the nature of the closure, which allows for producers to respond to exogenous shocks through changes in the employment level only, the figures reveal the short-run supply responses from the model, for a tariff decrease. With a few exceptions, the changes are similar, in sign, across the three regions although the percentage changes do vary, with negative changes of greater intensity in the North and Northeast and somewhat greater positive changes in the Center-South.

Table 9.2 summarizes the simulation results on some macro variables. The real GDP of Brazil is shown to increase by 0.120 % with all the regions positively affected, with real GRP increases ranging from 0.057% (Northeast) to 0.132% (Center-South). Regarding the regional distribution of income, the tariff reduction worsens the Northeast's relative position in the country, even though that is a Pareto-improvement situation (outcome of tariff policy is said to be Pareto superior to outcome without tariff change, as GRP improves in all the regions). Another indication of the potential regional concentration effects of the reduction in tariffs is the result achieved for the regional/national consumption ratio, which shows the Center-South increasing its share in national consumption at the expense of the other two regions.

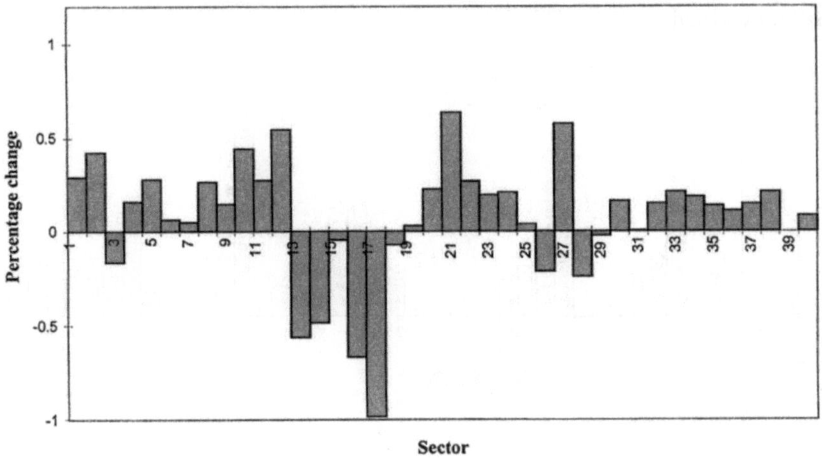

Figure 9.1 B-MARIA projected short-run employment effects of a uniform 25% tariff reduction: North

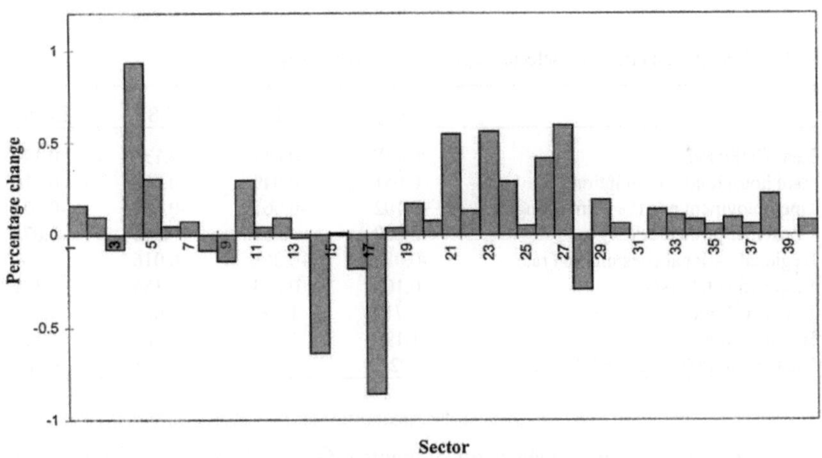

Figure 9.2 B-MARIA projected short-run employment effects of a uniform 25% tariff reduction: Northeast

In B-MARIA, household consumption in each region is assumed to be a function of household disposable income, in a Keynesian specification. Real household consumption in all the regions experiences an increase: North (0.051%), Northeast (0.049%), Center-South (0.126%). Since real wages are assumed fixed (nominal wages are indexed to the national CPI), this effect results directly from the increase in the activity level (employment effect). An examination of regional unemployment rates (which fall – see Table 9.2) confirms this result. Moreover, differential movements in the regional CPI also contribute to the consumption pattern verified.

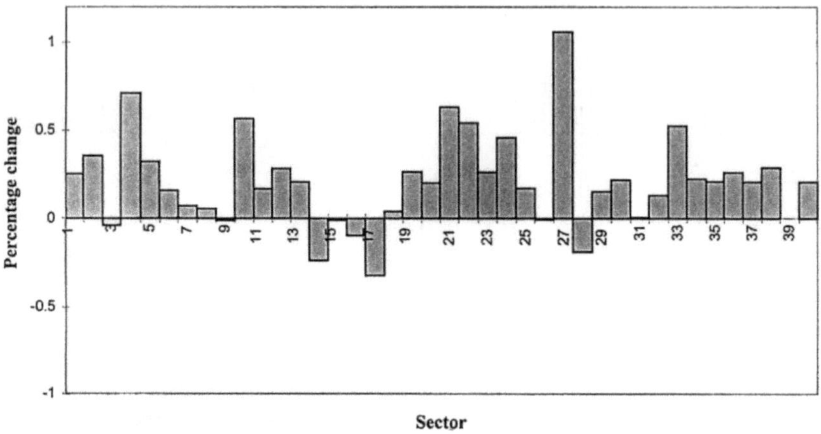

Figure 9.3 B-MARIA projected short-run employment effects of a uniform 25% tariff reduction: Center-South

Table 9.2 Short-run effects on selected regional and macro variables

	N	NE	C-S	Brazil
Real GDP/GRP	0.118	0.057	0.132	0.120
Real household consumption	0.051	0.049	0.126	0.110
Unemployment rate (% point change)	-0.102	-0.061	-0.151	-0.124
Consumer price index	-0.080	-0.57	-0.82	-0.078
Regional/national consumption rate	-0.058	-0.061	0.016	--
Employment: persons	0.105	0.063	0.156	0.129
Export volume	0.719	1.096	1.061	1.054
Import volume	1.199	1.705	1.106	1.124
Balance of trade (ordinary change)	2.9	106.9	504.5	614.3

One final comment on the short-run impacts refers to the results achieved for the balance of trade. In every region, the percentage change in the import volume is greater than the export volume. However, because of trade surpluses presented in the benchmark year, this movement of imports is not enough to produce a deficit in the balance of trade in the short-run. Even though a trend towards the reduction of trade surplus is suggested by the results, the model does not project balance of trade problems in the short-run.

9.4.2 Long-Run

The results described above refer to the short-run effects of the tariff reduction, which are important for macroeconomic management. As trade reform aims at improving the allocation of resources in the long-run, a simulation was carried out adopting the long-run closure, described in the previous section. In this exercise, the assumptions on interregional mobility of capital and labor are relaxed and a steady-state-type of solution is achieved, in which regional natural unemployment rates and regional aggregate rates of return are reestablished. Attention will be focused on the interregional flows of capital and labor and their effects on regional and macro variables.

Table 9.3 shows the long-run results of the simulation for selected regional and national variables. As the aggregate level of employment is now assumed to be exogenously determined by demographic variables, the national real wage is allowed to change to keep national employment at the base case level. Supply-side effects are restricted to the distribution of labor across sectors and regions, and to capital movements. At the national level, the increase in GDP by 0.166% above the base case level is made possible through the increase in the capital stock of the economy (0.320%) induced by the fall in the aggregate rental price of capital. Imported commodities are important inputs for capital creation and the fall in the prices of imports reduces the cost of producing capital. Contributing to the increase in GDP is also an increase in the productivity of the labor force as measured by the percentage change in employment using wage-bill weights; with the number of people employed fixed, the impact of the tariff cut has been to increase the output produced per worker by 0.109%.

Table 9.3 Long-run effects on selected regional and macro variables

	N	NE	C-S	Brazil
Real GDP/GRP	0.210	-0.579	0.299	÷0.166
Real household consumption	0.260	-0.468	0.379	0.233
Real investment	0.396	-0.493	0.451	0.295
Capital stock	0.386	-0.499	0.456	0.320
Regional government consumption	0.260	-0.468	0.379	0.208
Federal government consumption	0.233	0.233	0.233	0.233
Consumer price index	-0.039	0.107	-0.015	0.004
Regional/National consumption rate	0.027	-0.699	0.145	--
Employment: persons weights	0.131	-0.521	0.217	--
Employment: wage-bill weights	0.131	-0.521	0.217	0.109
Interregional export volume	0.316	-0.339	-0.126	--
Interregional import volume	0.210	-0.313	-0.067	--
International export volume	-0.231	-1.598	0.509	0.306
International import volume	1.480	1.026	1.269	1.268
Balance of trade (ordinary change)	-50.4	-246.8	-431.3	-728.5
Nominal wage	0.333	0.333	0.333	0.333
GDP/GRP deflator	-0.069	0.212	0.005	0.032
Population	0.131	-0.521	0.217	--

Regional unemployment and wage differentials are assumed constant in the simulation. B-MARIA accommodates the labor market assumptions by allowing population movements between regions so that labor supply is increased in regions experiencing employment expansion, and vice-versa. The impact of the trade liberalization policy favors employment in the North (0.131%) and Center-South (0.217%) at the expense of the Northeast (-0.521%), with a consequent transfer of population from the latter. The estimates of regional wage differentials considered in the benchmark data base show the average wage in the Center-South as 119.8% of the national average wage, while the wages in the North and Northeast are, respectively, 81.5% and 53.0% of the national average. Because labor is paid the value of its marginal product, the value of the marginal product of labor in the Northeast is, on the average, smaller than in other regions. Hence, as labor moves towards the North and Center-South, the economy experiences an increase in the productivity of labor (assuming the level of national employment constant).

The overall increase in the national supply is met by movements in the demand aggregates. Investment in each regional industry is assumed to deviate from the base case line together with deviations in the industry capital stock; as capital moves away from the regional industries in the Northeast, that region is harmed by lower levels of investment (-0.493%), while the recipient regions benefit from the positive flows of capital. An increase in capital stocks also provides a rise in the household income derived from capital earnings; in combination with the rising wage income from higher levels of employment and higher real wages, household disposable income in the North and Center-South goes up, inducing an increase in the real household consumption (led by an increase in the rate of household formation). In the Northeast, an opposite movement appears. This finding is also reflected in the declining share of the Northeast in national consumption, as shown by the estimates of the regional/national consumption rate. Again, the Center-South is the region that increases most its share in national consumption.

Regional government consumption of public goods is assumed to move with regional household consumption. In the case of federal public goods, a compensatory rule is imposed in which its consumption follows the national household consumption level. As the latter is a weighted average of the regional results, federal government redistributes resources towards the less favored regions.

B-MARIA projects balance of trade problems in the long-run generated by tariff reductions. Even though this result seems to go against the general goals of trade liberalization, it has to be kept in mind that the simulation only takes into account the reduction in tariff rates, which, as noted above, is only part of a successful trade liberalization program. Moreover, the results for the export volumes from the Center-South signal a potential increase in international competitiveness in the region.

In the long-run, producers are able to reevaluate their investment decisions, which was not possible in the short-run. The long-run movements in the rental values of capital and cost of capital define differential rates of returns in each sector, providing indicators of more profitable investment opportunities. As was mentioned before, current rates of return are defined by the ratio of the rental values of a unit of capital (that depend on the productivity of the current capital stock in each industry) and the cost of a unit of capital, based on its cost structure. B-MARIA assumes that if the percentage change in the rate of return in a regional industry grows faster than the national average rate of return, capital stocks in that industry will increase at a higher rate than the average national stock. For industries with lower-than-average increase in their rates of return to fixed capital, capital stocks increase at a lower-than-average rate, i.e., capital is attracted to higher return industries. Figures 9.4-6 depict the *short-run* movements of the industry-specific rates of return and the *long-run* investment in capital creation.

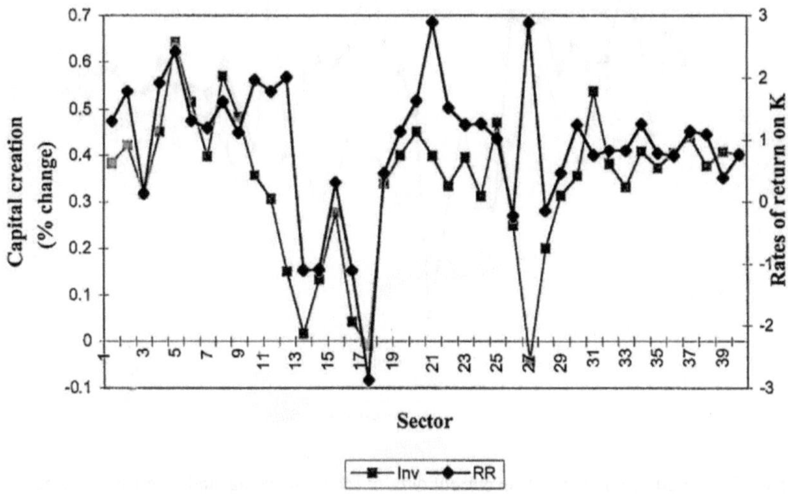

Figure 9.4 Short-run movements in current rates of return on capital and long-run movements in capital creation: North

Differential rates of return in the short-run play a major role in explaining investments in the model. The demand stimulus of lower prices for imported commodities may increase rates of return in regional industries, initially, drawing capital to those industries exhibiting above-average increases until the point that rates of return are driven back to their initial levels. Because the Northeast economy is relatively closed to international markets, the cost of capital in the region is relatively less affected by the tariff cut, influencing long-term investment decisions.[9]

Regarding the rental value of capital (and abstracting from the temporal dichotomy presented by the closures), changes might be thought to be related also to short-term changes in unemployment; as capital stocks are fixed, changes in the employment level determine changes in capital productivity by altering the industry capital-labor ratio. It was shown that, on the average, sectors in the Northeast had a smaller change in the unemployment rate (Table 9.2). The combined effects of higher costs to capital creation and lower productivity of capital, reinforced by internal multiplier effects, determine the generalized outflow of capital in the Northeast. This discussion is important to show how the short-run and long-

9 The average share of imports used in capital creation is only 0.22% in the Northeast, compared to 3.58% in the Center-South and 4.04% in the North.

run results are connected in the model. When capital is allowed to move, short-run rates of return represent a crude indicator of its future direction.

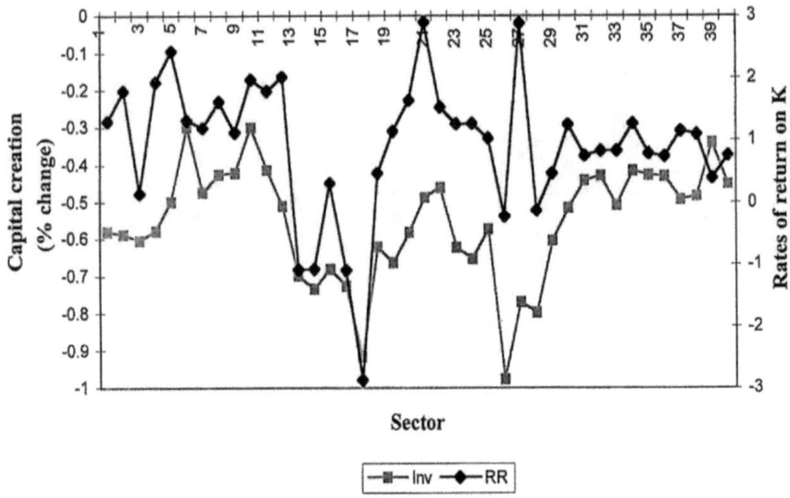

Figure 9.5 Short-run movements in current rates of return on capital and long-run movements in capital creation: Northeast

In the long-run, producers are not restricted anymore to respond to production decisions through labor changes, as was the case in the short-run. Employment changes are less strong, in the sense that the changes in employment are no longer necessarily greater than the changes in activity in absolute terms. The generalized flow of capital away from the Northeast towards the North and Center-South is induced by differential industry rates of return in the first round, and multiplier effects, derived from the increasing demand for investments in the recipient regions and decreasing demand for investment in the Northeast. The effects of the simulation on the industrial activity are depicted in Figures 9.7-9. As a general result, the industries that become exposed to higher levels of foreign competition, more precisely those of the chemical complex, are adversely affected in the three regions. The activity in these sectors is hampered by a switch of expenditure in favor of foreign produced goods (import volume changes for these commodities are among the highest).

Primary factors movements play a major role in the results achieved for the long-run. The Northeast is the most harmed region (all the sectors reveal output results below the base case level). As capital moves away from the region, multiplier effects operate, further deteriorating the regional economy. The increase in the demand for investment in the other regions operates in the opposite direction. The increasing demand for capital goods generates a round of expenditures in these regions (North and Center-South). The capital goods industries are characterized by using a greater amount of regional commodities (especially construction) and facing lower elasticity of substitution of the Armington type (both for substitution between foreign and the domestic composite, and between goods from different regions in the country). In combination, these effects contribute to the generation of higher internal multipliers in the regions. Labor migration to these regions is another source of stimuli to the regional economies. By increasing the demand for consumption goods, which are to a great extent provided by regional sources, the new migrants induce the expansion of those sectors, which sell predominantly to domestic markets and face little foreign import competition. Finally, the higher level of domestic absorption allows higher activity levels in the non-traded-commodity sectors, such as services.

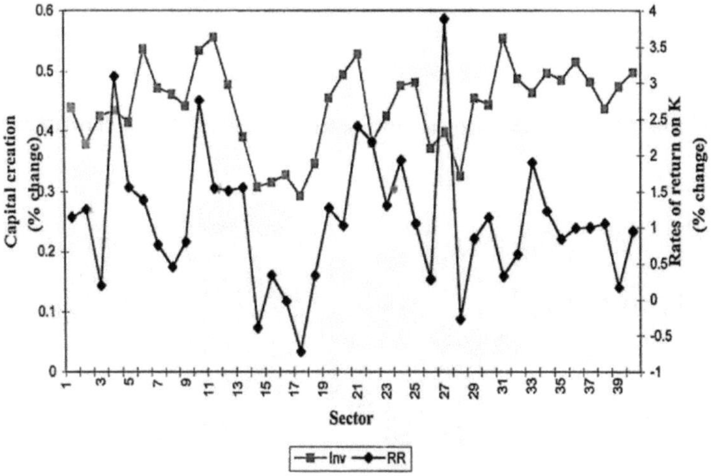

Figure 9.6 Short-run movements in current rates of return on capital and long-run movements in capital creation: Center-South

These facts are reflected in the sectoral activity results; the weak performance of industries in the Northeast is shown dramatically in Figure 9.8. In the North (Figure 9.7), capital-good industries lead growth in the region, with sectors producing consumer goods and services also achieving positive outcomes. The expansion of the mining sector is mainly induced by the increase in its exports (0.377%), the only regional sector with a favorable record in the international arena. Finally, the Center-South (Figure 9.9) presents the best performance overall. With the exception of the chemical complex industries and the other food products sector (which faces one of the highest tariff rates and includes typically luxury goods), all the sectors are positively affected. Because of the higher internal multipliers in the region, the relative performance of the non-capital-good-producing sectors is better than in the North. Moreover, the region also benefits from increased international competitiveness, as suggested by the rise in its exports: mining, 0.103%; coffee, 0.184%; sugar, 0.560%, and non-traditional exports, 0.625% on average.

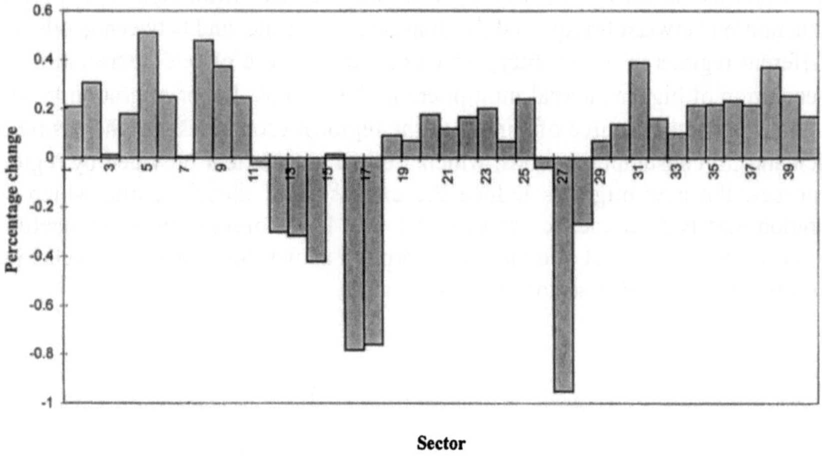

Figure 9.7 B-MARIA projected long-run activity effects of a uniform 25% tariff reduction: North

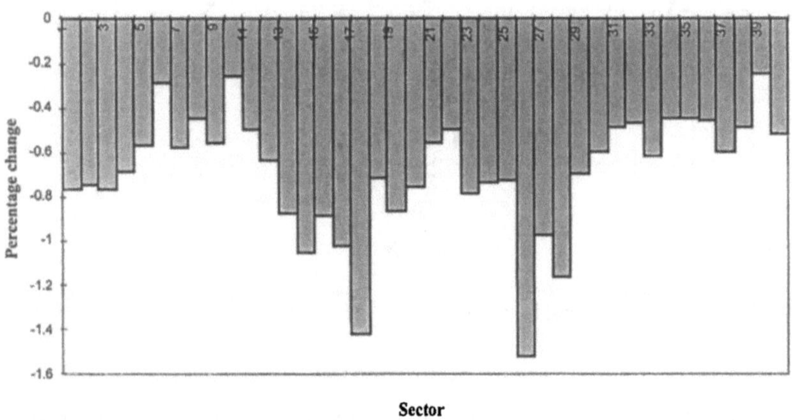

Figure 9.8. B-MARIA projected long-run activity effects of a uniform 25% tariff reduction: Northeast. Same sector numbering as in Figure 9.7.

The changes in the basic price of commodities can be explained in three different ways, for each region (Figure 9.10). These price deviations from the base case levels influence the results for the GRP deflator presented above, in which the values for each region differ both quantitatively and qualitatively (North, -0.069%; Northeast, 0.212%; Center-South, 0.005%). The high value for the Northeast is the result of strong supply shortages. With lower levels of primary factors, the regional industries, given Leontief technology, decrease their demand for intermediate inputs. However, the supply restrictions that arise from the outflow of capital and labor are stronger than the fall in demand (most of the inputs for current production are from regional sources). Therefore, the relative excess demand (or, more precisely, supply shortage) in most sectors induces prices to increase so that market-clearing conditions are reached. The initial cost reduction due to the lowering of tariffs on imported inputs is counterbalanced by the price-increasing mechanism, generating higher production costs in the region.

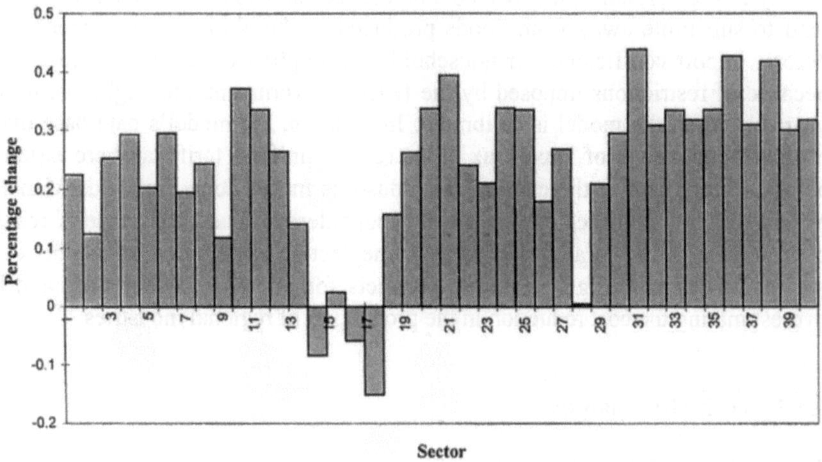

Figure 9.9 B-MARIA projected long-run activity effects of a uniform 25% tariff reduction: Center-South

In the case of the Center-South, prices of intermediate inputs are forced to rise because of the increased demand generated by capital and labor appropriation in the regional industries. However, in many industries, the rising supply is more than enough to offset the increased demand for intermediate inputs, and prices fall. This is especially the case for most of the traded-goods sectors that also benefit from the fall in the price of intermediate foreign inputs. In the non-traded-goods sectors (mainly services), a rise in prices is verified; since they do not face international competition, the tariff effect on intermediate inputs is not observed. Finally, the North, led by the performance of the electronic equipment sector, presents an overall fall in prices of basic commodities.

In summary, the results achieved by the model when there is a tariff reduction show that there is a potential concentration of economic activities in the more competitive sectors of the Center-South. In the short-run, an increase in economic activity is observed in all regions; however, the less developed region (Northeast) experiences lower levels of economic growth. In the long-run, regional imbalance is reinforced as a result of the Northeast's negative performance compared to the other regions.

The positive performance of the North should be taken with caution. The results for the region are heavily determined by the sectoral results of the electrical and electronic equipment industries of Manaus; these sectors are positively affected by their sales increases to the Center-South (the volume of the North's interregional exports to the Center-South increased by 0.340%). It is well known that the electric-electronic complex of the Zona Franca de Manaus benefited for a long time from non-tariff barriers as well (Diniz and Santos, 1996). Included in the trade liberalization policies is the removal of such barriers; hence, it is very likely that the region will be hampered by these policies. By facing cheaper, better-quality foreign products, not available before, consumers in the Center-South will tend to substitute away from goods produced in the North. As was mentioned before, import coefficients for household consumption of electronic equipment, because of restrictions imposed by the federal government, are negligible in the year for which the model is calibrated. In addition, the model's data base over-estimates the weight of import tax in the region; uniform tariff rates are assumed across sectors, and in the case of the industries in the Zona Franca de Manaus, exemptions on imported inputs are not considered. Thus, the model's results overestimate the sectoral performance by neglecting substitution effects between the regional and foreign electronic products for internal consumption, and by overestimating the cost reduction in the production of regional industries.

9.4.3 Sensitivity Analysis

The analysis of the effects of tariff reduction described above revealed that substitution effects play a major role in the final results. As tariff rates go down, imported commodities become relatively cheaper and domestic agents tend to substitute away from the domestic goods: for instance, producers tend to use more imported inputs and consumers tend to purchase the cheaper imported goods. The strength of these substitution effects is captured in the model by a set of parameters usually called "Armington elasticities". In B-MARIA, there are six groups of such parameters for each commodity; these are divided into two subgroups defining elasticities of substitution between commodities from different domestic sources (interregional Armington elasticities), and defining substitution between imported commodities and a domestic composite (international Armington elasticities). As there are no econometric estimates for these parameters for the Brazilian economy, the model is calibrated by assuming that the value of the parameter for domestic/foreign products is equal to 0.5 for capital goods (sectors 1, 6-12, 29, 31), 4.0 for consumer goods (18-28), and 2.0 for the remaining sectors (basically, services).[10] The values specified for the interregional elasticites are twice as large as those for the international elasticities (1.0 for capital goods, 8.0 for consumer goods, and 4.0 for the remaining sectors). The arbitrary choice of the values is based on the international literature that addresses the estimation of this type of elasticity but it also carries some personal judgment. It is implicitly assumed that consumer goods are more easily substitutable than capital goods, in the sense that, for instance, if the prices of imported shoes become cheaper it is more likely that consumers will buy more imported shoes. However, if the price of a machine declines, given technological restrictions, the strength of substitution is

[10] Sectors are defined in the Appendix.

smaller. Moreover, it assumed that substitution between goods from different regions inside the country is stronger than substitution between domestic and imported commodities.

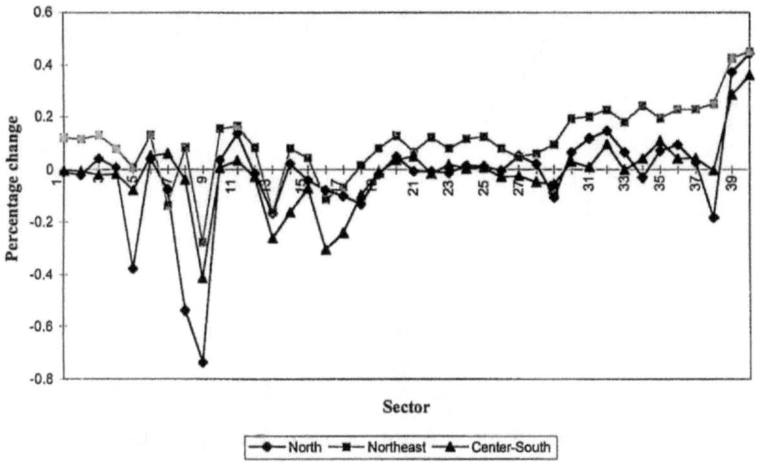

Figure 9.10 B-MARIA projected long-run effects of tariff cut on the basic prices of commodities

In order to evaluate how these parameters affect the results of the model, the values of the international Armington elasticities were set equal to 2.0 to all commodities, and the interregional Armington elasticities were set to be, again, twice as high. Figures 9.11-13 indicate the intensity of changes by comparing industrial activity levels in the three regions for the original and the new sets of Armington elasticities. The change in the parameters' values is felt more strongly by the industries in the North, which show a downward shift in their activity level. Capital goods facing relatively stronger substitution effects (the parameter affecting these commodities are four times higher) are deeply harmed, with the results reversing qualitatively in these industries. Since the increased demand for capital goods in the North had led the regional growth in the original simulation, the region now experiences a decrease in total GRP, as shown in Figure 9.14.[11] The results for the Northeast go in the opposite direction, led by better performance of the consumer goods industries that now face relatively weaker competition, as measured by the Armington elasticities. Finally, the Center-South also presents a qualitative reversal in the results for the capital goods industries. However, the consumer goods industries improve slightly, together with the traditional exports sectors. Overall, the results show stronger differential effects in the North (negative) and Northeast (positive), with the positive results for the Center-South counterbalancing the negative ones.

---

[11] It was previously shown that the electric-electronic complex in Manaus compensates the adverse effects of high international competition in the region by increasing its sales to capital creation in domestic markets.

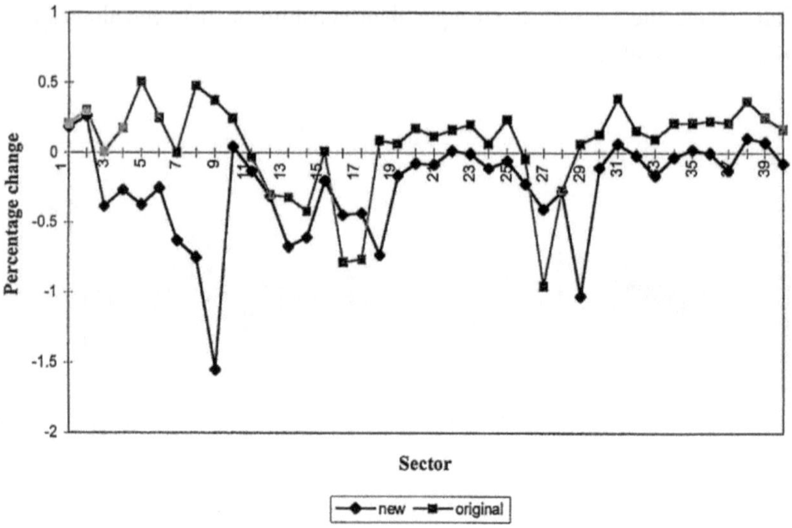

Figure 9.11 Long-run activity effects of the tariff simulation using different sets of Armington elasticities: North

9.5 Conclusion

The simulation exercises described in section 9.4 were intended to examine some of the aspects of the changing economic environment in Brazil in the 1990's. The quantitative results supported a thorough qualitative analysis of the evolution of Brazil's productive structure after the economic liberalization process of the 1990's. The goal was to use a formal analytical framework to capture the role of interindustrial and interregional relations in the economic development process through the evaluation of the differential regional impact of economic policies. Attention was directed to the short-term and long-term analysis of the effects of trade liberalization.

The choice of the CGE method to address the regional impact of trade policies is justified by its ability to address the issue of regional inequality and structural changes based on the theoretical framework of the dynamics of regional development. It borrows from the input-output framework the capability of handling details at the sectoral level, providing a detailed picture of the regional economies. Input-output linkages also incorporate the static dimension of regional development by revealing the strength of interregional spread effects through the specification of existing trade flows in the economy. However, the CGE approach goes one step further; regional competitiveness is captured by incorporating price effects in those models. Differential regional prices play a major role in the theories of regional development, in which regional competitive advantage is addressed through differential production costs and activities allocation (Myrdal, 1957; Hirschman, 1958; Williamson, 1965; Krugman, 1991).

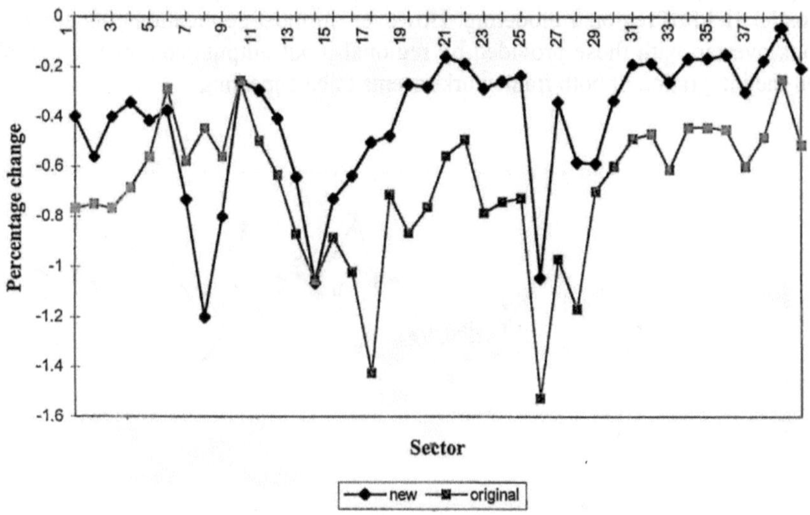

Figure 9.12 Long-run activity effects of the tariff simulation using different sets of Armington elasticities: Northeast

In summary, the discussion of regional and sectoral impacts of alternative strategies of regional development in the present macroeconomic context of the Brazilian economy has shown that the open policies of the 1990's and the national strategies for increasing international competitiveness are very likely to increase regional imbalances in the country. The results suggest that the interplay of market forces in the Brazilian economy favors the more developed region of the country. In other words, the trickling-down effects generated by market forces are still very unlikely to overtake the polarization effects emanating from the Center-South. If regional equity is part of the country's development agenda, an active regional policy by the central government is still needed, in order to reduce regional economic disparities, and specifically to address the problems of the North and Northeast, traditionally backward areas reliant on low technology activities. The improvement of the economic infrastructure in those regions, as well as the establishment of dynamic competitive advantages, through a consistent human capital policy, are necessary to attenuate the adverse regional effects of the development strategy pursued by the public authorities.

The use of the interregional CGE modeling approach has proved to be very relevant to the Brazilian case. Its ability to handle detail – both in terms of its disaggregation level and in terms of its theoretical specification – is useful for the analysis of policy issues and their effects on the Brazilian economy. However, the field of interregional CGE modeling is relatively recent and a line defining how these models' results differ from predictions obtained from alternative methods has yet to be drawn. The CGE approach provides improvement upon traditional fixed-price methods in the sense that the latter represent a limiting case within the CGE framework (Harrigan and McGregor, 1989; McGregor et al., 1997). By incorporating price substitution effects, it also provides unique contributions to the broader field of regional modeling. However, in most applications, its contributions overlap with those provided by regional input-output econometric models, and the integration of both frameworks seems to be appealing.

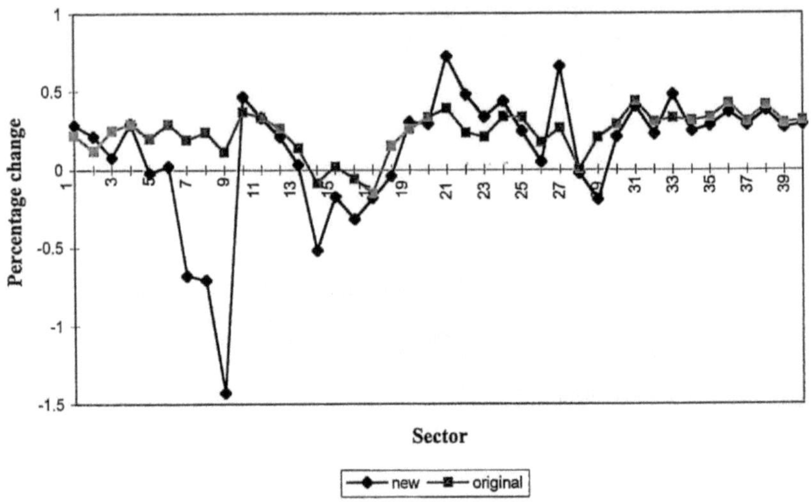

Figure 9.13 Long-run activity effects of the tariff simulation using different sets of Armington elasticities: Center-South

Data limitations, especially the lack of estimates for the specified elasticities, and some unresolved theoretical issues in region specification which are important in the regional development literature, such as imperfect competition in regional markets and increasing returns to scale, provide room for the future development of regional CGE models. To be sure, most of the relevant theoretical issues have already been tackled in national CGE models (see Ginsburgh and Keyzer, 1997) and regional scientists with a solid formation in economic theory will not face strong technical restrictions in the incorporation of the regional dimension to such models. However, because of data limitations, the implementation of interregional CGE models carries a heavy burden which cannot be overcome unless more information at the regional level is available. That is a common issue of the broader field of regional modeling and researchers will have to face this bottleneck until the institutions responsible for data collection improve their process of collection and presentation of regional statistics.

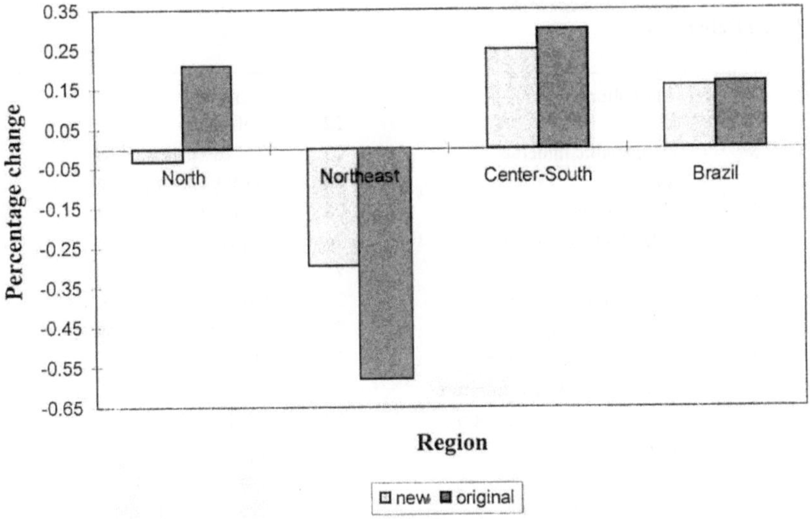

Figure 9.14 Long-run output effects of the tariff simulation using different sets of Armington elasticities: GRP/GDP

Acknowledgement

Haddad would like to recognize the important financial support received from CNPq – Brazil.

Appendix

Sector Definitions

1	Agriculture	21	Footwear
2	Mining	22	Coffee
3	Nonmetallic minerals	23	Processed vegetables
4	Steel	24	Meat packing plants
5	Nonferreous metals	25	Dairy products
6	Other metals products	26	Sugar
7	Machinery	27	Vegetable oil mills
8	Electrical equipment	28	Other food products
9	Electronic equipment	29	Other manufacturing
10	Transportation equipment	30	Utilities
11	Wood products and furniture	31	Construction
12	Paper products and printing	32	Trade
13	Rubber	33	Transportation
14	Chemicals	34	Communication
15	Petroleum refining	35	Financial institutions
16	Other chemicals	36	Personal services
17	Pharmaceuticals	37	Business services
18	Plastics	38	Real estate
19	Textiles	39	Public administration
20	Clothing	40	Community services

References

Agénor, P.R. and P.J. Montiel, 1996, *Development Macroeconomics*, Princeton University Press, Princeton.

Baer, W., E.A. Haddad and G.J.D. Hewings, 1998, "The Regional Impact of Neo-Liberal Policies in Brazil", *Revista de Economia Aplicada*, 27:233-256.

Baer, W. and W. Maloney, 1997, "Neoliberalism and Income Distribution in Latin America", *World Development*, 25:311-327.

BNB, 1992, *Matriz de Insumo-Produto do Nordeste 1980 e 1985: Metodologia e Resultados*, Banco do Nordeste do Brasil, S.A., Fortaleza.

Bruno, M., 1987, "Opening Up: Liberalization with Stabilization", in. R. Dornbusch and L. Helmer (eds.), *The Open Economy: Tools for Policymakers in Developing Countries*, EDI Series in Economic Development, Oxford University Press.

Diniz, C. C. and F.B.T. Santos, 1996, "Manaus: Uma "Satellite Platform" na Região Amazônica", ANPEC, *XXIV Encontro Nacional de Economia,* Dezembro.

Dinsmoor, J. and E.A. Haddad, 1996, *Brazil: The Use of State Fiscal Data Bases for Financial Projections, Region I Technical Note*, Inter-American Development Bank, August.

Dixon, P.B. and B R. Parmenter, 1994, "Computable General Equilibrium Modelling", Preliminary Working Paper no. IP-65, IMPACT Project, Monash University, Clayton, July.

Dixon, P.B., B.R. Parmenter, A.A. Powell and P.J. Wilcoxen, 1992. *Notes and Problems In Applied General Equilibrium Economics*, North-Holland, Amsterdam.

Dixon, P. B., B.R. Parmenter, J. Sutton and D.P Vincent, 1982, *ORANI: A Multisectoral Model of the Australian Economy,* North-Holland, Amsterdam.

FIBGE, 1995, *Matriz de Insumo-Produto Brasil – 1985*, Departamento de Contas Nacionais, Fundação Instituto Brasileiro de Geografia e Estatística, Rio de Janeiro, February.

Gazel, R.C., 1994, *Regional and Interregional Economic Effects of the Free Trade Agreement between the US and Canada*, Unpublished Ph.D. dissertation, University of Illinois at Urbana-Champaign.

Ginsburgh, V. and M. Keyzer, 1997, *The Structure of Applied General Equilibrium Models*, MIT Press, London.

Guilhoto, J.J.M., 1986, *A Model for Economic Planning and Analysis for the Brazilian Economy*, Unpublished Ph.D. dissertation, University of Illinois at Urbana-Champaign.

Guilhoto, J.J.M., 1995, *Um Modelo Computável de Equilíbrio Geral para Planejamento e Análise de Políticas Agrícolas (PAPA) na Economia Brasileira,* ESALQ, Piracicaba, Tese de Livre Docência, June.

Guilhoto J.J.M. and M.A.R. Fonseca, 1990, "As Principais Correntes de Modelagem Econômica e o Caso Brasileiro", *Anais do XII Encontro Brasileiro de Econometria, Brasília*, December 03-06.

Haddad, E.A., 1997, *Regional Inequality and Structural Changes in the Brazilian Economy*, Unpublished Ph.D. dissertation, University of Illinois at Urbana-Champaign.

Haddad, E.A and G.J.D. Hewings, 1997, "The Theoretical Specification of B-MARIA", Discussion Paper REAL 97-T-5, Regional Economics Applications Laboratory, University of Illinois at Urbana-Champaign.

Haddad, P.R., 1997, "A Questão Regional em Três Ciclos de Expansão do Brasil", *Idéias e Debates*, Instituto Teotônio Vilela, Brasília.

Harrigan, F. and P.G. McGregor, 1989, "Neoclassical and Keynesian Perspectives on the Regional Macro-Economy: A Computable General Equilibrium Approach", *Journal of Regional Science*, Vol. 29, 4:555-573.

Hewings, G.J.D, 1985, *Regional Input-Output Analysis*, Sage Publications, Beverly Hills.

Hirschman, A.O., 1958, *The Strategy of Economic Development*, Yale University Press, New Haven.

Horridge, J.M., B.R. Parmenter and K.R. Pearson, 1993, "ORANI-F: A General Equilibrium Model of the Australian Economy", *Economic and Financial Computing*, Vol. 3, No. 2.

Krugman, P., 1991, "Increasing Returns and Economic Geography", *Journal of Political Economy*, 99:483-499.

Liew, L.H., 1984, "'Tops-Down' versus "Bottoms-Up" Approaches to Regional Modeling", *Journal of Policy Modeling*, Vol. 6, 3:351-367.

McGregor, P.G., J.K Swales and Y.P. Yin, 1997, "Spillover and Feedback Effects in General Equilibrium Interregional Models of the National Economy: A Requiem for Interregional Input-Output?", *Strathclyde Papers in Economics*, 97/8, The University of Strathclyde in Glasgow.

Moreira, A.R.B. and A. Urani, 1994, "Um Modelo Multissetorial de Consistência para a Região Nordeste", *Texto para Discussão 352*, IPEA, Brasília, October.

Myrdal, G., 1957, *Rich Lands and Poor: The Road to World Prosperity*, Harper and Brothers, New York.

Naqvi, F. and M.W. Peter, 1995, "Notes on the Implementation of MONASH-MRF: A Multiregional Model of Australia", Working Paper no. OP-82, IMPACT Project, Monash University, Clayton, April.

Naqvi, F. and M.W. Peter, 1996, "A Multiregional, Multisectoral Model of the Australian Economy with an Illustrative Application", *Australian Economic Papers*, June.

Patridge, M.D. and D.S. Rickman, 1998, "Regional Computable Equilibrium Modeling: A Survey and Critical Appraisal", *International Regional Science Review*, 21:205-250.

Sachs, J.D. and A. Warner, 1995, "Economic Reform and the Process of Economic Integration", in W. C. Brainard and G L. Perry (eds.), *Brookings Papers on Economic Activity*, Brookings Institution, Washington D.C.

Sampaio, M.C.S., 1987, "Proteção, Crescimento e Distribuição de Renda no Brasil – Uma Abordagem de Equilíbrio Geral", *Revista Brasileira de Economia,* Vol. 41, 1:99-116.

Savedoff, W.D., 1990, "Os Diferenciais Regionais de Salários no Brasil: Segmentação Versus Dinamismo da Demanda", *Pesquisa e Planejamento Econômico*, 20.

Shoven, J.B. and J. Whalley, 1984, "Applied General-Equilibrium Models of Taxation and International Trade: An Introduction and Survey", *Journal of Economic Literature*, Vol. 22, 3:1007-1051.

SUDAM, 1994, *Matriz de Insumo-Produto do Norte 1980 e 1985: Metodologia e Resultados*, Superintendência de Desenvolvimento da Amazônia, Belém.

Whalley, J. and I. Trela, 1986, *Regional Aspects of Confederation,* Vol. 68 for the Royal Commission on the Economic Union and Development Prospects for Canada, University of Toronto Press, Toronto.

Williamson, J., 1965, "Regional Inequality and the Process of National Development: A Description of the Patterns", *Economic Development and Cultural Change*, 13:3-84.

10 The Economic System of Small-to-Medium Sized Regions in Japan

Se-il Mun and Komei Sasaki
Graduate School of Information Sciences, Tohoku University
Katahira 2, Aoba-ku, Sendai 980-8577, Japan

10.1 Introduction

In Japan, since the end of World War II, the three largest metropolitan areas (Tokyo, Kinki, and Chukyo) have constantly experienced population growth and, in particular, the Tokyo metropolitan area has been attracting positive net population in-migration. In brief, population and economic activities have continued to concentrate in a few of the larger areas. The central government has attempted to alter this tendency to concentrate so as to disperse population and economic activities from central metropolitan areas to peripheral, less-dense areas through transportation system improvements, industry-related infrastructure investment, lower taxes and subsidies. However, this effort has not been very successful because such policies have not been effective in modifying the results brought about by market forces. In other words, planners intending to change the spatial structure of the economy need to investigate carefully the market forces prevailing in the existing system of regions. The present research is motivated by this conclusion.

Urban economists have developed models to explain how the size of each city is determined in a system of cities with emphasis on the role of agglomeration economies (e.g., Abdel-Rahman, 1990; Henderson, 1987; Kanemoto, 1980). However, they have not considered spatial factors such as the existing location of cities and distances or transport costs between cities. Therefore, such models are not capable of explaining what type of city has developed at each location, how large it is, and how inter-city transport improvements affect its size and scope.

A new theory of spatial agglomeration has emerged and been flourishing since Krugman's, 1991, work on regional increasing returns, which incorporates transport cost in the model of interregional trade with scale economies. Various extensions of the model have been made recently, e.g., endogenous determinants of cities' locations (Fujita and Krugman, 1995), incorporation of multiple industrial sectors (Fujita, Krugman and Mori, 1995), intra-city land use (Tabuchi, 1996; Helpman, 1998), and inter-city transport networks (Mun, 1997). Most models, however, treat only two-city economies or cities locating in a one dimensional space. Furthermore, most of the existing studies on systems of cities have been confined to theoretical analysis or numerical simulation with hypothetical parameters. It has been recognized that the properties of spatial economies with increasing returns are indeed complex. For example, Tabuchi, 1996 showed that a reduction in transport cost may cause either a concentration or a dispersion of activities, depending on the initial conditions and parameter values. Thus effects of policy changes need to be evaluated by empirical analysis.

In this paper, a system of small and medium sized regions in Japan is analyzed empirically and numerically using theoretical methods. The "Tohoku" area (hereafter, the T-area) in the north-eastern part of Japan was divided into 37 regions for the purpose of this study. The largest region (whose central city is Sendai) in this area has a population of 1.29 million and the smallest one (central city is Nagai) has 72,500. The average regional population is 263,000 (see Table 10.1 and Figure 10.1).

Table 10.1 37 regions in the T-area

Region code	Representative city of region	Population	Class
1	Aomori	323604	II
2	Hachinohe	352240	II
3	Hirosaki	350603	II
4	Towada	286990	III
5	Goshogawara	169436	IV
6	Morioka	536579	I
7	Ofunato	82689	V
8	Hanamaki	198602	IV
9	Ichinoseki	154389	IV
10	Miyako	190400	IV
11	Kamaishi	106481	IV
12	Mizusawa	147788	IV
13	Sendai	1292282	I
14	Furukawa	223144	III
15	Ishinomaki	237353	III
16	Shiroishi	196143	IV
17	Kesen-numa	114468	IV
18	Tsukidate	185168	IV
19	Honjo	127327	IV
20	Akita	430784	I
21	Noshiro	109635	IV
22	Omagari	165848	IV
23	Kaduno	190946	IV
24	Yuzawa	202938	III
25	Yamagata	465910	I
26	Sakata	161458	IV
27	Yonezawa	180795	IV
28	Shinjo	102214	IV
29	Tsuruoka	166905	IV
30	Nagai	72567	V
31	Obanazawa	108541	IV
32	Fukushima	489514	I
33	Iwaki	433386	I
34	Koriyama	559599	I
35	Aidu-wakamatsu	336785	II
36	Haranomachi	133211	IV
37	Shirakawa	151563	IV

Note: Class divisions of regions based on population (N_i). I = $400000 < N_i$, II = $300000 < N_i \leq 400000$, III = $200000 < N_i \leq 300000$, IV = $100000 < N_i \leq 200000$, V= $N_i \leq 100000$

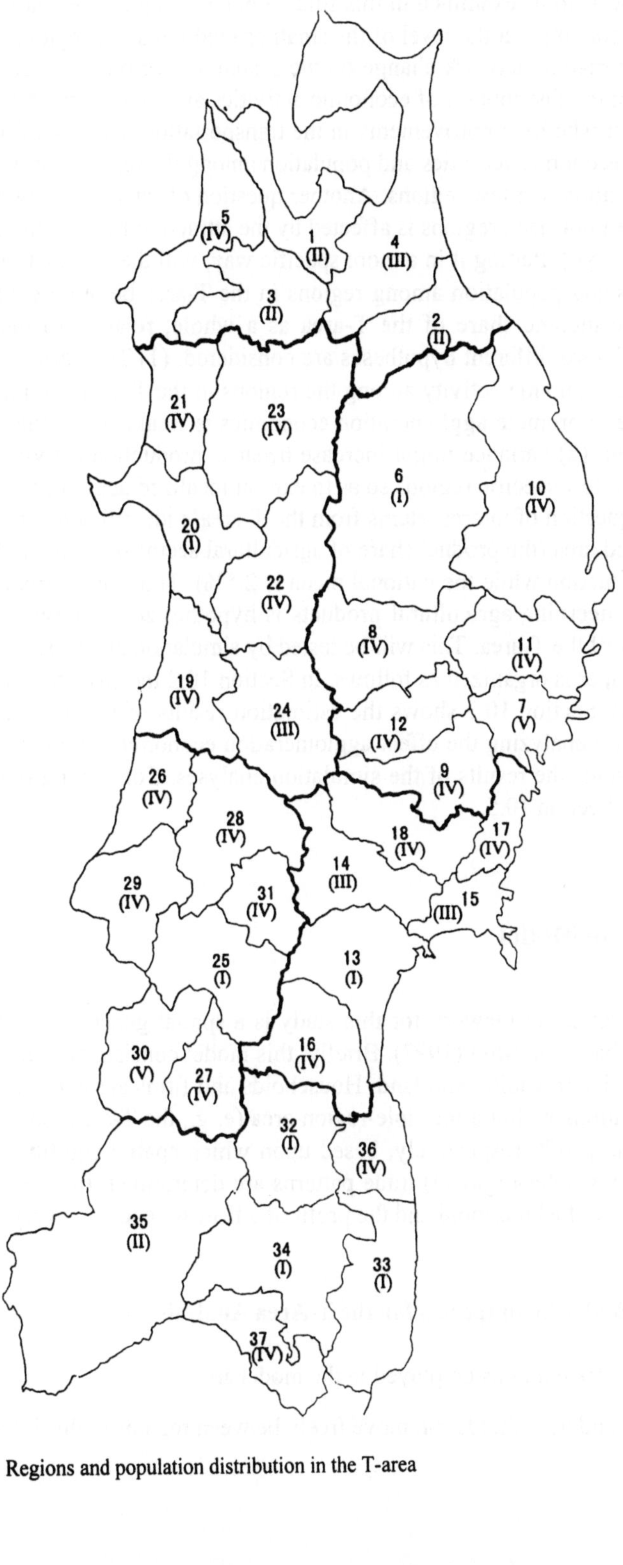

Figure 10.1 Regions and population distribution in the T-area

The subjects to be examined in this study are: the comparative analysis of agglomeration economies at the level of the small or medium sized region; and the effects of transportation network change on the system of regions. In particular, focus is placed on the distribution of economic activities among regions in the T-area. It is of interest whether improvements in the transportation network of the T-area will disperse economic activities and population among the regions, or will result in the agglomeration in a few regions. Another question of interest is how this system of small-medium sized regions is affected by the attraction force of the central area in Japan (Tokyo). Putting it in a more specific way, will the distribution of economic activities and population among regions in the T-area be more even or localized when the income share of the T-area as a whole, relative to the Tokyo area, declines? Two different hypotheses are considered: (1) The variance of the distribution of economic activity among the regions in the T-area declines because the incentive to promote agglomeration economies is weakened as the T-area's share is reduced; (2) variance might increase because production activities are concentrated in a few specific regions so as to exploit localized agglomeration economies. A final question of interest stems from the T-area's identification as an agriculture dominated area (the product share of agricultural sector is relatively high: 5.5% of total production while the national mean is 2.5%). Thus, the current liberalization policy concerning agricultural products is hypothesized to largely influence the economy of the T-area. This will be tested by simulation analysis.

The paper is organized as follows. In Section 10.2 the basic theoretical model is presented. Section 10.3 shows the estimation results of the model, in particular, empirically analyzing the effect agglomeration economies on production. Section 10.4 presents the results of the simulation analyses. Concluding remarks are presented in Section 10.5.

10.2 The Model

The theoretical framework for this study is a spatial general equilibrium model, which is based on Mun (1997). Briefly, this model consists of four markets: commodities, labor, capital and land. Households and firms are assumed to search for their locations within a multiple-region area (e. g., the T-area) so as to maximize utility and profit, respectively, based upon which spatial equilibrium prices and equilibrium (inter-regional) trade patterns are determined. In an equilibrium, the utility level of a household and the profit of a firm are equal in every region.

10.2.1 Model Assumptions for the T-Area Analysis

The basic assumptions employed in the model are:

1. Firms and households can move freely between regions in the T-area.

2. There is no migration between the T-area and rest of the world. The population in the T-area is fixed (i.e., a large area is closed).
3. Land in each region is used only for households' residences. Residents in a region have equal ownership over the land in that region, so land rental revenue in a region is equally distributed among the residents in that region.
4. M different commodities and services are produced in the T-area by employing labor and capital inputs (intermediate inputs are neglected).
5. In transporting a particular commodity between regions, a certain amount of that commodity is consumed as transport cost (i.e., ice-berg type transport cost is assumed).
6. Capital is mobile between the T-area and rest of the world. The current balance between the T-area and the rest of the world is in equilibrium, so that the difference between the T-area's imports and exports is covered by capital net inflow (or outflows) from (or to) the rest of the world.
7. Every resident in the T-area has equal ownership over a given amount of capital, $\bar{K}$, and the capital rent revenue is equally distributed among residents in the T-area.

10.2.2 Behavior of a Firm

A Cobb-Douglas type production function is assumed.

$$y_i^m = \delta^m G^m(N_i)(L_i^m)^{a^m}(K_i^m)^{1-a^m} \tag{10.1}$$

in which the superscript m $(m = 1,\cdots M)$ denotes a type of commodity (industrial sector) and the subscript i $(i = 1,\cdots I)$ a particular region. In (10.1), L and K are, respectively, labor and capital inputs, and $G^m(N_i)$expresses what are called urbanization economies and agglomeration economies are measured as a function of regional population, N_i. Under the specification in (10.1), input demand functions are derived as

$$L_i^m = \frac{a^m}{w_i} q_i^m y_i^m \tag{10.2}$$

$$K_i^m = \frac{(1-a^m)}{r} q_i^m y_i^m$$

$$i = 1,\cdots I, \quad m = 1,\cdots M$$

where: q_i^m is the f.o.b. price of commodity $m$produced in region i; w_i and r are, respectively, wage rate in region i and capital cost common in a nation. The maximized profit is zero in a competitive market, and if firms produce positive amounts, then the average cost is equal to the commodity supply price. If the f.o.b. price is lower than the average cost, then output of that commodity is zero in the region. Namely it holds that:

$$y_i^m \geq 0 \text{ when } q_i^m = C^m(N_i, w_i, r) \tag{10.3}$$

$$y_i^m = 0 \text{ when } q_i^m < C^m(N_i, w_i, r)$$

where C^m is the average cost.

10.2.3 Behavior of a Household

The utility function of a household is specified in the following form.

$$U = \alpha \ln h_i + \sum_{m=1}^{M} \beta^m \ln x_i^m \tag{10.4}$$

where h is residential lot size, and x^m is consumption of commodity m. In (10.4), it is assumed that $\alpha + \sum \beta^m = 1$. Income constraint of a household is

$$w_i + \frac{p_i^h H_i}{N_i} + \frac{r\overline{K}}{N} = \sum_{m=1}^{M} p_i^m x_i^m + p_i^h h_i \tag{10.5}$$

in which p_i^m is the c.i.f. price of commodity m in region i (which is different from q_i^m), and p_i^h is the land rent in region i. The second and third terms on the LHS in (10.5) are, respectively, the distributions from land and capital rental revenues. The variables H_i and N are, respectively, total land size in region i and total population in the T-area.

A household in region i plans consumption bundles so as to maximize (10.4) subject to (10.5). Noting that $h_i = \frac{H_i}{N_i}$, the following are optimum conditions.

$$x_i^m = \frac{\beta^m}{1-\alpha} \frac{1}{p_i^m} (w_i + \frac{r\overline{K}}{N}) \tag{10.6}$$

$$h_i = \frac{\alpha}{1-\alpha} \frac{1}{p_i^h} (w_i + \frac{r\overline{K}}{N}).$$

10.2.4 Interregional Trade

Suppose a consumer has a demand for commodity m. He (or she) does not care where that commodity is produced as long as the quality is the same. However, if the supply price of that commodity differs (depending on where it is produced), then he (or she) rationally chooses commodities with the lowest c.i.f. price. In the circumstance of perfect competition, c.i.f. price is the sum of f.o.b. price and transport cost. Thus, in the strict sense, it holds that

$$p_i^m = \min_j q_j^m(1+t^m d_{ij}) \tag{10.7}$$

where t^m is the transport cost of commodity m per unit per distance, and d_{ij} is the physical distance between regions i and j. However, in reality, commodities consumed in a particular region are shipped from regions even though the theoretical c.i.f. price from those regions is not the lowest. Furthermore, there are many actual patterns of "cross-hauling" between regions although, theoretically, they cannot take place. Such counter-theoretical phenomena are observed mainly because the classification of industrial sectors is not fine enough to ensure the homogeneity of product in a particular sector. Thus, from the standpoint of better explaining as modeling reality, the following probabilistic approach is employed[1].

It is assumed that, in consuming a commodity, each consumer chooses a firm which produces that commodity on the basis of her own preference towards a particular firm. The choice depends not only on the "theoretical" c.i.f. price level but on each consumer's preference.[2] The cost incurred by a consumer in region i for purchasing a commodity m from a firm f in region j is expressed as

$$q_{jf}^m(1+t^m d_{ij}) + \varepsilon_{jf}^m$$

In this expression ε_{jf}^m is a consumer specific preference term and varies among consumers in region i. Thus ε_{jf}^m is regarded as being distributed among consumers according to a specific density function. In this situation, the probability that a randomly drawn consumer in region i chooses firm f in region j is represented as

$$P_r\left\{q_{jf}^m(1+t^m d_{ji}) + \varepsilon_{jf}^m \le q_{kf'}^m(1+t^m d_{ki}) + \varepsilon_{kf'}^m\right\} \text{ for all } k \text{ and } f' \tag{10.8}$$

if ε_{jf}^m obeys the Weibull function with parameters $(0, \lambda^m)$, then the probability in (10.8) is obtained as

$$S_{jfi}^m = \frac{\exp[-\lambda^m q_j^m(1+t^m d_{ji})]}{\sum_k n_k^m \exp[-\lambda^m q_k^m(1+t^m d_{ki})]} \tag{10.9}$$

where n_k^m is the number of firms in sector m in region k. Thus, the probability that some firm in region j is chosen is

---

1 This is the only difference from the model in Mun (1997).

2 The formulation of "Random cost" in this paper is based on Sasaki (1982).

$$S_{ji}^m = \frac{n_j^m \exp[-\lambda^m q_j^m (1+t^m d_{ji})]}{\sum_k n_k^m \exp[-\lambda^m q_k^m (1+t^m d_{ki})]}. \tag{10.10}$$

The data on number of firms is not available, so the total regional output y_i^m is used as a proxy on the presumption that the average size of a firm is the same among regions: i.e.,

$$S_{ji}^m = \frac{y_j^m \exp[-\lambda^m q_j^m (1+t^m d_{ji})]}{\sum_k y_k^m \exp[-\lambda^m q_k^m (1+t^m d_{ki})]}. \tag{10.11}$$

Using the estimate of S_{ji}^m, the shipped quantity of commodity m from region j to region i is estimated as

$$Z_{ji}^m = \left\{N_i x_i^m (1-\mu^m) + E_i^m\right\} S_{ji}^m \tag{10.12}$$

where μ^m is the import coefficient (common to all the regions) and E denotes the quantity exported outside the T-area from region i.

Finally, the supply purchase price in region i is defined as the average of c.i.f. price: i.e.,

$$p_i^m = \sum_j S_{ji}^m q_j^m (1+t^m d_{ji}). \tag{10.13}$$

10.2.5 Market Equilibrium

Equilibrium conditions in this system are represented as follows.

labor market:

$$\sum_{m=1}^{M} L_i^m = N_i \quad i = 1, \cdots I \tag{10.14}$$

$$\sum_{i=1}^{I} N_i = N$$

commodity market: demand:

$$N_i x_i^m (1-\mu^m) + E_i^m = \sum_{j=1}^{I} Z_{ji}^m \tag{10.15}$$

$$i = 1, \cdots I$$

$$m = 1, \cdots M$$

commodity market: supply:

$$y_i^m = \sum_{j=1}^{I} Z_{ij}^m (1 + t^m d_{ij}) \tag{10.16}$$

$i = 1, \cdots I$

$m = 1, \cdots M$

current balance with the rest of the world: assumption 6

$$r\left(\sum_{i=1}^{I}\sum_{m=1}^{M} K_i^m - \overline{K}\right) = \sum_{i=1}^{I}\sum_{m=1}^{M} q_i^m E_i^m - \sum_{i=1}^{I}\sum_{m=1}^{M} \mu^m p_i^m N_i x_i^m \tag{10.17}$$

households location:

$$U(h_i, x_i^1, x_i^2, \cdots x_i^m) = U \text{ for } i = 1, \cdots I \tag{10.18}$$

In this system, the numeraire is the price of capital service, r which is set to unity. Endogenous variables of the system are: $x_i^m, h_i, y_i^m, L_i^m, Z_{ij}^m, S_{ij}^m, w_i, q_i^m, p_i^m$, $p_i^h, N_i$and U. In the subsequent sections, the models in (10.1) through (10.13) are estimated using the cross-section data at 1990; and, on the basis of the estimated structures, the "theoretical value" of each endogenous variable is calculated.

10.3 Estimation of Production Function

The data required for the subsequent empirical analysis include population, employment, output of each industry, wage, land area, distances between each pair of regions, and interregional trade. Data on population and employment of each region are obtained from the Population Census. Data on industrial output and wage are obtained from Annual Report on Prefectural Accounts published by each prefecture. Data on land area are obtained from property tax register reports. The distance between each pair of regions was measured as the shortest time path between the central cities of regions along the road network shown in Figure 10.2.

A statistical model for estimating the production function in (10.1) is specified in the following way.

$$\frac{y_i^m}{L_i^m} = \delta^m G^m(N_i)\left(\frac{K_i^m}{L_i^m}\right)^{1-a^m} + e_i^m. \tag{10.19}$$

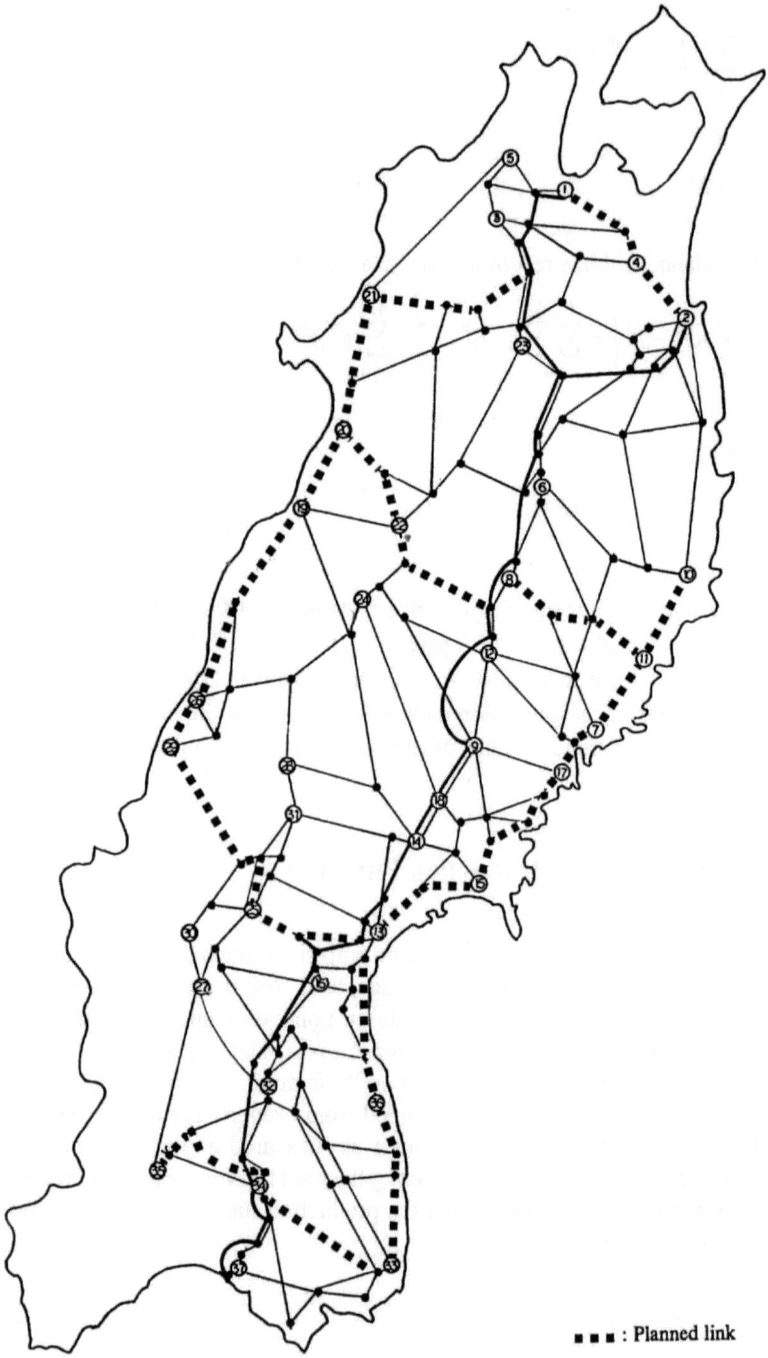

Figure 10.2 (Road) transportation network of the T-area

G^m represents the agglomeration economies. Agglomeration of population and economic activities in a specific location causes positive externalities for several reasons: cost of materials is lower because of collective purchasing; firms have better access to market trend information relating to the industry; scale economies in related sectors (such as transportation and repair services) work so as to lower the cost incurred by firms; and the search cost of labor is cheaper and the quality of labor higher relative to other regions.

There is extensive literature on empirical analysis of agglomeration economies in production (e.g., Sveikauskas, 1975; Sasaki, 1985; Nakamura, 1985; Henderson, 1986).

Our hypothesis, distinct from the analysis so far, is that there is some threshold level of regional population above which agglomeration economies start working. That is,

$$G^m(N_i) = \begin{cases} N_i^{\sigma_m} & \text{when } N_i \geq \overline{N}^m \\ \overline{N}^{m\sigma_m} & \text{when } N_i < \overline{N}^m \end{cases}$$

in which $\overline{N}^m$ is the threshold level of population for industry m. In the conventional specification of the agglomeration economies effect, it is implicitly assumed that externality effects work even for small population. However, such benefits as are cited above (e.g., lower purchase cost of materials, lower labor search cost, and better access to information) cannot be realized in a small population, i. e., in a region that is below the threshold.

Data on capital stock were not available. Instead, the value of K was estimated in the following way. Under a linear homogeneous production function and a competitive market, the total value added is fully distributed between capital and labor inputs. Then, defining the price of capital service to unity, the amount of capital stock was calculated as $K = qy - wL$.

In our model, the threshold level $\overline{N}^m$ itself is estimated. Estimation was repeatedly performed by changing the value of $\overline{N}^m$ by 100,000, and the estimated structure with the highest R^2 was selected. The value of $\overline{N}^m$ associated with the selected structure is regarded as the estimate of threshold population level. Such an estimation method is similar to the likelihood maximization approach.

Industries were classified into 20 sectors as shown in Table 10.2.

In fifteen of the twenty sectors, the value of R^2 reacted to changing $\overline{N}^m$ in an inverted-U shape, so that a maximum of R^2 can be obtained (Figure 10.3 shows the case of industry 7). The estimation results are shown in Table 10.3. Surprisingly and interestingly, the threshold level of population was 200,000 in every sector except one sector (the tenth industry). This implies that a population concentration of at least 200,000 is a prerequisite for agglomeration economies to work. Therefore it follows that, according to the population size in 1990, agglomeration economies do not occur in twenty-one of the thirty-seven regions in the T-area.

Table 10.2 Industry classification

Industry code	Industry
1	Lumber & wooden products
2	Textile mill products, Furniture & fixtures, Lether & lether products
3	Chemical, Petroleum refining & related products, Transport equipment
4	Plastic products, Electrical machinery
5	Food products
6	Precision machinery, Other manufacturing
7	Stone, clay, and glass products
8	Beverage, food & tobacco, Iron & steel, Metal products
9	Printing & publishing
10	Apparel & related products, Pulp, paper & allied products, Rubber products, Nonelectrical machinery
11	Nonferrous metal industry
12	Agriculture, forest & fishery
13	Mining
14	Construction
15	Wholesale & retail
16	Finance, insurance & real estate
17	Transport, communication
18	Electricity, gas supply
19	Service
20	Government

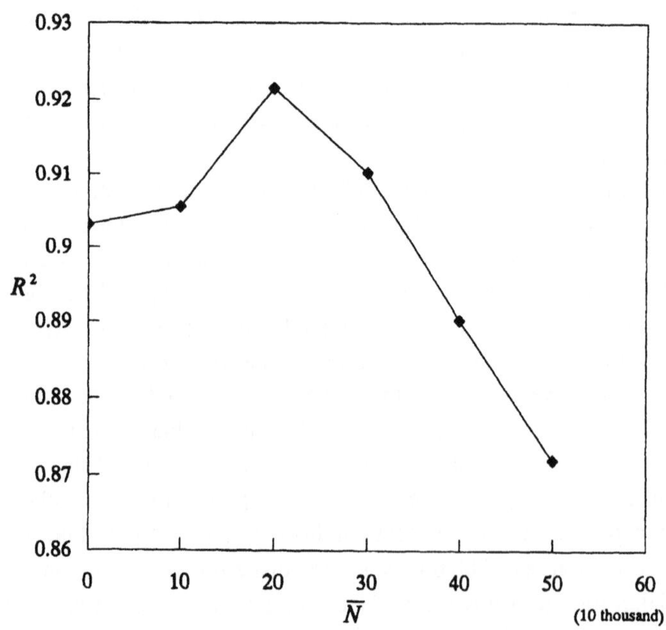

Figure 10.3 Estimation of population threshold level (for Industry 7)

Table 10.3 Estimation results of sectoral production functions

Industry code	δ^m		σ^m		$\overline{N}$	a^m		R^2
1	0.3827	(2.635)	0.1738	(6.033)	20	0.4864	(17.110)	0.9001
2	0.1581	(3.171)	0.2561	(5.498)	20	0.8069	(7.363)	0.7898
3	1.0706	(0.148)				0.2855	(48.128)	0.9871
4	0.6152	(0.947)	0.1451	(3.522)	20	0.5796	(18.233)	0.9217
5	0.2926	(0.840)				0.7489	(10.082)	0.7576
6	0.4932	(1.266)	0.1676	(3.743)	20	0.7087	(13.153)	0.8586
7	0.7755	(1.025)	0.1096	(5.675)	20	0.4118	(17.924)	0.9215
8	0.7032	(1.071)	0.1115	(4.034)	20	0.3589	(27.828)	0.9716
9	0.6071	(0.936)	0.1472	(3.410)	20	0.5914	(12.269)	0.9193
10	0.6242	(1.127)	0.1475	(4.480)	0	0.6029	(15.984)	0.9609
11	1.3145	(0.323)				0.5690	(10.977)	0.8959
12	3.9354					0.8180		
13	2.1510	(22.814)				0.2090	(61.345)	0.9823
14	0.6016	(1.512)	0.1389	(5.075)	20	0.5073	(18.876)	0.9309
15	0.1842	(3.561)	0.2429	(6.385)	20	0.8586	(5.780)	0.6869
16	1.0710	(0.695)	0.0339	(5.605)	20	0.1193	(78.872)	0.9957
17	0.5069	(1.836)	0.1573	(5.162)	20	0.5554	(15.139)	0.9118
18	0.9492	(0.782)	0.0333	(6.383)	20	0.0786	(184.520)	0.9990
19	0.4999	(2.130)	0.1646	(6.372)	20	0.6310	(15.747)	0.8885
20	0.5470	(2.261)	0.1226	(5.638)	20	0.2844	(46.106)	0.9860

A general tendency from an industry perspective is that the agglomeration economies do not work in the relative capital-intensive industries such as chemical and nonferrous sectors, and in primary sectors such as mining and agriculture.

As far as the agglomeration economy elasticity, σ_m, is concerned, it is relatively higher in the textile and wholesale & retail trade sector: that is, the production efficiency in these sectors rises with regional population growth. It might be counterintuitive that the elasticity is relatively lower in the sectors of finance, insurance and real estate (0.0339), and electric and gas utilities (0.0333), since these are typical industries with scale economies in production with average cost declining over a wide range of output. An interpretation of the observed small elasticities in these sectors is that the supply area of those industries is wider than the standard region in this study. So, "true" agglomeration economies are measured by population size in a large area including the contiguous regions of the region considered. That is, as the specification of G^m,

$$\left(\sum_{j \in R_i} N_j\right)^{\sigma_m}$$

is used, in which R_i is a set of regions in the supply area of firms in region i.

We explain briefly how other parameters were estimated. The parameters in a log-linear utility function (10.4) were estimated by the T-area average share of each commodity (or service) in total expenditure. The two parameters in the trade coefficient model, λ^m and t^m, were estimated by the non-linear least squared method using the data on inter-prefecture trade in the T-area.

Goodness-of-fit of the model in (10.19) is generally high: in thirteen sectors R^2 is above 0.9.

A final test was performed to evaluate how well the estimated structures fit the reality. Table 10.4 shows Mean Absolute Percentage Error (MAPE) of the main endogenous variables.

Table 10.4 Final test result

Variable	N_i	w_i	$\sum q_i^m y_i^m$ Manufacturing	$\sum q_i^m y_i^m$ Service industry
MAPE(%)	20.69	14.62	45.14	46.79
CORR	0.972	0.797	0.841	0.944

The result indicates that the estimated structures do not reproduce reality very accurately. A main reason is that the structures were estimated using cross-section data for only one year. In the last row, the simple correlation coefficient between the predicted and actual values is shown for each variable. This correlation coefficient is rather high, which implies that the pattern of actual variance of each variable among regions is better reproduced by the estimated structures, although the absolute discrepancy between the actual and predicted values is not small.

10.4 Simulation Analysis

10.4.1 Transportation System Change

Changes in the T-area transportation network will affect the system of regions within the T-area. A concrete question is whether population and production activities will be more concentrated in some regions or more dispersed across regions. In determining a trend toward concentration or dispersion, two opposing forces operate: agglomeration economies promote more concentration, while relatively lower factor prices such as wage rate and land rent in peripheral regions work to disperse activities.

A simulation analysis is performed making hypothetical changes in the transportation network. In the first simulation, the effect of uniform improvement in the transportation network is investigated. That is, expressing the new transport cost as $t^m = \theta t_0^m$, where t_0^m is the initial transport cost, and θ is gradually lowered from 1 to 0. As a measure of interregional disparity, the coefficient of variation of the regional population in the T-area, v_p, was calculated, and is illustrated in Figure 10.4. The value of v_p in Figure 10.4 decreases monotonically as transport cost is lowered (the value of the coefficient of variation in the "actual" distribution where $\theta = 1$ is 1.1252). That is, the dispersion-force rather than the concentration-force prevails in reaction to lowered transportation cost. This implies that the advantages of peripheral or small regions under conditions of lower wage rate and land rent are exploited more than the agglomeration economies of large regions.

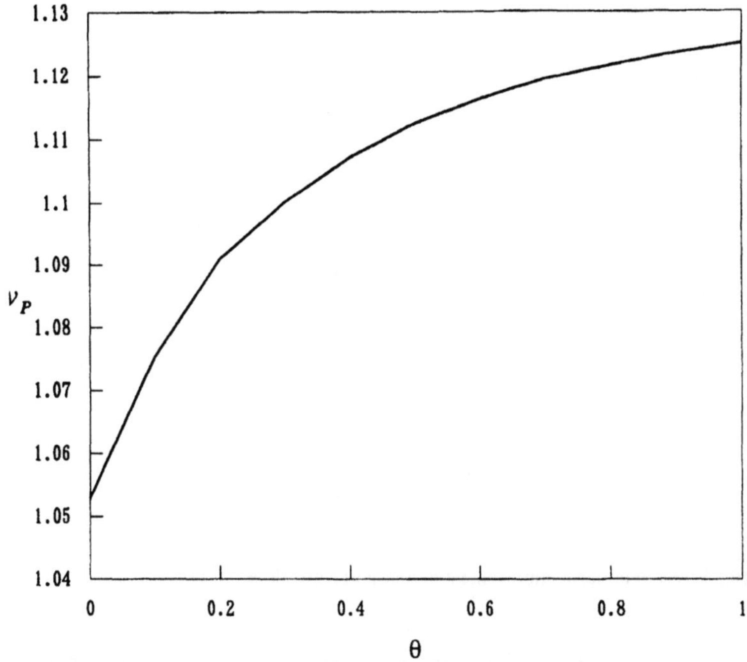

Figure 10.4 Effect of uniform improvement of transportation network

When transport cost is lowered by half (i.e., θ=0.5), the biggest decrease in population takes place in the Sendai region, i. e., the central region in the T-area. Among the ten largest regions accounting for 59% of the total population of the T-area, only three regions experience population increase due to homogeneous transportation system change. Wage rates are lowered in most larger regions, reflecting a decrease in labor demand. However, the rate of decrease tends to be larger in the smaller of the large regions (see Table 10.5 and Figure 10.5).

Table 10.5 Effect of uniform improvement of transport network (θ = 0.5). Coeff. of variation 1.1123

Class of region	Population bef.change	Population after change	After – before	After/ before	Number of regions	Increase	Decrease
I	4403491	4376268	-27223	0.9938	6	3	3
II	1327650	1331318	3668	1.0028	4	2	2
III	1493272	1493725	453	1.0003	6	2	4
IV	1916326	1929687	13361	1.0070	13	9	4
V	597550	607287	9737	1.0163	8	7	1

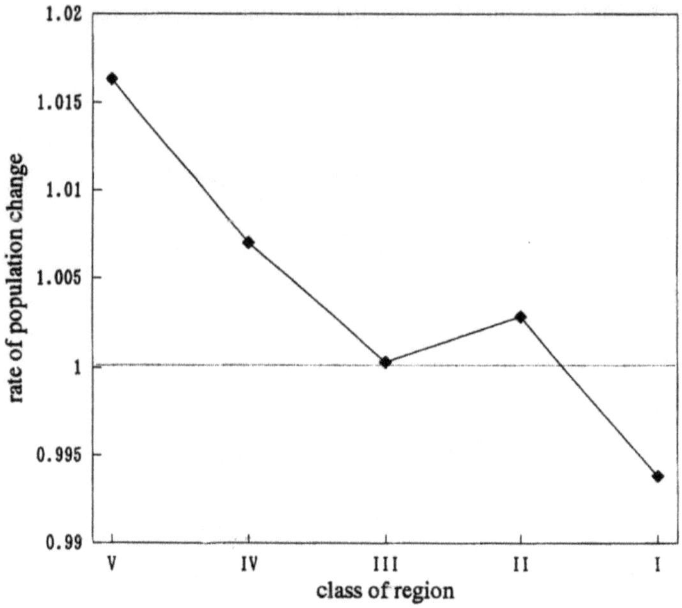

Figure 10.5 Effect of uniform improvement of transport network (θ = 0.5)

In the simulation analysis so far, the transportation network in the T-area was hypothesized to be uniformly improved. However, in reality, some links of the network are improved more than others. This part of the analysis examines the effects of such non-uniform transportation system change. It is assumed that transport cost is decreased by half along the seven specific links in the network (drawn with a thick dotted line in Figure 10.2), while no change occurs along the other links. This presumption is parallel to the existing plan of network improvement (in fact, most of the seven link improvements have been completed or are well on the way to completion). How is this system of regions affected by non-uniform improvement? Simulation results indicate that, in the largest three regions, population decreases, though not by a large amount: the rates of decrease are 0.34%, 1.34% and 1.80%. On the other hand, in some peripheral regions located on the improved links, population increases at a relatively high rate: higher than 10% in some regions. It is generally observed that population increases in the regions located at the nodes of improved links, and that population decreases in the regions on the non-improved links parallel to the improved links as a consequence of competition between links. Also, there is a tendency for the regions between two nodes of an improved link to lose population. Probably this is because the advantage of the location at nodes increases relatively. The overall result of the non-uniform network improvement is that the variance of population distribution is smaller, implying that the assumed transportation system change works to disperse activities within the T-area (see Table 10. 6 and Figure 10.6).

Table 10.6 Effect of improvement of specific links in transport network. Coefficient of variation 1.1190

Class of region	Population bef.change	Population after change	After - before	After/ before	Number of regions	Increase	Decrease
I	4403491	4386938	-16553	0.9962	6	2	4
II	1327650	1334377	6727	1.0051	4	1	3
III	1493272	1483800	-9472	0.9937	6	1	5
IV	1916326	1919704	3378	1.0018	13	5	8
V	597550	613464	15914	1.0266	8	5	3

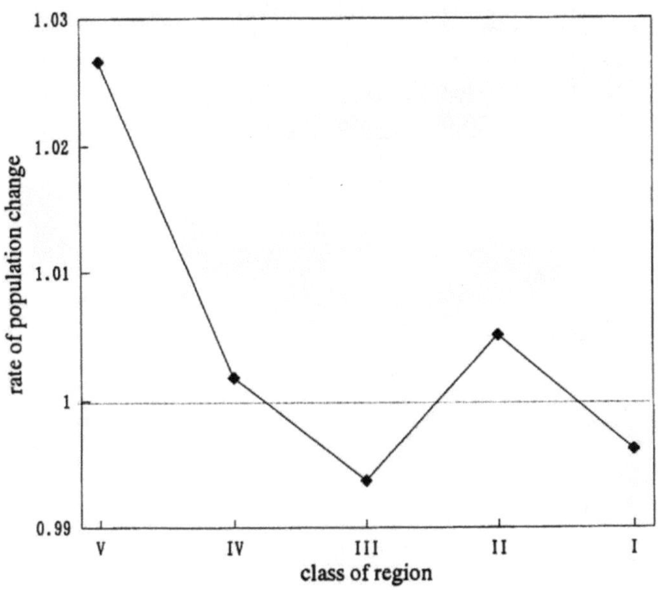

Figure 10.6 Effect of improvement of specific links in transport network

10.4.2 Relative Position of a Large Area

The third issue described in the Introduction concerns how the system of regions is affected by the relative position of the T-area in the nation. In order to examine this, the total population of the T-area was hypothetically changed in simulations on the assumption that the national population is fixed. Simulation was performed for the two cases where N is increased and decreased, respectively, by 50%. When the relative proportion of population in the T-area is increased (N is increased by 50%), every region in the T-area, naturally, observes population growth. However, in some larger regions (including the first, second and fourth largest ones), population increases by less than 50%, whereby the variance of the population distribution in the T-area is lowered. Furthermore, in some small or medium sized regions also, in the vicinity of those larger regions, the population growth rate is lower than 50% (see Table 10.7 and Figure 10.7[3]). This may be interpreted as follows: the population in some small or medium sized regions in the T area possibly exceeds the threshold level for agglomeration economies, whereby production efficiency is increased, and thus more economic activities are attracted to those regions.

Table 10.7 Effect of change in relative position of T-area (when N is increased by 50%)
Coefficient of variation 1.1203

Class of region	Population bef.change	Population after change	After - before	After/ before	Number of regions	Increase	Decrease
I	4403491	4394968	-8523	0.9981	6	3	3
II	1327650	1326871	-779	0.9994	4	2	2
III	1493272	1491135	-2137	0.9986	6	2	4
IV	1916326	1923304	6978	1.0036	13	9	4
V	597550	602000	4450	1.0074	8	7	1

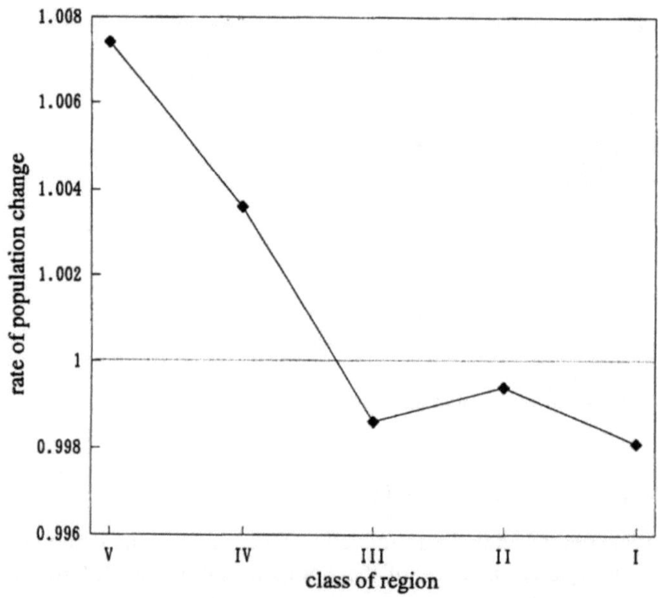

Figure 10.7 Effect of a change in relative position of T-area (when N is increased by 50%)

In contrast, when the T-area population is relatively small (i.e., N is decreased by half), the larger regions' share in the T-area is increased, and consequently the

3 In Table 18.7 and Figure 18.7, the simulated regional population is divided by 1.5 so as to make the result comparable with the other simulated values. Likewise in Table 18.8 and Figure 18.8, the simulated value is divided by 0.5.

variance of population distribution within the T-area is larger (see Table 10.8 and Figure 10.8). In summary, the relation between the relative population share of the T-area to the nation and variance within the T-area is monotonic: the larger the relative share, the smaller the intra-area variance.

Table 10.8 Effect of change in relative position of T-area (when N is decreased by 50%)
Coefficient of variation 1.1317

Class of region	Population bef.change	Population after change	After - before	After/ before	Number of regions	Increase	Decrease
I	4403491	4414152	10661	1.0024	6	3	3
II	1327650	1330457	2807	1.0021	4	2	2
III	1493272	1497322	4050	1.0027	6	4	2
IV	1916326	1908372	-7954	0.9958	13	6	7
V	597550	587984	-9566	0.9840	8	1	7

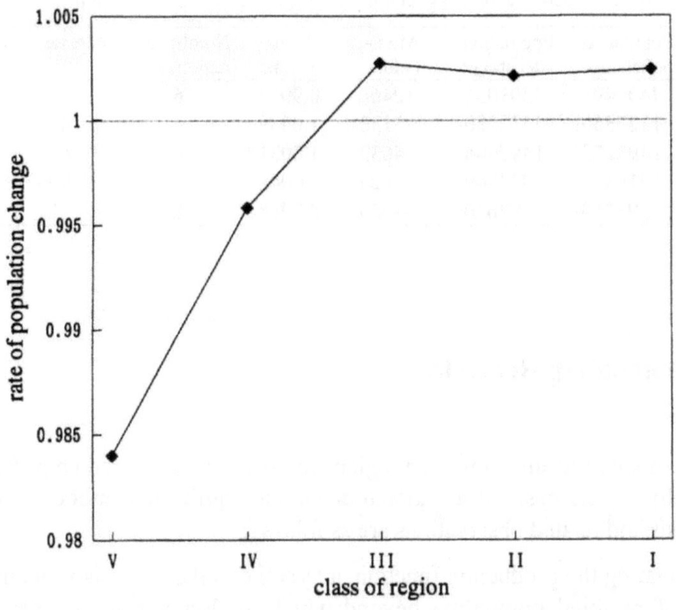

Figure 10.8 Effect of a change in relative position of T-area (when N is decreased by 50%)

10.4.3 Liberalization of Agricultural Product

The Japanese agricultural sector is currently faced with the liberalization of agricultural products in international trade. It is said that Japan has a comparative advantage not in agricultural production but in manufacturing. In parallel to this evaluation, per capita regional production is lower in areas where the agricultural sector's share of total production is higher. The T-area is a typical agricultural area. Thus, its economic system is hypothesized to be influenced, to a greater or less extent, by the future liberalization program. To investigate the effect of this, a naive simulation was carried out where agricultural production (as a result of liberalization) is assumed to be reduced by half in every region.

Surprisingly, the largest region (Sendai) loses population most (by about 1%) while population increases in most medium sized regions where agriculture is one of the major sectors. One interpretation is that the agricultural sector is relatively capital intensive in the sense that the parameter $(1-a)$ in the Cobb-Douglas production function in (10.19) is relatively large (0.818 as compared with 0.510, which is the average for all sectors). Thus, when other industries are located in a region as a result of reductions in the agricultural sector, there is a higher demand for labor in order to maintain the same total regional production as before. However, it is noted that, in nine of the ten smallest regions (where the agricultural sector's share was originally very high), population decreases as a result of curtailed agricultural production (see Table 10.9 and Figure 10.9).

Table 10.9 Effect of liberalization of agricultural product. Coefficient of variation 1.1163

Class of region	Population bef.change	Population after change	After - before	After/ before	Number of regions	Increase	Decrease
I	4403491	4391031	-12460	0,9972	6	3	3
II	1327650	1332780	5130	1.0039	4	4	0
III	1493272	1497904	4632	1.0031	6	6	0
IV	1916326	1923949	7623	1.0040	13	9	4
V	597550	592630	-4920	0.9918	8	1	7

10.5 Concluding Remarks

A system of small-to-medium sized regions in Japan was analyzed empirically and numerically on the basis of a regional economic equilibrium model. The major conclusions and related observations are as follows:

1. In estimating the production function, a hypothesis that there is some threshold level of regional population beyond which agglomeration economies start working was tested for each industrial sector. In most sectors, the existence of the threshold level was confirmed, and surprisingly, the threshold level of population was 200,000 in every sector.
2. As for the effect of transportation network improvement, two opposing forces operate: the concentration effect due to agglomeration economies, in large regions, and the dispersion effect due to lower factor prices in peripheral regions. Our simulation analysis shows that the dispersion-force rather than concentration-force prevails in response to lowered transport cost.
3. In order to examine how the system of regions is affected by the relative position of a large area in the nation, the total population in the studied area was

hypothetically changed with the national population unchanged in the simulation analysis. The relation between the relative population share of the studied area to the nation and variance within the area was found to be monotonic and the larger the relative share, the smaller the intra-area variance.

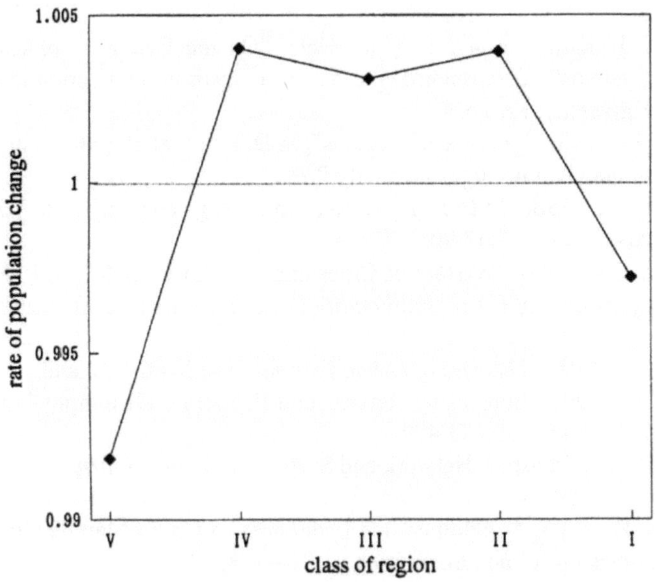

Figure 10.9 Effect of liberalization of agricultural product (when agricultural output is reduced by half)

4. In the simulation to assess the effect of the future liberalization program (where agricultural production is assumed to be reduced by half in every region), surprisingly, the largest region loses population most, and population increases in most medium sized regions where agriculture is one of the major sectors. This is because the agricultural sector is relatively capital intensive, and thus, when other industries replace the agricultural sector in a region, there is a higher demand for labor in order to maintain the same total regional production as before.

Acknowledgement

The original version of this paper was presented at the international workshop on "Theories of Regional Development-Lessons for Policies of Regional Economic Renewal and Growth" at Uddevalla, Sweden on 14th-16th June, 1998. We wish to thank the participants of the workshop for useful comments.

References

Abdel-Rahman, H.M., 1990, "Agglomeration Economies, Types, and Sizes of Cities", *Journal of Urban Economics,* 27:25-45.

Fujita, M. and P. Krugman, 1995, "When is the Economy Monocentric? *von Thünen* and Chamberlin Unified", *Regional Science and Urban Economics*, 25: 505-528.

Fujita, M., P. Krugman and T. Mori, 1995, "On the Evolution of Hierarchical Urban Systems", Discussion Paper No. 419, Institute of Economic Research, Kyoto University, Kyoto.

Helpman, E., 1998, "The Size of Regions", in D. Pines, et al. (eds.), *Topic in Public Economics*, Cambridge University Press.

Henderson, J.V., 1986, "Efficiency of Resource Usage and City Size", *Journal of Urban Economics,* 19:47-90.

Henderson, J.V., 1987, "System of Cities and Inter-City Trade", in P. Hansen, et al. (eds.), *System of Cities and Facility Location*, Harwood Academic Publishers.

Kanemoto, Y., 1980, *Theories of Urban Externalities*, North-Holland.

Krugman, P., 1991, "Increasing Returns and Economic Geography", *Journal of Political Economy,* 99:483-499.

Mun, S., 1997, "Transport Network and System of Cities", *Journal of Urban Economics*, 42:205-221.

Nakamura, R., 1985, "Agglomeration Economies in Urban Manufacturing Industries", *Journal of Urban Economics,* 17:108-124.

Sasaki, K., 1982, "Travel Demand and Evaluation of Transportation Change: A Reconsideration of the Random Utility Approach", *Environment and Planning* A, 14:169-182.

Sasaki, K., 1985, "Regional Difference in Total Factor Productivity and Spatial Feature", *Regional Science and Urban Economics*, 15:489-516.

Sveikauskas, L., 1975, "The Productivity of Cities", *Quarterly Journal of Economics,* 89:393-413.

Tabuchi, T., 1996, "From Urban Agglomeration to Dispersion", Working Paper No.34, Faculty of Economics, Kyoto University, Kyoto.

11 Agglomeration, Enterprise Size, and Productivity[1]

Edward J. Feser
Department of City and Regional Planning, University of North Carolina at Chapel Hill, BC 3140, New East Building, Chapel Hill, NC 27599-3140, USA

11.1 Introduction

Much research on agglomeration economies, and particularly recent work that builds on Marshall's concept of the industrial district, postulates that benefits derived from proximity between businesses are strongest for small enterprises (Humphrey, 1995; Sweeney and Feser, 1998). With internal economies a function of the shape of the average cost curve and level of production, and external economies in shifts of that curve, a small firm enjoying external economies characteristic of industrial districts (or complexes or simply urbanized areas) may face the same average costs as the larger firm producing a higher volume of output (Oughton and Whittam, 1997; Carlsson, 1996; Humphrey, 1995). Thus we observe the seeming paradox of large firms that enjoy internal economies of scale co-existing with smaller enterprises that should, by all accounts, be operating below minimum efficient scale. With the Birch-inspired debate on the relative job- and innovation-generating capacity of small and large firms abating (Ettlinger 1997), research on the small firm sector has shifted to an examination of the business strategies and sources of competitiveness of small enterprises (e.g., Pratten, 1991; Nooteboom, 1993). Technological external scale economies are a key feature of this research (Oughton and Whittam, 1997).

One argument is that smaller firms utilize superior flexibility and innovativeness to compete with larger producers. Borrowing from Stigler's (1951) model of market size and industry structure, Scott (1988) argues that changing market conditions favor the vertical disintegration of larger producers and the greater utilization of outsourcing. Though Stigler (1951) emphasized that growth in the size of the market makes disintegration possible, Scott considers uncertainty in demand and growing sophistication of consumers as favoring a more flexible, vertically disintegrated production regime. Since greater outsourcing puts pressure on the management of external transactions, business partners are encouraged to seek proximate locations. The result is a re-agglomeration of economic activity with the shift toward more flexible production modes. Pratten's (1991) study of small firms would seem to lend some credence to this scenario. He found that small firms co-

---

[1] Part of the research in this paper was conducted while the author was a research associate at the Center for Economic Studies, U.S. Bureau of the Census. Research results and conclusions expressed are those of the author and do not necessarily indicate concurrence by the Bureau of the Census or the Center for Economic Studies.

exist with larger ones because the former serve market niches appropriate for their scale of production. As Carlsson (1996) notes, Pratten's study suggests that there is, in effect, a division of labor between the two types of producers, one that shows signs of being characterized as much by complementarity as competitiveness.

None of this suggests that only small firms will benefit from local externalities. Under the Scott framework, larger firms gain economies of scope and scale by outsourcing certain functions once the market is sufficient to support the independent production of those functions (giving rise to de facto externalities). Urban or industry scale might be sufficient proxies for those types of effects, provided demand is localized. Likewise, both large and small firms would be expected to benefit from other types of external economies associated with interfirm proximity, including a network of suppliers, pools of skilled labor, and knowledge spillovers, though small firms may depend to a greater degree on such advantages.

In this paper, I examine the degree to which local business externalities differ in magnitude and type among large and small enterprises in two U.S. manufacturing sectors (farm and garden machinery, SIC 352, and measuring and controlling devices, SIC 382). I begin by specifying a four factor micro-level production function with oft-cited sources of agglomeration economies (local input supply, labor pools, knowledge spillovers) treated as technology parameters. The inclusion of dummy variables representing varying definitions of plant size (and type, i.e., single or multi establishment unit) permit an investigation of differences in output elasticities indicating the magnitude of any agglomeration economies or spillovers for small versus large plants.

11.2 A Test for Local External Economies among Small and Large Firms

To determine the influence of interfirm proximity on production efficiency, I assume that plant-level manufacturing activity may be described by the following four factor function:

$$Y_i = f(K_i, L_i, E_i, M_i, A_{im} \ldots A_{in}) \quad (11.1)$$

where Y, K, L, E, and M are output, capital, labor, energy, and materials, respectively, and A_m(m, . . ., n) represent different sources of external economies affecting the level of technical efficiency across production units. The latter include proximity to input supplies, proximity to producer services, nature and size of the local labor pool, proximity to research universities, and degree of local innovative activity. Estimation of (11.1) is implemented for the two study sectors for 1992 utilizing the flexible translog production system outlined by Kim (1992), whereby a standard translog production function is estimated jointly with a set of non-linear cost share equations. The principal feature of the approach is that it imposes fewer *a priori* restrictions on the production structure than typical production function based studies of agglomeration economies. Space constraints preclude a full exegesis of the modeling framework here; it is described in full in Feser (1997).

To test whether geographic proximity influences efficiency differently for small versus large and branch versus single establishment enterprises, two dichotomous variables are included in the basic production function and interacted with the proximity measures. The first, *SINGLE*, takes a value of one if the plant is a single establishment firm and zero otherwise. The second, *SMALL*, takes a value of one if the plant employs fewer workers than a specified employment size threshold, and zero otherwise. While defining firms as 'small' or 'large' on the basis of employment is imperfect at best, employment is a common means of identifying small and medium sized firms in the literature and is also the primary means of targeting small business programs in the United States. Because there is some dispute in the literature over what constitutes a small firm (see, for example, Harrison, 1994), I tested three definitions of "small" based on the first, second, and third sample quartiles for each industry. Those quartiles are 13, 27, and 70 for the farm and garden machinery sample and 15, 31, and 88 for the measuring and controlling devices sample. The third sample quartile roughly corresponds with the standard adopted by the OECD (at 100 workers) in studies of the small firm sector. Of course, any threshold is arbitrary. In the absence of better theory, the comparison of alternative standards is more informative than the use of any single definition.

11.2.1 Input and Proximity Measures

The estimated model includes four basic sets of variables: conventional inputs, measures of enterprise proximity, controls, and indicators of plant size and type. The construction of the conventional inputs (*K, L, E, M*), output (*Y*), and cost shares (S_k, S_l, S_e, S_m) is described in detail in Feser (1997) as well as a technical appendix available upon request. The principal data source for the conventional input measures is the Longitudinal Research Database (LRD) of the U.S. Census. The LRD contains confidential plant-level data from the yearly *Survey of Manufactures* and quincentennial *Census of Manufactures*.

The sources of local business externalities are based on Marshall's (1961) analysis of industrial districts: availability of inputs (including manufactured inputs and producer services), the presence of specialized labor pools, and knowledge spillovers. Where appropriate, each variable is constructed to take explicit account of industry mix (i.e., the specific types of inputs and services typically utilized by study industry firms), relative demand (i.e., the importance of particular inputs, services, and labor in the production process), and proximity (weighted distance between enterprises using alternative decay specifications). The knowledge spillover variables are designed to measure the spatial influence of innovation activity and public knowledge infrastructure (research universities). In sum, the proximity measures embody a number of plausible assumptions derived from available theory. Although they are undoubtedly imperfect in many respects, they represent reasonable first attempts to detect subtle effects that are easily missed by simpler measures.

Local manufactured inputs and producer services. The measures of the availability of local manufactured inputs and producer services utilize an income/agglomeration potential framework. Each plant i in industry k purchases intermediate inputs from p ($p, \ldots, q$) industries with plants located at points j. The share of total intermediate input purchases by plant i in industry k from each supplier industry p is given by r_{kp}. Then, a measure of total potential intermediate input supply in the region surrounding plant i($A_{T,i}$) that accounts for the distance between plant i and sources of supply, the size of sources of supply, and the relative mix of inputs available to plant i, is given by the index

$$A_{T,i} = \sum_{p}\sum_{j} E_{pj} h_{ij} r_{kp} \tag{11.2}$$

where E_{pj} is employment in industry p at point j. The factor h_{ij} is the distance between plant i and sources j specified in the general decay form:

$$h_{ij} = (m - d_{ij})/(m - \alpha d_{ij}) \tag{11.3}$$

where m is the maximum allowable distance and alpha is the decay parameter. As α approaches minus infinity, h_{ij} resembles the simple inverse of distance.

Note that (11.2) assumes that a given plant in industry k purchases the same relative mix of inputs as the average in its industry. A measure of producer services availability, $A_{S,i}$, may be derived analogously, where the E_{pj} refer to employment at points j for p producer services sectors.

To implement $A_{T,i}$ and $A_{S,i}$ an appropriate distance measure and an assumption about the form of decay over some relevant distance are required, in addition to spatially and sectorally disaggregated data on industry size and data on input purchasing patterns. Assuming that each establishment is located at the centroid of its county, the primary measure of size (E_{pj}), where j is the county centroid, is employment as reported in 1992 *County Business Patterns* (CBP).[2] Great circle arc distances between counties are used as weights, with distance decaying slowly at first and then more rapidly up to a maximum distance of fifty miles (beyond which the distance weight falls to zero). Fifty miles was selected since transportation studies have shown that very little commuting occurs beyond this distance. The r_{kp} are derived from the 1987 *Benchmark Input-Output Accounts of the United States*, released by the Bureau of Economic Analysis in 1994.

The adopted decay profile (Figure 11.1), which is generated by setting alpha equal to 0.75, simply represents one plausible specification of the distance-related intensity of interaction likely among neighboring enterprises.[3] Feser (1997) describes results of the basic model without firm size/type effects under alternative

[2] Where employment information is suppressed in CBP, the reported data on the number of plants in specific size categories were used to construct estimates by assuming the category's midpoint employment for each establishment.

[3] A common default is the simple inverse of distance. But the inverse assumes extreme decay and all variables essentially revert to individual county-level measures.

distance decays but without size/type effects. The use of alternative decays effectively provides useful information about the relative spatial influence of alternative proximity factors. But as a practical matter, comparing size/type effects and different decay specifications multiplies significantly the number of models and estimated parameters that require interpretation. Such an exercise would constitute a separate study in itself. The problem points to the need for further studies of business externalities that specify the decay as an inherent part of the model, rather than as an initial assumption that must be subjected later to detailed sensitivity tests.

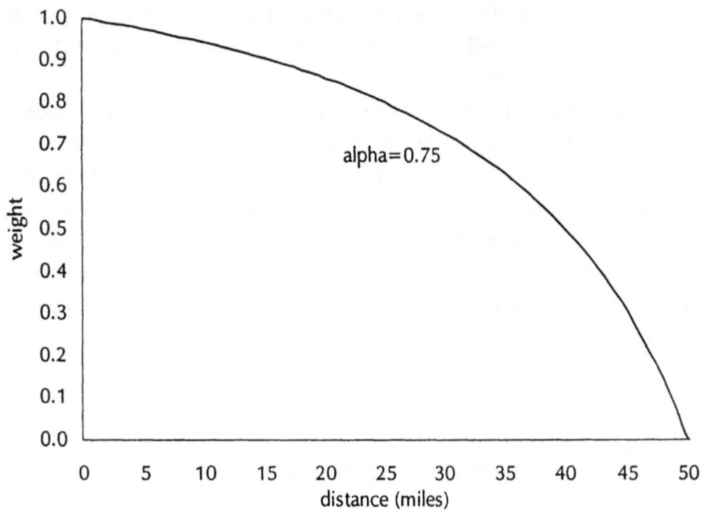

Figure 11.1 Distance decay specification

Local manufacturing demand. The measuring and controlling devices industry is itself a key supplier to a wide range of final market manufacturing industries, from aerospace (aircraft, missiles, and space vehicles) to industrial and home appliances (refrigeration and heating equipment, electric housewares and fans, and household refrigerators and freezers). Just as manufacturers may derive benefits from proximate location to their own suppliers, they may also benefit from proximity to their major customers. To test this conjecture a demand-side proximity variable for plant *i* in SIC 382 is included, $A_{D,i}$, that is constructed in a similar fashion to the intermediate input and producer services supply variables.

Specialized labor pool. According to Marshall, in an industrial district mutual learning takes place through a concentration of workers engaged in similar tasks, "a habit of responsibility, of carefulness and promptitude in handling expensive machinery and materials becomes the common property of all." (Marshall, 1961, p.205). To construct a labor pool variable that measures the specialized skill base of each study plant's location, I first grouped national 3-digit SIC industries into clusters based on similarities in occupational staffing patterns. Those clusters were then used to create a labor market specialization variable specific to each study plant's fifty mile location shed. Although ideally industries would be grouped together based on similar worker skill needs, detailed occupation may serve as a reasonable proxy.

The *Occupational Employment Survey* conducted every three years by the Bureau of Labor Statistics reports the number of workers employed in 400 occupations for each 3-digit manufacturing sector. Each column in this 400 x 140 data matrix represents the labor requirements vector – in terms of detailed occupation – for a specific manufacturing sector. A form of oblique factor analysis was performed on the matrix to reduce the number of columns (or variables) to a set of principal components. The method is typically used to reduce to a manageable number a large number of closely related variables. Here it is used to divide industries (the variables) into mutually exclusive groups that may be interpreted as sharing similar labor force requirements.

The procedure reduced the 140 industries to 24 clusters. SIC 352 was assigned to a cluster with 19 other industries, while SIC 382 was joined with 5 other industries. Those "labor requirements" clusters represent groups of industries that arguably employ and draw from a joint worker pool. Cities and regions specialized in a particular cluster's industries are therefore where cluster members are likely to enjoy the most significant labor pooling economies. Given p cluster industries, a measure of labor specialization in the fifty mile shed surrounding plant i in a given study industry is defined as:

$$A_{R,i} = \frac{\sum_p \sum_j \frac{E_{pj} / E_{j,m}}{E_p / E_m} h_{ij}}{\sum_j h_{ij}}, \qquad h_{ij} = 1 \textit{ if } d_{ij} \leq 50, \textit{ otherwise } h_{ij} = 0 \tag{11.4}$$

where E_{pj} is employment in industry p at location j, $E_{j,m}$ is total manufacturing employment at location j, E_p is U.S. employment in industry p, E_m is total U.S. manufacturing employment, and h_{ij} is a zero-one indicator variable based on the distance, d_{ij}, between county centroids. Equation (11.4) is simply a location quotient averaged across cluster industries for each fifty mile commuting or activity shed. Since the variable does not account for regional differences in educational quality, the share of the population 25 years old and older with some college education (*ATTAIN*) is included in the model as well. The variable is calculated for all counties with centroids in plant i's 50 mile shed.

Innovative activity and knowledge infrastructure. Knowledge spillovers are expected to be most prevalent in locations with high rates of public and private sector innovation. A public sector innovation rate for plant i's location is defined as:

$$A_{U,i} = \sum_j U_j h_{ij}$$

where U_j is total research expenditures by universities in location j and h_{ij} is distance between the locations i and j, specified in hyperbolic decay form. As before, locations are defined on the basis of county centroids. Data for U_j (628 doctoral granting institutions in 1993) are from the National Science Foundation's 1993 *Survey of Scientific and Engineering Expenditures at Universities and Colleges*. Only the research disciplines most relevant to the study industries are included in the expenditure figures.[4]

The degree of private sector innovative activity at plant *i*'s location, $A_{P,i}$ is measured by the number of utility patents granted to residents, per capita, in the plant's fifty mile shed. U.S. patent grants by county are from the U.S. Patent and Trademark Office. These data represent the most geographically disaggregated information on patenting activity available at the time of the study. Because of the way they are reported (residence of the first named inventor and undistinguished by patent class), $A_{P,i}$ is only a broad indicator of real inventive activity in particular localities. It is, nevertheless, an improvement over other possible alternatives, including employment in high technology sectors or state-level private sector research and development expenditures.

11.2.2 Controls

The control for educational attainment has already been noted. Two additional controls are included in the model: the competitive structure of local industry and a proxy for urban diseconomies. The former is based on Chinitz's (1961) hypothesis that external economies may be most prevalent in places less dominated by a few large producers. The degree to which industry in plant *i*'s location is competitively organized, *CRATIO*, is defined as the share of total manufacturing sales made by the four largest *firms* in the plant's *commuter zone*. (Sales by firm are from the LRD.) Commuting zones, first created by the U.S. Department of Agriculture's Economic Research Service in 1980 and updated in 1990, constitute mutually exclusive groups of counties defined on the basis of Census journey-to-work data. To ensure that the proximity indicators do not indirectly measure generalized disadvantages associated with dense, urban places, a population density variable is included to control for possible congestion effects and other urban diseconomies. The variable, *DENSITY*, is defined as the number of persons per square mile in plant *i*'s 50 mile shed (population figures are from the Bureau of Economic Analysis).

11.2.3 Descriptive Statistics

Table 11.1 lists the model variables, along with means and standard deviations.[5]

---

4 Disciplines identified as relevant to the farm and garden machinery industry are mechanical engineering, industrial engineering, electrical engineering, other engineering, other physical sciences, and agricultural sciences. For the measuring and controlling devices sector, relevant disciplines are aerospace engineering, electrical engineering, mechanical engineering, industrial engineering, other engineering, astronomy, physics, other physical sciences, and computer science.

5 One legitimate question regarding the set of proximity measures is whether they measure distinct underlying concepts. In fact, no pairwise correlations among the indicators are

The SIC 352 and 382 samples are comprised of 863 and 2,609 observations, respectively. Due to data limitations, the overall establishment size distributions of those samples do not match the population distributions for the two sectors. Each census year, to reporting burdens, the Census Bureau exempts from filing requirements a significant number of the smaller enterprises in its sampling frame (most affected businesses employ 1-2 workers, although some may employ as many as five workers). For the purposes of the preparation of summary published reports, data for those establishments are imputed and are therefore of questionable quality for micro-level analyses. Since the validity of the translog production function depends strongly on the quality of the underlying data, administrative records are excluded from this study. The exclusion of the smallest enterprises is important for the interpretation of the firm size findings, particularly their external validity to the industries a whole. Yet there are still many small manufacturers in the samples; one-quarter of enterprises in the SIC 352 and 382 samples employ fewer than 13 and 15 workers, respectively.

Table 11.1 Descriptive statistics, study industries, 1992

		SIC 352, n=863		SIC 382, n=2,609	
Variable	Description	Mean	Std Dev	Mean	Std Dev
Y	Output (000s)	16,667	74,559	12,553	36,314
K	Capital (000s)	5,821	31,992	4,717	17,307
L	Labor (000s manhours)	216	576	276	724
E	Energy (million Btus)	16,882	79,943	8,494	26,368
M	Materials (000s)	7,774	32,947	4,160	12,830
S_k	Capital cost share	.0492	.0246	.0955	.0567
S_l	Labor cost share	.3882	.1244	.4938	.1271
S_e	Energy cost share	.0148	.0116	.0124	.0106
S_m	Material cost share	.5478	.1355	.3983	.1400
A_T	Local material pool	455	806	3,128	2,775
A_S	Local service pool	2,620	5,808	15,876	16,953
A_R	Specialized labor pool	25.78	15.16	7.28	5.76
A_D	Local market demand (SIC 382)	------	------	2,659	3,916
A_U	Proximity to R&D universities	7,910	15,501	60,416	68,200
A_P	Patents per 100,000 population	15.09	11.64	26.84	12.00
CRATIO	4-firm local concentration variable	.4515	.2181	.2663	.1567
ATTAIN	Share popul. 25+ with some college	.1747	.0448	.2338	.0485
DENSITY	Population per square mile	190	359	850	791
SMALL	1 if establishm. is 'small', 0 otherwise	------	------	------	------
SINGLE	1 if plant is single estab., 0 otherwise	------	------	------	------

particularly high. The correlation between the material input and producer services measures for SIC 352 is the highest at 0.70. Moreover, most correlations are positive, as is consistent with the underlying theory. These results suggest that not only does each factor measure a different concept, but the potential for excessive multicollinearity in the estimated model is not as extreme as might be expected given the variables' similar construction.

11.3 Estimation and Results

To generate estimates of the link between enterprise proximity and productivity, a fully specified translog production system excluding plant size and type variables was first estimated jointly using iterated seemingly unrelated regressions (IZEF), a maximum likelihood-equivalent procedure (stage 1). After eliminating insignificant controls and cross-terms on the proximity factors (i.e., testing for factor augmentation and Hicks-neutrality of specific technology indicators), the revised models were re-estimated under different assumptions regarding production technology: homotheticity, homogeneity, and constant returns to scale (stage 2). Those tests suggested – though were not entirely conclusive – that constant returns are present in both sectors, particularly farm and garden machinery. To avoid imposing an overly restrictive assumption, the plant size and type variables were included and homogeneity imposed to generate the final estimates (stage 3). Homogeneity is less restrictive than constant returns yet is still more efficient (given the results) than non-homotheticity.

The final revised production function for SIC 352 is written as:

$$
\begin{aligned}
\mathrm{Ln}Y = {} & \alpha_0 + \alpha_k \ln K + \alpha_l \ln L + \alpha_e \ln E + \alpha_m \ln M + \zeta \ln DENSITY + \\
& \kappa \ln ATTAIN + \frac{1}{2}\beta_{kk}(\ln K)^2 + \beta_{kl} \ln K \ln L + \beta_{ke} \ln K \ln E + \\
& \beta_{km} \ln K \ln M + \frac{1}{2}\beta_{ll}(\ln L)^2 + \beta_{le} \ln L \ln E + \beta_{lm} \ln L \ln M + \\
& \frac{1}{2}\beta_{ee}(\ln E)^2 + \beta_{em} \ln E \ln M + \frac{1}{2}\beta_{mm}(\ln M)^2 + \xi_T \ln A_T + \xi S \ln A_S + \\
& \gamma_{ks} \ln K \ln A_S + \gamma_{ls} \ln L \ln A_S + \gamma_{es} \ln E \ln A_S + \gamma_{ms} \ln M \ln A_S + \\
& + \xi_r \ln A_R + \xi_U \ln U + \xi_P \ln A_P + \gamma_{kP} \ln K \ln A_P + \gamma_{lP} \ln L \ln A_P + \\
& \gamma_{eP} \ln E \ln A_P + \gamma_{mP} \ln M \ln A_P + \lambda_f SINGLE + \lambda_{Tf} \ln A_T SINGLE + \\
& \lambda_{Sf} \ln A_s SINGLE + \lambda_{Rf} \ln A_R SINGLE + \lambda_{Uf} \ln A_U SINGLE + \\
& \lambda_{Pf} \ln A_P SINGLE + \lambda_z SMALL + \lambda_{Tz} \ln A_T SMALL + \lambda_{Sz} \ln A_S SMALL + \\
& \lambda_{Rz} \ln A_R SMALL + \lambda_{Uz} \ln A_U SMALL + \lambda_{Pz} \ln A_P SMALL + \varepsilon
\end{aligned}
$$

while the revised function for SIC 382 is:

$$
\begin{aligned}
\mathrm{Ln}Y = {} & \alpha_0 + \alpha_k \ln K + \alpha_l \ln L + \alpha_e \ln E + \alpha_m \ln M + \iota \ln CRATIO + \kappa \ln ATTAIN + \\
& \frac{1}{2}\beta_{kk}(\ln K)^2 + \beta_{kl} \ln K \ln L + \beta_{ke} \ln K \ln E + \beta_{km} \ln K \ln M + \frac{1}{2}\beta_{ll}(\ln L)^2 + \\
& \beta_{le} \ln L \ln E + \beta_{lm} \ln L \ln M + \frac{1}{2}\beta_{ee}(\ln E)^2 + \beta_{em} \ln E \ln M +
\end{aligned}
$$

$$\frac{1}{2}\beta_{mm}(\ln M)^2 + \xi_T \ln A_T + \xi_S \ln A_S + \xi_R \ln A_R + \gamma_{kR} \ln K \ln A_R +$$

$$\gamma_{lR} \ln L \ln A_R + \gamma_{eR} \ln E \ln A_R + \gamma_{mR} \ln M \ln A_R + \xi_U \ln U +$$

$$\gamma_{kU} \ln K \ln A_U + \gamma_{lU} \ln L \ln A_U + \gamma_{eU} U \ln E \ln A_U + \gamma_m \ln M \ln A_U +$$

$$\xi_P \ln A_P + \gamma_{kP} \ln K \ln A_P + \gamma_{lP} L \ln A_P + \gamma_{eP} \ln E \ln A_P + \gamma_{mP} \ln M \ln A_P +$$

$$\xi_D \ln A_D + \lambda_f SINGLE + \lambda_{Tf} \ln A_T SINGLE + \lambda_{Sf} \ln A_S SINGLE +$$

$$\lambda_{Rf} \ln A_R SINGLE + \lambda_{Uf} \ln A_U SINGLE + \lambda_{Pf} \ln A_P SINGLE +$$

$$\lambda D_f \ln A_D SINGLE + \lambda_z SMALL + \lambda_{Tz} \ln A_T SMALL + \lambda_{Sz} \ln A_S MALL +$$

$$\lambda_{Rz} \ln A_R SMALL + \lambda_{Uz} \ln A_U SMALL + \lambda_{Pz} \ln A_P SMALL +$$

$$\lambda_{Dz} \ln A_D SMALL + \varepsilon$$

Parameter estimates of the revised production functions along with asymptotic standard errors for the second quartile enterprise size specifications are reported in Tables 11.2 and 11.3. Note that in the case of the farm and garden machinery sample, there was no evidence in early stage regressions that the degree to which local industry is competitively organized determines productivity differentials among manufacturers in the industry. *CRATIO*, or the share of area sales held by the largest four local manufacturing firms, was consistently insignificant across all earlier estimated models and therefore does not appear in the revised models reported in Table 11.2. Likewise, the insignificant population density variable (*DENSITY*) was eliminated in the course of earlier SIC 382 model revisions. All early-stage models are reported in Feser (1997).

For the farm and garden machinery industry, educational attainment (*ATTAIN*) and population density (*DENSITY*) do appear to influence productivity. Plants located in denser urban environments, other things equal, tend to be less efficient – an indication that urban diseconomies are binding for this industry. Surprisingly, in the revised model that excluds size/type effects, a doubling of the share of population 25 and over with some college education reduces output by 13 percent, holding all else constant (the impact varies from 13 to 15 percent in the models reported in Table 11.2). A possible explanation for this finding is that the farm and garden machinery producer located in a region with a better educated workforce must pay a wage premium for comparably skilled workers.

In the case of measuring and controlling devices, the local four-firm concentration ratio (*CRATIO*) and degree of educational attainment (*ATTAIN*) variables are found to strongly influence productivity. Location in a region dominated by – in Chinitz's (1961) words – an "oligopolistically organized" local industrial base is associated with a significant efficiency penalty. A doubling of the local concentration ratio reduces output by approximately 13 percent, holding conventional input levels constant. On the other hand, a doubling of the rate of college attainment increases output, other things equal, of the average SIC 382 manufacturer by around 12 percent. The finding with respect to educational attainment is not particularly surprising, given the heavy dependence of this sector on a highly trained workforce.

Checking the parameters of the estimated models as reported in Tables 11.2 and 11.3, it appears that there are few statistically significant differences between small versus large plants and branch versus single establishments in terms of spatial economies. For SIC 352, the individual parameter estimates in Table 11.2 provide no evidence of significant interaction effects for any spatial variable with respect to the single/branch establishment dummies. There is only weak evidence of differences in spatial economies by firm size for the material and producer services pool variables. (The results are similar for the unreported first and third quartile plant size estimations.) When both sets of interaction terms are included in the estimated models however, the results get somewhat stronger, with most interaction effects with respect to plant size significant. The findings for SIC 382 are even weaker, suggesting few differences by plant size or type in this sector.

More easily interpreted than the parameter estimates are the calculated output elasticities in Tables 11.4 and 11.5. The tables, which report elasticities at the sample means, compare the sensitivity of the results to the definition of the *SMALL* variable. They also report the elasticities for the models excluding plant size/type effects (the overall sample estimates). Comparing the overall elasticities with those for plants of alternative sizes and types provides the best evidence of any differences in local externalities among large and small enterprises.

11.3.1 Farm and Garden Machinery

Looking first at the elasticities of proximity for the full sample, location in regions with high relative rates of innovation as proxied by patenting activity represents the strongest efficiency effect for the overall farm and garden machinery sample. Each doubling of the patent rate is associated with an increase in output of nearly 4 percent, holding input levels constant. At the same time, the elasticity of output with respect to proximity to university R&D is effectively zero.

A second influence on productivity among plants in the farm and garden machinery sector is proximity to their required mix of producer services. The magnitude of the services efficiency effect is very close to that of the patent rate. Doubling the producer services pool index increases output, other things equal, by 3.5 percent. Since the spatial variables in index form may be difficult to interpret, it is important to note that the producer services variable increases with geographic proximity, the size of the services sector, and as the mix of the services sector matches the unique requirements of the farm and garden machinery sector. Key producer services sectors for the industry are radio and television broadcasting, engineering, and advertising. The local availability of those services may allow manufacturers to outsource certain services-related functions while also enjoying face-to-face contact with suppliers.

Table 11.2 Production function with local externalities: SIC 352. Plant type and size effects, homogeneity assumed, small plants: <27 employees: Parameter estimates and asymptotic standard errors

	Plant type effects			Plant size effects			Plant types and size effects		
Parameter	Estimate	s-e.	t-statistic	Estimate	s-e.	t-statistic	Estimate	s-e.	t-statistic
α_0	2.0687	0.2575	8.03	1.5289	0.2436	6.28	1.7066	0.2865	5.96
α_K	0.0246	0.0025	9.77	0.0256	0.0026	9.75	0.0256	0.0026	9.72
α_L	0.8567	0.0133	64.57	0.8763	0.0141	62.21	0.8722	0.0145	60.34
α_M	0.1541	0.0175	8.83	0.1638	0.0188	8.70	0.1635	0.0190	8.62
α_E	-0.0074	0.0019	-3.93	-0.0073	0.0019	-3.79	-0.0072	0.0019	-3.78
DENSITY	-0.0456	0.0230	-1.98	-0.0415	0.0230	-1.80	-0.0441	0.0231	-1.91
ATTAIN	-0.1302	0.0646	-2.02	-0.1507	0.0643	-2.35	-0.1496	0.0646	-2.32
β_{KK}	0.0402	0.0006	65.70	0.0411	0.0007	61.80	0.0410	0.0007	60.30
β_{KL}	-0.0117	0.0006	-19.08	-0.0120	0.0006	-19.13	-0.0119	0.0006	-19.03
β_{KM}	-0.0286	0.0006	-49.28	-0.0292	0.0006	-46.74	-0.0291	0.0006	-46.24
β_{KE}	0.0001	0.0003	0.19	0.0001	0.0003	0.19	0.0001	0.0003	0.17
β_{LL}	0.1753	0.0034	51.27	0.1791	0.0035	50.95	0.1781	0.0036	49.70
β_{LM}	-0.1611	0.0031	-52.40	-0.1647	0.0032	-51.96	-0.1638	0.0032	-50.67
β_{LE}	-0.0024	0.0005	-5.23	-0.0025	0.0005	-5.16	-0.0024	0.0005	-5.13
β_{EE}	0.0115	0.0004	32.20	0.0118	0.0004	31.67	0.0117	0.0004	31.45
β_{EM}	-0.0091	0.0004	-21.39	-0.0094	0.0004	-21.21	-0.0093	0.0004	-21.18
β_{MM}	0.1988	0.0030	66.93	0.2032	0.0031	65.03	0.2022	0.0032	62.80
ζ_T	0.0103	0.0181	0.57	-0.0189	0.0140	-1.35	-0.0062	0.0193	-0.32
ζ_S	0.0389	0.0383	1.02	0.0952	0.0363	2.62	0.0695	0.0417	1.67
γ_{KS}	-0.0002	0.0003	-0.61	-0.0003	0.0004	-0.84	-0.0003	0.0004	-0.83
γ_{LS}	0.0003	0.0005	0.66	0.0004	0.0005	0.84	0.0004	0.0005	0.82
γ_{ES}	0.0011	0.0002	4.69	0.0011	0.0002	4.52	0.0011	0.0002	4.51
γ_{MS}	-0.0043	0.0026	-1.65	-0.0055	0.0029	-1.91	-0.0055	0.0029	-1.87
ζ_U	0.0036	0.0038	0.96	-0.0025	0.0027	-0.91	0.0023	0.0039	0.59
ζ_P	0.0878	0.0344	2.55	0.0898	0.0263	3.42	0.0937	0.0351	2.67
γ_{KP}	-0.0011	0.0005	-2.07	-0.0011	0.0006	-1.99	-0.0011	0.0006	-1.99
γ_{LP}	0.0139	0.0032	4.32	0.0135	0.0033	4.04	0.0135	0.0033	4.03

Table 11.2 cont.

	Plant type effects			Plant size effects			Plant types and size effects		
Parameter	Estimate	s-e.	t-statistic	Estimate	s-e.	t-statistic	Estimate	s-e.	t-statistic
γ_{EP}	-0.0012	0.0004	-2.85	-0.0012	0.0004	-2.78	-0.0012	0.0004	-2.77
γ_{MP}	-0.0115	0.0029	-3.92	-0.0112	0.0031	-3.67	-0.0112	0.0031	-3.66
ζ_R	0.0081	0.0320	0.25	0.0398	0.0229	1.74	0.0265	0.0335	0.79
λ_F	-0.1256	0.1857	-0.68				-0.2225	0.1923	-1.16
λ_{TF}	-0.0127	0.0192	-0.66				-0.0205	0.0202	-1.01
λ_{UF}	-0.0051	0.0044	-1.16				-0.0082	0.0046	-1.77
λ_{RF}	0.0050	0.0364	0.14				0.0166	0.0381	0.43
λ_{SF}	0.0298	0.0293	1.02				0.0446	0.0309	1.44
λ_{PF}	-0.0164	0.0344	-0.48				-0.0103	0.0370	-0.28
λ_Z				0.4234	0.1626	2.60	0.4599	0.1690	2.72
λ_{TZ}				0.0314	0.0166	1.89	0.0374	0.0175	2.14
λ_{UZ}				0.0048	0.0039	1.25	0.0071	0.0041	1.75
λ_{RZ}				-0.0447	0.0308	-1.45	-0.0455	0.0324	-1.41
λ_{SZ}				-0.0464	0.0253	-1.83	-0.0588	0.0267	-2.20
λ_{PZ}				-0.0305	0.0285	-1.07	-0.0259	0.0307	-0.84
N		863			863			863	
Adj. R^2's Output		0.968			0.968			0.968	
K Share		0.837			0.837			0.837	
L Share		0.711			0.711			0.711	
M Share		0.798			0.799			0.799	

Table 11.3 Production function with local externalities: SIC 382. Plant type and size effects, homogeneity assumed, small plants: <31 employees. Parameter estimates and asymptotic standard errors

	Plant type effects			Plant size effects			Plant type and size effects		
	Estimate	s.e	t-statistics	Estimate	s.e	t-statistics	Estimate	s.e	t-statistics
α_0	2.8616	0.1141	25.09	2.7264	0.1080	25.25	2.8316	0.1213	23.35
α_K	0.0285	0.0037	7.72	0.0285	0.0037	7.61	0.0286	0.0037	7.71
α_L	0.8438	0.0101	83.71	0.8529	0.0106	80.50	0.8449	0.0107	79.15
α_M	0.1375	0.0078	17.60	0.1389	0.0079	17.54	0.1378	0.0079	17.54
α_E	-0.0016	0.0012	-1.29	-0.0015	0.0012	-1.24	-0.0015	0.0012	-1.26
CRATIO	-0.1317	0.0467	-2.82	-0.1329	0.0468	-2.84	-0.1349	0.0468	-2.88
ATTAIN	0.1284	0.0346	3.71	0.1173	0.0346	3.40	0.1271	0.0347	3.67
β_{KK}	0.0746	0.0008	99.26	0.0755	0.0008	93.71	0.0747	0.0008	91.43
β_{KL}	-0.0342	0.0008	-42.02	-0.0347	0.0008	-42.20	-0.0342	0.0008	-41.60
β_{KM}	-0.0408	0.0006	-62.90	-0.0411	0.0007	-60.24	-0.0408	0.0007	-60.03
β_{KE}	0.0003	0.0002	1.45	0.0003	0.0002	1.44	0.0003	0.0002	1.43
β_{LL}	0.1693	0.0020	84.27	0.1714	0.0021	82.69	0.1694	0.0021	80.78
β_{LM}	-0.1308	0.0016	-82.74	-0.1323	0.0016	-80.96	-0.1309	0.0016	-79.38
β_{LE}	-0.0043	0.0003	-15.57	-0.0043	0.0003	-15.47	-0.0042	0.0003	-15.48
β_{EE}	0.0086	0.0002	39.76	0.0087	0.0002	39.30	0.0086	0.0002	39.19
β_{EM}	-0.0047	0.0002	-20.28	-0.0047	0.0002	-20.22	-0.0047	0.0002	-20.21
β_{MM}	0.1762	0.0015	113.71	0.1782	0.0017	106.90	0.1764	0.0017	103.95
ζ_T	0.0156	0.0171	0.91	0.0106	0.0152	0.70	0.0193	0.0179	1.08
ζ_S	0.0181	0.0153	1.18	0.0119	0.0133	0.90	0.0149	0.0160	0.93
ζ_U	0.0034	0.0044	0.78	0.0031	0.0038	0.80	0.0013	0.0044	0.30
γ_{KU}	-0.0001	0.0002	-0.40	-0.0001	0.0002	-0.41	-0.0001	0.0002	-0.40
γ_{LU}	0.0014	0.0005	2.77	0.0014	0.0005	2.82	0.0014	0.0005	2.78
γ_{EU}	-0.0002	0.0001	-3.29	-0.0002	0.0001	-3.31	-0.0002	0.0001	-3.29
γ_{MU}	-0.0011	0.0004	-2.71	-0.0011	0.0004	-2.75	-0.0011	0.0004	-2.71
ζ_P	0.0018	0.0257	0.07	0.0156	0.0234	0.67	0.0043	0.0264	0.16
γ_{KP}	-0.0039	0.0012	-3.33	-0.0040	0.0012	-3.33	-0.0039	0.0012	-3.32
γ_{LP}	0.0117	0.0031	3.78	0.0119	0.0031	3.79	0.0117	0.0031	3.77
γ_{EP}	-0.0007	0.0004	-1.82	-0.0007	0.0004	-1.81	-0.0007	0.0004	-1.82

Table 11.3 cont.

		Plant type effects			Plant size effects			Plant type and size effects		
		Estimate	s.e	t-statistics	Estimate	s.e	t-statistics	Estimate	s.e	t-statistics
γ_{MP}		-0.0071	0.0025	-2.80	-0.0072	0.0026	-2.81	-0.0071	0.0026	-2.78
ζ_{R}		0.0248	0.0120	2.06	0.0385	0.0109	3.55	0.0299	0.0125	2.39
γ_{KR}		-0.0009	0.0006	-1.54	-0.0009	0.0006	-1.59	-0.0009	0.0006	-1.54
γ_{LR}		0.0072	0.0015	4.96	0.0073	0.0015	5.00	0.0072	0.0015	4.95
γ_{ER}		0.0013	0.0002	7.15	0.0013	0.0002	7.14	0.0013	0.0002	7.15
Y_{MR}		-0.0076	0.0012	-6.37	-0.0077	0.0012	-6.40	-0.0076	0.0012	-6.35
ζ_{D}		-0.0263	0.0127	-2.07	-0.0178	0.0113	-1.58	-0.0229	0.0133	-1.72
λ_{F}		-0.0874	0.1016	-0.86				-0.0986	0.1098	-0.90
λ_{TF}		-0.0358	0.0212	-1.69				-0.0256	0.0234	-1.09
λ_{SF}		0.0026	0.0193	0.13				-0.0038	0.0211	-0.18
λ_{DF}		0.0054	0.0157	0.34				0.0110	0.0171	0.64
λ_{UF}		0.0118	0.0047	2.50				0.0068	0.0053	1.27
λ_{PF}		0.0360	0.0278	1.29				0.0320	0.0313	1.02
λ_{RF}		0.0048	0.0122	0.39				0.0152	0.0132	1.16
λ_{Z}					0.0587	0.1016	0.58	0.0877	0.1095	0.80
λ_{TZ}					-0.0318	0.0208	-1.53	-0.0200	0.0229	-0.87
λ_{SZ}					0.0099	0.0190	0.52	0.0114	0.0207	0.55
λ_{DZ}					-0.0104	0.0154	-0.68	-0.0138	0.0167	-0.83
λ_{UZ}					0.0168	0.0047	3.60	0.0117	0.0053	2.22
$\lambda_{\underline{PZ}}$					0.0085	0.0277	0.31	-0.0060	0.0311	-0.19
$\lambda_{\underline{RZ}}$					-0.0168	0.0116	-1.45	-0.0216	0.0125	-1.73
N			2609			2609			2609	
Adj. R^2's	Output		0.958			0.958			0.958	
	K Share		0.780			0.780			0.780	
	L Share		0.691			0.691			0.691	
	M Share		0.829			0.828			0.829	

Table 11.4 Output elasticities by plant type and size, farm and garden machinery, estimates and asymptotic t-statistics

Proximity index	*Full Sample*						*Small: <13 Employees*				*Small: <27 Employees*				*Small: <70 Employees*			
	Est.	t-Stat.	Est.	t-Stat.	Est.	t-Stat.	Est.	t-Stat.	Est.	t-Stat.	Est.	t-Stat.	Est.	t-Stat.	Est.	t-Stat.	Est.	t-Stat.
	All		**Single**		**Branch**		**Small**		**Large**		**Small**		**Large**		**Small**		**Large**	
Manufact. input pool	0,000	0,05	-0,002	-0,23	0,010	0,57	0,026	1,62	-0,009	-0,85	0,012	1,09	-0,019	-1,35	0,005	0,51	-0,018	-0,93
Producers svcs pool	0,038	1,89	0,046	2,16	0,017	0,56	-0,017	-0,59	0,053	2,48	0,018	0,76	0,064	2,62	0,028	1,34	0,074	2,30
Special. labor pool	0,013	0,86	0,013	0,75	0,008	0,25	-0,010	-0,38	0,026	1,44	-0,005	-0,23	0,040	1,74	0,004	0,20	0,047	1,37
Research universities	0,000	-0,02	-0,001	-0,65	0,004	0,96	0,004	0,94	-0,001	-0,53	0,002	0,86	-0,002	-0,91	0,002	0,79	-0,006	-1,53
Local innovation rate	0,035	2,20	0,030	1,73	0,047	1,48	-0,016	-0,53	0,045	2,57	0,019	0,93	0,050	2,24	0,030	1,72	0,047	1,45
							Single		**Branch**		**Single**		**Branch**		**Single**		**Branch**	
Manufactured input pool																		
Small							0,025	1,52	0,041	1,72	0,011	0,92	0,031	1,46	0,002	0,21	0,031	1,40
Large							-0,013	-1,09	0,003	0,17	-0,027	-1,68	-0,006	-0,32	-0,034	-1,49	-0,006	-0,28
Producers services pool																		
Small							-0,012	-0,41	-0,053	-1,39	0,024	1,04	-0,020	-0,58	0,038	1,73	-0,018	-0,54
Large							0,067	2,88	0,026	0,87	0,083	3,02	0,039	1,26	0,107	2,88	0,051	1,47
Specialized labor pool																		
Small							-0,010	-0,37	-0,034	-0,77	-0,002	-0,11	-0,019	-0,48	0,006	0,34	-0,009	-0,22
Large							0,033	1,51	0,009	0,29	0,043	1,60	0,027	0,79	0,050	1,14	0,034	0,91
Research universities																		
Small							0,003	0,77	0,010	1,72	0,001	0,45	0,009	1,94	0,000	0,05	0,011	2,17
Large							-0,003	-1,30	0,003	0,89	-0,006	-1,78	0,002	0,59	-0,012	-2,44	-0,002	-0,43
Local innovation rate																		
Small							-0,015	-0,49	0,002	0,05	0,018	0,85	0,028	0,70	0,028	1,55	0,033	0,77
Large							0,039	1,96	0,056	1,77	0,044	1,66	0,054	1,68	0,042	0,96	0,047	1,34

Note: Elasticities are evaluated at sample means.

Table 11.5 Output elasticities by plant type and size, measuring and controlling devices, estimates and asymptotic t-statistics

Externality/Spillovers							*Small: <15 Employees*				*Small: <31 Employees*				*Small: <88 Employees*			
							Est.	t-Stat.	Est.	t-Stat.	Est.	t-Stat.	Est.	t-Stat.	Est.	t-Stat.	Est.	t-Stat.
	All		**Single**		**Branch**		**Small**		**Large**		**Small**		**Large**		**Small**		**Large**	
Manufact. Input Pool	-0,006	-0,49	-0,02	-1,43	0,016	0,91	-0,023	-1,04	0,001	0,04	-0,021	-1,35	0,011	0,70	-0,006	-0,45	-0,005	-0,22
Producers Svcs Pool	0,019	1,88	0,021	1,64	0,018	1,18	0,015	0,79	0,019	1,63	0,022	1,53	0,012	0,90	0,022	1,86	0,013	0,68
Specializ. Labor Pool	0,013	1,86	0,013	1,65	0,008	0,79	0,016	1,36	0,013	1,62	0,005	0,57	0,022	2,36	0,009	1,24	0,024	1,72
Local Demand Pool	-0,023	-2,75	-0,021	-2,00	-0,026	-2,07	-0,026	-1,69	-0,022	-2,28	-0,028	-2,45	-0,018	-1,58	-0,025	-2,65	-0,018	-1,13
Research Universities	0,007	2,78	0,012	3,76	0,000	-0,01	0,014	2,83	0,005	1,69	0,016	4,48	0,000	-0,13	0,008	2,83	0,004	0,76
Local Innovation Rate	-0,006	-0,38	0,01	0,48	-0,03	-1,19	0,003	0,10	-0,010	-0,58	-0,005	-0,23	-0,013	-0,67	0,002	0,10	-0,027	-0,98
							Single		**Branch**		**Single**		**Branch**		**Single**		**Branch**	
Manufactured Input Pool																		
Small							-0,028	-1,24	0,005	0,18	-0,026	-1,62	-0,001	-0,03	-0,018	-1,21	0,023	1,08
Large							-0,016	-0,97	0,017	0,96	-0,006	-0,30	0,019	1,08	-0,036	-1,28	0,005	0,22
Producers Services Pool																		
Small							0,017	0,84	0,013	0,50	0,022	1,51	0,026	1,17	0,022	1,74	0,029	1,39
Large							0,022	1,47	0,018	1,16	0,011	0,61	0,015	0,93	0,005	0,20	0,012	0,60
Specialized Labor Pool																		
Small							0,016	1,34	0,011	0,68	0,007	0,78	-0,008	-0,59	0,011	1,35	0,000	0,01
Large							0,014	1,46	0,009	0,82	0,029	2,48	0,013	1,21	0,030	1,79	0,019	1,34
Local Demand Pool																		
Small							-0,024	-1,53	-0,030	-1,47	-0,026	-2,15	-0,037	-2,05	-0,023	-2,12	-0,035	-2,23
Large							-0,020	-1,62	-0,025	-1,97	-0,012	-0,79	-0,023	-1,72	-0,007	-0,33	-0,019	-1,16
Research Universities																		
Small							0,014	2,83	0,003	0,44	0,016	4,41	0,010	1,63	0,011	3,60	-0,002	-0,32
Large							0,011	2,91	-0,001	-0,16	0,005	1,04	-0,002	-0,52	0,015	2,26	0,002	0,34
Local Innovation Rate																		
Small							0,009	0,31	-0,026	-0,67	0,002	0,07	-0,030	-0,90	0,010	0,53	-0,023	-0,77
Large							0,008	0,36	-0,026	-1,16	0,008	0,27	-0,024	-1,04	0,002	0,04	-0,031	-1,11

Note: Elasticities are evaluated at sample means.

Among the other proximity variables, the parameter on the specialized labor market variable is positive though not significant at conventional levels. Local industrial specialization in sectors using like labor does not appear to confer a productivity advantage to farm and garden machinery producers. There is also no evidence that proximity to supplies of manufactured inputs is associated with greater technological efficiency. That result would seem to suggest that a farm and garden machinery plant's proximity to producers of its key intermediate inputs does not imply a higher level of efficiency, as is predicted by conventional agglomeration theory (as well as more recent contributions in industrial organization and economic geography).

How do the findings differ when controls for plant size and type are introduced? Perhaps the most significant trend in Table 11.4 is that across all three size definitions, economies associated with proximity to producer services are strongest for large, single establishment plants. In fact, as the definition of the small plant changes from the first to third employment quartile, the magnitude of the economy with respect to producer services increases for the large establishments. That may mean that access to producer services is most important for larger manufacturers that, by definition, demand more services. The service needs of smaller manufacturers may not be significant enough to make proximity to producer services suppliers a critical determinant of efficiency. In other words, perhaps the smaller manufacturers are not necessarily outsourcing services to a greater degree than their larger counterparts, as might be predicted by the flexible specialization literature. In the case of the local rate of innovation, large plants under all three definitions derive the greatest productivity advantage; indeed, there is also weak evidence of a difference between branch and single establishments, with the calculated elasticities higher for the former. Therefore, for SIC 352, the results are not uniform in one direction: small, single establishments do not necessarily benefit to a greater degree from proximity to services, inputs, and spillovers.

11.3.2 Measuring and Controlling Devices

The importance of some types of externalities and spillovers might be expected to be stronger for the SIC 382 sample than SIC 352. The ratio of capital to labor for the SIC 382 sample suggest more labor-intensive manufacturing that is perhaps characterized by more customized or batch mode production. It is the customized manufacturer that is often hypothesized to depend much more strongly on spatial externalities and spillovers than the standardized producer that internalizes many important business functions. The measuring and controlling devices sector is also made up of producers that supply components to several highly technology-oriented sectors. The flexible specialization literature argues that rapid shifts in demand and markets force producers to depend to a greater degree on outsourcing for goods and services in order to maintain flexibility, and, therefore, that external economies are becoming much more prevalent in industries operating under such demand conditions. Manufacturers in SIC 382 supply industries such as automobiles, military weapons, guidance systems, and medical equipment that are continually innovating and introducing new products. The sector would appear to face, certainly more than the farm and garden machinery industry, the types of uncertain demand pressures described by Scott (1988) and others.

The results in Tables 11.3 and 11.5 suggest that the measuring and controlling devices plant located in a competitively organized region with a highly educated labor force, well-developed producer services, universities conducting basic research and development, and labor markets specialized in related industries is more productive, all else equal, than plants located in areas with a less favorable array of characteristics. The rate of local innovation as proxied by per capita patent grants does not influence efficiency in overall terms, although it does appear to positively augment some factors and negatively augment others. At the same time, proximity to sources of final market demand appears to negatively affect productivity; on average, the coefficient for the demand variable indicates a reduction in output of 2.4 percent for each doubling of the demand pool, holding input levels constant. Although there is a greater variety of positive spatial economies for SIC 382, the magnitudes of the individual effects tend to be smaller than those for SIC 352. For example, a doubling of the local producer services pool index is associated with an increase in output of 1.9 percent, compared with 3.5 percent for SIC 352. Proximity to producer services yields the largest effect in percentage terms for the measuring and controlling devices sector, followed by specialization in the labor market and proximity to research universities.

Finally, consider more closely the efficiency effect with respect to proximity to R&D universities. Results from the model indicate an increase in output of less than 1 percent with each doubling of the university R&D index. Although the impact appears small, the statistical evidence with regard to this variable is among the strongest of any spatial factor. The finding, coupled with the efficiency effect with respect to local educational attainment, may provide some information about the nature of the impact of universities on regional economies. Given the high technology nature of the measuring and controlling devices sector, scientific and engineering occupations are more common among the top fifteen occupations in this sector than in the farm and garden machinery industry. Perhaps some of the effect of the educational attainment variable may be attributed to the role of R&D universities, where most basic training of scientists and engineers takes place. On the other hand, the low magnitude of the weighted R&D variable would suggest that the university's most important role, at least for this sector, may be an educational one. That is, effectively training the next generation of workers may yield greater gains for the regional economy compared to conducting R&D that eventually leads to improved productivity and manufacturing performance.

Overall, the results suggest spatial externalities may differ by plant type and size primarily only with respect to one variable: proximity to university R&D. The elasticities of output with respect to university R&D reported in Table 11.5 are uniformly higher for small, single establishments under all three plant size definitions. For example, when small plants are defined as those with fewer than 15 employees, a doubling of the university R&D index increases output for smaller producers by 1.4 percent, holding all conventional input levels constant. By comparison, the increase for the larger plants under the same scenario is only 0.5

percent.[6] The elasticities for branch plants, particularly large branch plants, are generally near zero and statistically insignificant. There is also some evidence that diseconomies associated with proximity to sources of final market demand are strongest for smaller establishments, although the results vary by the definition of the *SMALL* dummy variable.

Finally, note that the models are estimated with spatial variables defined under the default distance decay for the study ($\alpha = 0.75$ in equation 11.3). Spatial effects may arguably manifest themselves at different spatial scales. It is possible that the plant type/size results might vary under alternative distance decay profiles; neglecting such variation could lead to false findings regarding type and size differences for specific spatial effects. Yet based on an analysis of the influence of alternative decays (πreported in Feser 1997), there appears to be enough consistency in the basic results across different decays, particularly in qualitative terms, to draw reasonable general conclusions about the direction and strength of any employment size/type effects.

11.4 Summary and Implications

A number of theories of regional growth and change suggest that small to medium sized firms may depend on local externalities to remain competitive with larger producers. Size of firms, in most cases, is defined in employment terms. The hypothesis implies that smaller enterprises garner a more significant relative advantage from proximity to other producers than larger firms. In a micro-level test for two U.S. manufacturing sectors, this study found only weak evidence of such a dynamic, and then in the case of only one particular type of externality (knowledge spillovers).

References

Carlsson, B., 1996, "Small Business, Flexible Technology and Industrial Dynamics", in P.H. Admiraal (ed.), *Small Business in the Modern Economy*, Blackwell, Oxford.

Chinitz, B., 1961, "Contrasts in Agglomeration: New York and Pittsburgh", *American Economic Review,* 51:279-289.

Ettlinger, N., 1997, "An Assessment of the Small-Firm Debate in the United States", *Environment and Planning A*, pp. 419-442.

Feser, E.J., 1998a, "Enterprises, External Economies, and Economic Development", *Journal of Planning Literature,* 12:283-302.

Feser, E.J., 1997, *The Influence of Business Externalities and Spillovers on*

[6] The results refer to the model with only size effects (the *SMALL* dummy variable) included.

Manufacturing Performance, Unpublished PhD dissertation, University of North Carolina at Chapel Hill.

Harrison, B., 1994, *Lean and Mean*, Free Press, New York.

Humphrey, J., 1995, "Introduction", *World Development,* 23:1-7.

Kim, H.Y., 1992, "The Translog Production Function and Variable Returns to Scale", *Review of Economics and Statistics,* 74:546-551.

Marshall, A., 1961, *Principles of Economics: An Introductory Volume,* 9th edition, Macmillan, London.

Nooteboom, B., 1993, "Firm Size Effects on Transaction Costs", *Small Business Economics,* 5:283-295.

Oughton, C. and G. Whittam, 1997, "Competition and Cooperation in the Small Firm Sector", *Scottish Journal of Political Economy,* 44:1-30.

Pratten, C., 1991, *The Competitiveness of Small Firms*, Cambridge University Press, Cambridge.

Scott, A.J., 1988. *Metropolis: From the Division of Labor to Urban Form,* University of California Press, Berkeley.

Stigler, G. J., 1951, "The Division of Labour is Limited by the Extent of the Market", *Journal of Political Economy,* 59:185-193.

Sweeney, S. H. and E.J. Feser. 1998, "Plant Size and Clustering of Manufacturing Activity" *Geographical Analysis,* 30:45-64.

Part IV

Functional Regions, Clustering and Local Economic Development

12 The Learning Region and Territorial Production Systems

Denis Maillat and Leïla Kebir
Institute for Regional and Econonic Research, University of Neuchâtel,
Pierre-à-Mazel 7, 2000 Neuchâtel, Switzerland

12.1 Globalisation, Territorialisation and Learning Economy

With the growing importance of non-material resources in the dynamics of development today, more emphasis is placed on constructed resources (skills, know-how, qualifications, but also methods of doing and acting) than on natural resources. Indeed, within the framework of globalisation, nations and firms are obliged to base their competitive advantage rather on their non-material resources and on the abilities of the actors to co-operate and to develop synergies amongst each other (untraded interdependencies). As with these constructed resources, this knowledge is not learnt once and for all, the various actors (firms, organisations, regions) must attend to keeping them up to date, to reproduce and to process them. This is why learning processes become so important, since it is thanks to them that new knowledge emerges and existing knowledge is transmitted.

Thus, maintaining competitive advantage follows from the creation of non-material resources which are constructed through learning processes. Here, we agree with the arguments of Lundvall and Johnson (1994, p.24) who maintain that "knowledge is the fundamental resource in our contemporary economy and learning is the most important process". This statement stems from three important and interrelated phenomena which characterise the contemporary economy:

> the first relates to the development of information, computer and telecommunications technologies (ICT); the second to the movement towards flexible specialisation and the third to changes in the process of innovation

which require more and more interactions between the actors (Lundvall and Johnson, 1994, p.25; Gregersen and Johnson, 1997, p.479). The ability to react swiftly, to have at one's disposal the right resources at the right time, and to find competent partners fast is essential. In such a context, at the level of production and of innovation "knowing how to communicate, to co-operate and to interact becomes much more important than before" (Lundvall and Johnson, 1994, p.25), because it is a question of constantly creating new knowledge or of developing new skills in order to maintain or expand competitive advantages.

The learning economy provides the theoretical bases for understanding such a context in which change takes priority over the allocation of resources and in which knowledge and learning processes occupy a central position. In fact, the learning economy is a dynamic concept which highlights the capability to learn and to expand the knowledge base. It refers not only to the importance of the science and technology systems, universities, research organisations, in-house R&D departments and so – but also to the learning implications of the economic structure, the organisational forms and institutional set-up (Lundvall and Johnson, 1994, p.26).

Thus, Lundvall and Johnson (1994) suggest an approach to the economy based on learning processes and change rather than on allocation mechanisms leading to an equilibrium.

> The learning economy is, and has to be, a mixed economy in a very fundamental sense. In such economies, there are important roles for the public sector and for a different kind of policy. But its very basic institutions its firms and markets are mixed. Its markets are embedded in habits, rules and norms and are organised for communication and exchange of qualitative, non-price type of information. Its firms show a diversity of different organisational forms which influence communication between different persons and departments. Its continually changing institutional set-up forms the environment for interactive learning-by-producing and learning-by-searching processes which are the main mechanisms for recombining and introducing new knowledge in the economy (p.41).

It follows that

> it is the ability of firms as well as regions and countries to learn, change and adapt rather than their allocative efficiency which determines their long run performance (Maskell and Malmberg, 1995, p.3).

From this point of view, the economy is in a permanent learning situation (Perrin, 1995; Morgan, 1995; Simmie, 1997).

The learning concept squarely straddles evolutionism (given the cumulative and on-going nature of learning) on the one hand and institutionalism (given the role attributed to the institutional framework) on the other hand. The paradigm of the learning economy is in contradiction with the idea of perfect rationality on which orthodox economic science is built. In the latter, the relationship is defined by the traded object and behaviour is defined strictly (rules of trading in the market model of the neo-classical theory). In the case of the learning paradigm, the purpose of the relationship is to find a solution together to a problem in common; this is interactive co-operation. Unfortunately, interactive co-operation is not self-evident. A propitious environment (institutional set-up) is required in which it and the learning processes can emerge. In particular

> for learning to occur, there must be conversations between and among partners. But since working conversations that create new knowledge can only emerge where there is trust, trust and confidence proved to be essential inputs (Paquet, 1994, p.3).

In fact, these conditions can usually be found in territorial production systems which offer proximity conditions which are propitious for co-operation between different types of private and public actors. This is even truer

> when the technology changes rapidly and radically – when a new technological paradigm develops - the need for proximity in terms of geography and culture becomes even more important. A new technological paradigm will imply that established norms and standards become obsolete and that old codes of information cannot transmit the characteristics of innovative activities. In the absence of generally accepted standards and codes able to transmit information, face-to-face contact and a common culture background might become of decisive importance for the information exchange (Lundvall, 1988, p.355).

The problem is that not all production systems are territorial. Therefore, certain regions cannot be embedded in the economic model described above. Are they condemned to not develop ? The objective of the following text is to show, with the help of the concept of the learning region, the processes which can transform a region to become and remain "learning". In a first step we will consider the concept of the learning region, a relatively new concept, but a promising one, mainly because it may well steer regional policy. We will then show that many regions do not possess the characteristics of the learning region and that one must take care when using this concept. Lastly, we will demonstrate that to be a learning region it does not suffice to promote education or research institutions. New resources must be activated at the territorial level. This activation, however, depends on the nature of the territorial production systems.

12.2 The Learning Region

If one accepts that the modern, post-fordist economy "is entering a new age of knowledge creation and continuous learning and that the territorial context plays an important part, it follows that regions are becoming focal points for knowledge creation and learning" (Florida, 1995, p.528).

In fact,

> the shift to knowledge-intensive capitalism goes beyond the particular business and management strategies of individual firms. It involves the development of new inputs and a broader infrastructure at the regional level on which individual firms and production complexes of firms can draw. The nature of this economic transformation makes regions key economic units in the global economy (p.531).

Thus, even if one may have believed at times that globalisation signified the end of regions, one has to admit that they have a determining role to play. Indeed, globalisation does not entail the disappearance of territories, but quite on the contrary the emergence of new forms of territorialisation, especially when one considers that certain types of knowledge and information are exchanged more easily and more profitably face-to-face than through long distance relationships.

> There is some irony in the fact that the information and communication technologies driving the new paradigm simultaneously increase the flexibility of firms in co-ordinating production on a global basis, yet accentuate the importance of regional concentrations of related firms and industries (Wolfe, 1997, p.9).

This phenomenon owes its existence mainly to the growing importance of constructed resources (know-how, skills, qualifications, manner of acting). In fact, what globalisation causes is a multiplication of territorial production systems which compete with each other. Competition between these systems does not depend only on the cost of the factors of production, but on a complex set of factors which stimulate and generate innovation permanently. To participate in this new landscape, regions must become learning regions, i.e.: they must adopt the "principles of knowledge creation and continuous learning" (Florida, p.532).

Not many definitions of this type of region exist. To home in on the issue, one may refer to Walther (1998, p.4) who, writing about Germany, states

> ...they are debating how the partners in training and qualifications in local communication areas and in the regions (the Länder) can best form cooperating networks to deal with the complexity and rapidity of industrial change. Their idea of a learning region is not just one of getting together all the high-level expertise in a given place. Their concern is to build the maximum number of links between those who are active in the local training market, whoever they are and whatever their status.
>
> Their focus is on the employment and innovation potential of small firms; and their effect is to establish partnerships between small firms, training providers, social partners, and public authorities. The goal is to establish at one and at the same time an approach to improving skill and qualifications, and an effort to identify new jobs.
>
> At the same time, the ultimate purpose of these learning regions is to consolidate and augment the local economic infrastructure. In fact, the concept of learning region finally makes sense in the desire of companies themselves to develop a capacity for training and the means to deliver it. Thus, they become true learning enterprises (pp.4-5).

Similar arguments can be found in the "learning region" project of IRES (1998).

> Our project aims, by means of an integrated action towards different subjects that operate in a local area characterised by a growth deficit, to develop an endogenous capacity of managing economic and social innovation process. This will also be achieved through the transfer of tasks and skill developed in other areas characterised by consolidated growth. In this way, we will encourage the creation of a network for the diffusion of competencies, technological know-how and the ability of managing local policies, concerning technological modernisation, work management and organisation, development of local enterprise

> schemes, prevention against the risk that the employed labour force's competencies become obsolete (p.1).

In these two examples, learning is understood as the improvement of knowledge or of qualifications and the learning region is one that implements an education and training policy involving various actors. In this case, the learning concept is merely a new cloak for the development (sale) of an education and training policy at the regional level. Actually, the concept of the learning region needs to be more than this if one wishes to hitch it to the concept of the learning economy. It does not suffice to merely shift the responsibility for training and education policies or for innovation policy from the national to the local level to be able to speak of a learning region. It means above all to show how a region can fit into globalisation by setting up various learning processes. This is the approach taken by Asheim (1995) who describes the learning region as the achievement of transformation of an industrial district. Such a transformation can be achieved thanks to the modernisation of the economic structure of industrial districts which makes it possible to strengthen their competitive advantages.

> Such learning regions will be able to avoid a lock-in of development, caused by localised path-dependency, through the formation of dynamic flexible organisation both at an intra and inter-firms level. In a learning economy, the competitive advantage of firms or regions is based on innovations and innovation process. In this way, a learning region would be in a position of transcending the contradiction between functional and territorial integration, which in the past made the industrial districts so successful, but at the time so vulnerable to change in the global capitalist economy (p.18).

From this perspective, the learning region is the achievement of a process which is neither specified nor described. Moreover, the concept is applied to a special case of a territorial production system: the industrial district. To our thinking, one must be more general in order to give the concept of the learning region more body.

This is what Florida (1995) attempts to do. According to him, the learning regions

> function as collectors and repositories of knowledge and ideas, and provide an underlying environment or infrastructure which facilitate the flow of knowledge, ideas and learning. Learning regions are increasingly important sources of innovation and economic growth, and are vehicles for globalisation (p.528).

As far as we know this is the most adroit definition of the learning region.
Florida specifies his concept of the learning region by offering a comparison with "mass production regions". As far as he is concerned,

> learning regions provide the crucial inputs required for knowledge-intensive economic organisation to flourish: a manufacturing infrastructure of interconnected vendors and suppliers; a human infrastructure that can produce knowledge workers, facilitates the development of team orientation, and which is organised around long-life learning; a physical and communication infrastructure which facilitates and supports constant sharing of information, elec-

tronic exchange of data and information, just-in-time delivery of goods and services, and integration into the global economy; and capital allocation and industrial governance systems attuned to the needs of knowledge-intensive organisations (p.534).

This is a clear, but descriptive definition which we will try to improve on by not merely enumerating the characteristics of a learning region's functioning, but by highlighting the processes which turn a region into a learning region. One must give the learning concept dynamic substance. Indeed, one may not forget that learning is a process of acquisition and transformation of knowledge which allows for permanent adaptation in the face of the uncertainty of the environment.

This being so, we will characterise the learning region as a dynamic and evolving region. It is dynamic because each actor, be it an individual, a firm, institution or a network, is in continuous interaction with his environment (directly or indirectly). It is evolving because each actor who is part of it is an "apprentice" in an experimental situation.

It is characterised by the combination of three types of processes: a process of territorial implementation of innovation, a process of territorialisation of firms, and complex learning processes. These processes do not only occur within the given region (the "apprentice" actors interact with each other) but also through the relationships the region fosters with the rest of the world (the actors interact also with agents or actors outside the given region).

12.2.1 The Learning Region: A Territory of Innovation

As we have seen, the concept of the learning region is part of the paradigm of the learning economy which considers the economy as a process of communication and of cumulative causality, and not as a system of equilibrium. This being so, the learning region is composed of a production system which is characterised by its on-going ability to adjust and innovate.

The authors who are interested in learning regions start out with idea that innovation is linked to territories in which production systems are located. In general, innovation is understood in the widest sense, i.e. any "profitable change" regardless of whether it is technological (a new product or an improved existing product), organisational (management) or institutional (change of rules, change of habits). In the context of globalisation and of competition between territorial systems it is accepted that innovation has become a permanent necessity and that therefore the actors cannot act alone. They must co-operate in order to benefit from their complementary competencies.

In the current post-fordist stage, the notion of territory has made a come-back because the concentration of firms on a given territory is seen as the context which is most propitious for the emergence of co-operation during the process of innovation. Hence, proximity is of the essence as it favours research and encounters between reliable partners whose propensity for co-operation is known. It limits the errors in the choice of partners and it reduces uncertainty. It procures considerable time gains and diminishes transaction costs. It is interesting to note that because the process of innovation is discontinuous there is a need for frequent exchange of specific information between the actors. Thus, the phenomena of co-operation, and to a certain extent, of non-appropriation are the rule. In the learning region, the economic model is no longer focused on the allocation of resources and exchanges, it is a system which integrates learning and change. One could even say that the externalities linked to a territory no longer suffice to explain the significance of proximity. In fact, the externalities are the result of involuntary actions. What counts is the partners' voluntary, deliberate will to co-operate, to achieve projects together (Schmitz, 1997). The actors interact with each other by exchanging specific information required for the smooth functioning of the innovation process and by working together on projects in common.

This explains why learning regions must be organised to ensure exchange and contacts between the various partners (firms, research centres, interface organisations, etc.). Thus, the institutional framework (rules and regulations, habits, trust) play an essential part. Indeed, it must be appropriate for the implementation of the interaction process between the actors.

Lastly, the conditions which favour a climate that is propitious for interaction cannot be created in a day. In fact, the learning processes are based on the knowledge that the partners acquire from each other and on the trust they are willing to show each other. It takes time for such behaviour to establish itself and to be mobilised within the framework of innovation networks.

12.2.2 The Learning Region: Space Territorialisation of Firms

Territorial production systems possess the characteristics which favour innovation and learning. In fact, globalisation, deregulation of the world economy and the diminished power of states have forced firms to organise themselves and to co-operate at the territorial level in order to reduce the degree of uncertainty arising from the state of insecurity and from the growing complexity of the economic environment (Asheim, 1995, p.10). One can thus argue in the present context that

> domestic co-operation rather than domestic competition is the key determinant of global competitive advantage. For domestic industry to attain and sustain global competitive advantage requires continuous innovation, which in turn requires domestic co-operation. Domestic rivalry is an important determinant of enterprise strategies. But the substance of these competitive strategies, specifically whether they entail continuous innovation or cut-throat price-cutting depends on how and to what extent the enterprises in an industry co-operate with one another (Lazonick, 1993, p.4).

Two models of inter-firm co-operation emerge: the global, functionally integrated production systems dominated by large firms and the territorial production systems composed of SMEs. In the first case, co-ordination and co-operation are explicit; they depend on a hierarchy. In the second case, co-ordination and co-operation are implicit; they depend on the milieu (Maillat, 1998). It is obvious for the SMEs that proximity appears to be the most important element in setting up co-operation networks. This need for interactivity at the territorial level, however, appears to be becoming increasingly important for the large firms which find opportunities for co-operation in the territorial systems which enhance their flexibility and brings them in contact with new opportunities. It is interesting to note that today large companies, in their constant wish to adapt, no longer rely on themselves only, but also seek new opportunities by integrating themselves in the horizontally organised territorial production systems.

> The movement away from tall hierarchies with vertical flows of information towards more flat organisations with horizontal flows of information is one aspect of the learning economy (Lundvall and Johnson, 1994, p.39).

The evolution of firms towards horizontal exchange and relations re-enforces the role played by proximity. The latter is above all "an enabling factor in stimulating inter-firm learning networks involving long term commitment" (Asheim, 1995, p.15). Indeed, the authors who deal with learning regions often refer to industrial districts when emphasising the importance of the part played by proximity and by horizontal-type organisation.

It goes without saying that one of the objectives of learning regions or of those wishing to become learning regions is to embed the branches of big companies in their production systems, to territorialise them, so that they participate in the development of horizontal co-operation. What counts is to make them so much a part of the territorial system that their active participation in it becomes a contribution towards its dynamism. This is the argument put forward by Pratt (1997) for whom the learning region is

> a particular structured combination of institutions strategically focused on technological support, learning and economic development that may be able to embed branch plants in the regional economy, and hence cause firms to upgrade in situ rather than to relocate away from the region (p.128).

12.2.3 The Learning Region: A Territory for Learning

The learning regions are in constant change thanks to the active part played by the learning processes. Four types of processes are essential: interactive learning, organisational learning, institutional learning and learning by learning.

Interactive learning. Interactive learning is the way in which interaction between actors is established when productive activities are co-ordinated or when an innovation process is set up. To be more precise, by interactive learning we mean the process of interaction which integrates and pools the knowledge required for the production systems to run smoothly, the knowledge possessed individually by all of the actors (individuals, firms, institutions). These interactions occur between actors when productive activities are co-ordinated or when a process of innovation is implemented (through the sharing of experience, the passing on of information, etc.).

In the learning economy, interaction between actors is the rule. Indeed, in this type of economy,

> the organisational modes of firms are increasingly chosen in order to enhance learning capabilities: networking with other firms, horizontal communication patterns and frequent movements of people between parts and department (Lundvall and Johnson, 1994, p.26).

In order to function effectively the actors therefore have to learn interaction.

Even though interaction may also occur vertically (users/producers) in the learning regions, the authors mainly refer to horizontal relations. Two types of territorially-based networks may result from such interaction: the "trade networks" geared mainly to the trading of market goods and services and the "knowledge networks" "where focus is on the flow of information and exchange of knowledge irrespective of its connection to the flow of goods" (Gelsing 1992, p.117). The "trade networks" remain strictly linked to the production processes whereas the "knowledge networks" make it possible to go beyond these processes and to thus create a context which is propitious for innovation for they carry information and specific knowledge which do not emerge from trading (untraded interdependencies).

These "knowledge networks" are important for the setting up and the smooth functioning of the innovation networks which in turn are based on the actors' ability to interact (Maillat, Crevoisier, Lecoq, 1995). In fact, innovation results from an interactive learning process

> between firms and the basic science infrastructure, between the different functions within the firm, between producers and users at the inter-firm level and between firms and their wider institutional milieu (Morgan 1995, p.4).

The role of these territorial networks may differ depending on the nature of innovation. Thus, Lorenzen (1996) feels that the

> vertical user-producers relations with exchange of information provide the essential vehicle for incremental product innovation.... horizontal relations can be seen as promoting process innovations through inter-firm provision of information about technological opportunities and because other firms interested in adapting successful process innovations can increase 'the market' and thus the scope for these innovations (p.8).

It is also through the processes generated by interactive learning that knowledge is exchanged. Here, the distinction must be made between two types of knowledge: explicit (codified) knowledge and tacit knowledge. Explicit knowledge can be easily formalised, written down on paper or on computer. It can easily be passed on from one person to another (Journé, 1996, p.13). Tacit knowledge is bound up with each individual's know-how. In general, transmitting such knowledge requires a prolonged contact between the holder of the know-how and the apprentices (Journé, 1996, p.13). Depending on whether knowledge is explicit or tacit, the mode of exchange will differ (see Table 12.1). In fact, explicit (codified) knowledge can be exchanged like a commodity in the classical sense or it can be copied. It is the prevalent conventions which determine the degree of accessibility to it. On the other hand, tacit knowledge is difficult to exchange and less accessible (Lorenzen, 1996). In this case the market is not an adequate trading instrument, because we here find ourselves in the non-market, non-trading area. Thus, it is by the interactive learning processes that knowledge is exchanged, renewed and developed.

It is important, however, to emphasise that the more the knowledge used in a territorial production system is tacit, the more particular it is to the system in question, and the less easily it can be transmitted to other systems. In order to benefit from it, and to make use of it, one must belong to the territorial system possessing it, because the transmission of the knowledge will depend on the interaction the actors will or will not have learned to practise.

Table 12.1 Types of knowledge and learning by interacting

Knowledge type	Intra-firm learning	Pool	Learning by interaction	Channel	Code keys
Explicit	-education -formal training	formalising -records, -procedures	-transmitting information signals	-communication technologies	-formal qualifications
'Less' explicit	-apprenticeship -trial-and-error	following -routines -scripts	-sharing experience	-discussions -conferences -visits	-skills/ experience
Tacit	socialisation	understanding –norms -mental models	-communicating understanding -sharing norms	-employee mobility -informal personal relations	-culture

Source: Lorenzen, (1996, 9)

Institutional learning. As in the learning economy, the formal institutions (governmental organisations, development agencies, associations, laws, etc.) and the informal ones (values, routines, codes of conduct, customs, trust, etc.) play an essential part in the functioning of learning regions. They allow for a certain predictability of the actors' behaviour and hence diminish uncertainty. Because of their stability they shape an environment which is propitious to learning:

> Institutions reduce uncertainties, co-ordinate the use of knowledge, mediate conflicts and provide incentives systems. By serving these functions institutions provide the stability necessary for change (Johnson, 1992, p.26).

But institutions also introduce a certain inertia into the system. In fact, institutions which at a certain period of time stimulate innovation may, as time goes by, no longer be adequate. Depending on the circumstances and the problems to be resolved they must therefore be renewed or transformed in order to be adapted to the changes in the environment. In order to do so, the actors in the regions must establish institutional learning which is not aimed at adding institutions to the existing institutional base but rather at eliminating obsolete institutions, at transforming the inadequate ones and at creating new ones if necessary. Thus, it is indispensable, in view of the usual inertia of the existing institutions, that special attention be given to the processes which achieve the transformation of the inadequate institutions.

Institutional learning means the capability of institutions to question themselves, to adapt their structures and their objectives, to regenerate themselves in line with the changes of the environment. Johnson (1992) specifies that

> institutions have a strong impact on technological change. However partly as a consequence of the technical change they shape, a tension between technology and institutions and a pressure for institutional change is often provoked. At the same time institutions are normally quite rigid and do not change easily. The capability of national economies to cope with this problem, i.e. to learn about, adapt and change their institutional frameworks to engage in 'institutional learning' is important for the development of their international competitiveness (p.23).

Whilst institutional learning is important at the national level, it is without doubt even more important at the regional level. In fact, at the regional levels, institutions, namely the formal institutions, are less numerous and less diversified than at the national level. Thus, it is essential that they be dynamic and constantly evaluated, even incited to change and to adapt themselves. As to the informal institutions, namely the habit to co-operate, to trust one another, to renew tacit knowledge, they may well disappear and seriously endanger the actors' capability to interact. Such a situation may occur in a territorial production system when opportunistic behaviour takes priority over the principles of reciprocity.

Organisational learning. Organisational learning can be defined as

> the capacity of an organisation to learn how to do what it does, where what it learns is possessed not by individual members of the organisation but by the aggregate itself. That is, when a group acquires the know-how associated with its ability to carry out its collective activities, that constitutes organisational learning (Scott et al., 1995, p.438).

It allows the actors of the organisation to better co-ordinate their action. It is in some way the process by which

> the newcomer in an organisation will understand, assimilate the role assigned to him and reach this predictability of behaviour which is indispensable to organisational co-ordination (Midler, 1934, p.340).

This kind of learning

> occurs when individuals find disparities between or, in contrast, confirmation of their observations and the "theories-in-use", i.e.: the different ways of

> understanding, thinking or acting taught or produced in the firm (Thuderoz, 1997, p.79).

For this to happen each individual's experience and discoveries must be encoded in the common language of the organisation failing which only the individual and not the entire organisation will learn.

> What is at stake in organisational learning is not the individuals' private knowledge, but the "collectivised" knowledge which they mobilise in their action within the organisation (Midler, 1994, p.342).

Thus, organisational learning is indispensable to the functioning of learning regions. This learning process is at the basis of learning organisations, i.e.: of organisations

> that promote the learning of all its members and have the capacity of continuously transforming itself by rapidly adapting the changing environments by adopting and developing innovations (Pedler et al., 1991; Weinstein 1992, in Asheim 1995).

This ability to change arises largely from the fact that firms have understood the importance of developing the process of horizontal co-operation which makes it possible to enhance the whole system's learning capacities. In fact, a region cannot be a learning region if its organisations, or more generally, its actors, are not "learning".

In fact, organisational learning does not only affect what happens in an organisation, but also inter-organisational relations. This is important at the level of a territorial production system because the actors must learn to fix and to abide by the rules of co-operation and of competition which govern both trading and non-trading exchanges. They must, however, also learn to adapt these rules according to the evolution of technologies, of markets and of the changing characteristics of production processes or according to the evolution of the functioning of the production system itself (flexible production, alternation between the phenomena of internalisation and externalisation). In such situations, organisational learning is indispensable in order to maintain the territorial production system's coherence and its capacity to transform itself.

Learning by learning. Learning by learning is the process of improving skills linked to learning (Le Bas and Zuskovitch, 1993, p.158). There is in some way learning within learning. Indeed, the more one learns, the more one hones one's capacities for assimilating the techniques which facilitate learning.

The easier the latter becomes, the more the actors wish to learn. This type of learning actually acts like the motor of the system: learning stimulates learning which incites the actors to surpass themselves.

> When economies learn how to learn the process tends to accelerate... Some aspects of this institutional development may be described as evolution of a "learning culture" in which people regard long formal education, repeated re-

> education and retraining, and even long-life education, as a necessary and normal aspect of economic life (Gregersen and Johnson, 1997, p.481).

The result is the establishment of cumulative processes which speed up learning, but also the creation of irreversible and perpetuating situations. For the learning region this is essential because irreversible and perpetuating situations are localised and bind the actors to the territory.

The counterpart of learning by learning is forgetting. In a way, forgetting is an integral part of learning in that it also brings about change in the store of knowledge. In a general way, every scientific or technological change implies the forgetting of knowledge and know-how. The process of forgetting may, ex ante, be divided into two groups: creative forgetting and just forgetting.

Just forgetting presupposes the complete destruction of knowledge, of know-how, both of which do not in any way reappear in a different activity.

Creative forgetting is necessary before the innovation can be disseminated in the economy.

> Old habits of thought, routines and patterns of co-operation, with as well as between firms, have to be changed before technical change can begin to move ahead along new trajectories (Johnson, 1992, p.29).

Creative forgetting is simply the momentary oblivion of knowledge, of know-how before it is re-channelled in a new shape, in a new field of activity.

12.3 The Learning Process and Different Types of Regions

As we have seen, the learning region is a territory characterised by territorial innovation processes, by territorialisation of firms and of learning. It is a dynamic and evolutive region. It is a dynamic region inasmuch as its actors know how to interact. They know how to work with each other, to co-operate, to transmit knowledge, to work out common projects. Thus, interactive learning gives the region its dynamic dimension. It is evolutive inasmuch as its actors are in a permanent state of learning whether they are individuals, firms or institutions. Thus, the individuals who make up the region augment their knowledge and their skills within the framework of their activities. As they are individuals motivated by the constant wish to learn and to improve their knowledge (learning by learning), they deliberately place themselves in this situation at all times (by doing, by using, by searching). Thus, they never cease to learn and to evolve. Firms, made up of learning actors, themselves follow a learning process, i.e.: a process which goes beyond the individual processes. In this way, they adjust as frequently to the changes in the environment as is necessary. They have learned to learn. They are constantly seeking novelties regarding products, production processes, markets and technologies alike. In the same way, institutions follow a process of institutional learning and never stop evolving in order to lay down a framework and rules which favour action by the various actors. The individuals who make up or control the institutions know how to prevent them from holding on to their natural inertia. Naturally, all regions cannot be both dynamic and evolutive all the time. Nonetheless, these two dimensions make it possible to determine a typology depending on the different types of learning which characterise a learning region.

The typology is determined by retaining the dynamic dimension and the evolutive dimension (see Figure 12.1). The dynamic dimension is characterised by interactive learning; the evolutive dimension is characterised by organisational and institutional learning. To depict this one may compare the presence or absence of the respective learning modes in the region. Thus, one obtains four cases:

Case 1: ***Region without innovation but with interaction between the actors.*** This is a region in which the actors fortified by the past no longer obey the logic of learning. They exploit what has been handed down by the past by producing a successful commodity or service. Depending on the production's type of logical sequence, they are forced to trade, but only with regard to known transactions. The purpose of these interactions is to neither exchange skills nor to develop new ones. In these conditions, there is no will to innovate. There are no evolutive processes. This is, for example, the case of a declining, industrial district. The actors are incapable of renewing themselves.

Case 2: ***Regions without innovation nor interaction between actors.*** In this case, the actors are not "apprentices"; they merely execute orders. There are no networks. In this type of context, the non-material resources of the region are not renewed, therefore it is difficult for innovation to emerge. Here one recognises the regions used by fordist firms which practise spatial division of labour or the regions which as a result of a crisis have experienced the disappearance of their main activities.

Case 3: ***The innovating region with little or no interaction between the actors.*** The actors in the firms are still learning: they are evolving and innovating. They do not, however, interact with other firms. Thus, knowledge, know-how are neither transmitted nor integrated in the territorial production system. As a consequence, the actors do not contribute to the development of the region's non-material resources. If a firm or a research centre leaves the region, they take with them their know-how and their skills. This is for example the case in a region characterised by the presence of large, vertically structured firms or of branches of large firms which are able to innovate, but only internally.

Case 4: ***The innovative region with interaction between the actors.*** In this case, the actors are learning and interacting with each other. The resulting synergy strengthens the development of non-material resources and stimulates innovation. The actors thus increase their capacity to adapt to changes in the environment. This is the case of the learning region.

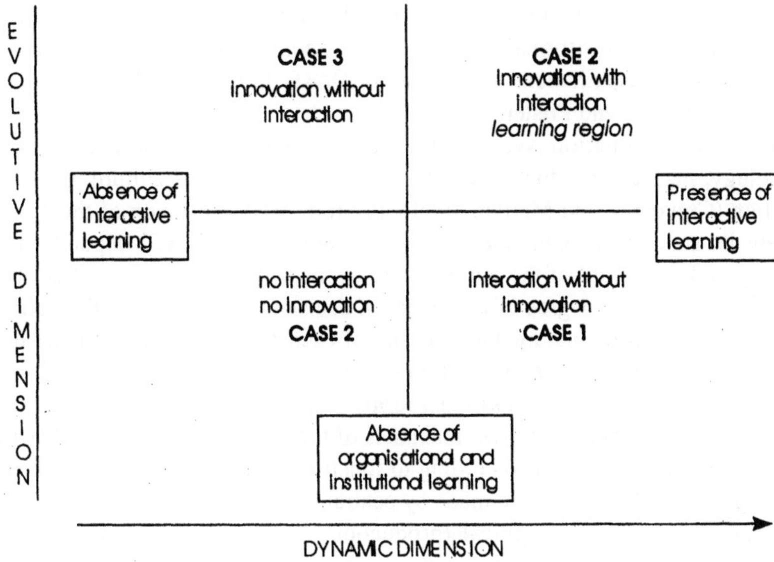

Figure 12.1 Typology of regions
Source: Maillat, Kebir, IRER

This typology clearly shows that all regions are not learning regions. The latter have very particular characteristics which cannot be imposed. One cannot decree a region to be a "learning" region; it has to evolve into one. What counts is to highlight the processes through which a region achieves this and which characteristics it has to acquire to succeed. Learning is not a factual state, but the result of a process. To qualify a region as a learning region is the same as saying that the actors of the system are engaged in learning processes which enable the development of knowledge, of know-how and of all the other skills required by innovation, namely the creation and dissemination of knowledge. It is also a question of showing interest in the nature of territorial systems which enables the development of such processes.

12.4 The Learning Region and Territorial Production Systems

12.4.1 The Importance of Territorial Production System

In order to contend with global competitiveness, productive organisations tend increasingly to become territorialised. This means that firms change from an organisation with hierarchic structures (with vertical exchange relations) to a "flatter" organisation (with vertical exchange relations between firms) (Lundvall and Johnson, 1994; Perrat, 1992). They then fit in the various exchange relations of the territorial production systems. We recall that these relations can be material, non-material, formal informal, trading or non-trading; they relate to the flow of goods and services, financial flows, labour flows, flows of technology and of knowledge (Nemeti and Pfister, 1994).

Territorial production systems thus create spaces for relations between technology, markets, productive capital, know-how, technical culture and representation (Crevoisier and Maillat, 1989). The territorial dimension of the system, namely its capacity to generate endogenous development, will depend on the intensity and the nature of the horizontal relations between the actors, their interdependence and their degree of autonomy in decision-making and in the defining of projects. So it is the nature and intensity of exchange relations which are decisive in qualifying a territorial production system.

The various, possible organisational forms of territorial production systems can be qualified by drawing from two main sets of logic: functional logic and territorial logic. Firms functioning according to functional logic have a vertical, hierarchic organisation (decisions are made by central management). Distribution of the various functions is geographical (production, marketing, sales, etc.) so as to diminish production costs (labour cost, transport costs, level of taxation, granted subsidies, etc.) Territorial location is only a back-up for these firms, they are not embedded in the territory concerned. Thus, the role of the territory is merely passive. On the other hand, territorial logic implies strong ties between the firms and the territorial location. The objective of territorial logic is to territorialise the firm, i.e.: to embed it in the territorial production system. In this case, firms are organised in horizontal networks, the milieu conducts the system (Maillat, 1998). They foster relations of co-operation/competition which generate synergies and complementarities which necessary for their functioning. The role of the territory is active, firms feel they belong to the territory. With the help of the characteristics of these two opposite sets of logic one can identify the various types of territorial production systems.

12.4.2 The Various Types of Territorial Production

Figure 12.2 illustrates the various types of territorial production systems based on two criteria. The first is related to the intensity of exchange relations between firms in the region (i.e.: at the territorial level). This criterion stands for the complementarities and the horizontal relations between the actors. The second criterion relates to the degree of integration in the added value chain. Are the activities which contribute to the production of a commodity or a service (research, development, production, marketing, selling) totally or partially integrated in the firm located in the region or are they split up (distributed) among many independent firms which are also located in the region. The typology is defined by contrasting on the one hand the presence or the absence of exchange relations taking place in the region with, on the other hand, the presence or the absence of internal integration (in the firm) in the region's added value chain. Four different types of territorial production systems can be established.

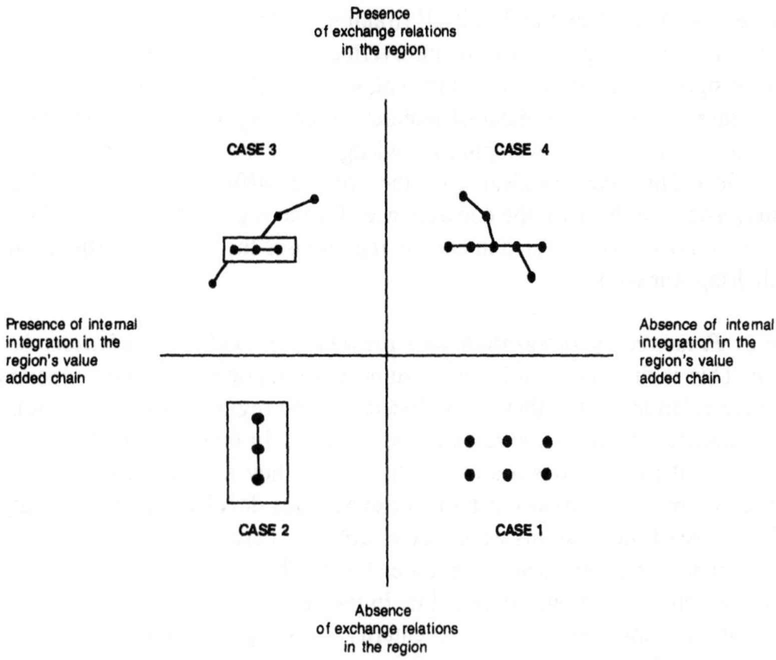

Figure 12.2 Typology of territorial production systems
Source: Maillat and Grosjean (1999)

Case 1: ***Absence of integration and absence of exchange relations in the region.*** The territorial production is made up of firms which act independently, i.e.: without establishing any links with each other in the region. There are no inter-firm relations. This is the typical case of regions in which large fordist firms have set up branches according to traditional localisation criteria (labour cost, access to infrastructure, to raw materials, to transport networks, etc.). In a given region, the branches concerned are dependent on a hierarchy located elsewhere (the parent company, for instance), namely for all decisions having to do with development or innovation. This kind of dependence does not favour endogenous development because it does not engender any interactive learning processes in the region, nor the development of resources specific to the territory. The latter serves as a passive support.

Case 2: ***Presence of integration and absence of exchange relations in the region.*** This is the case of the large firm, located in a given territory, the functions of which are entirely internalised. Everything is produced by the firm, from conception to packaging. Such a firm has practically no important relation with the other firms present in the region. Such a firm imposes itself on the territory and the region. It has a form of power over the region and because of it, it shapes it to its purpose, excluding competitors, as the case may arise, and developing those activities which are beneficial to itself (training centres, centres of recreation and leisure). These are typical cases of introverted or paternalistic firms. Even when there are signs of endogenous development, since the firm takes its decisions in the region, there is no development of interactive learning, nor of externalisation of knowledge, nor even of developing knowledge other than that which is necessary for the firm. This becomes clear when the firm has difficulties or when it decides to move away. In this case the consequences for the region are disastrous because no other activity is able to step in and the region loses the firm's specific resources which disappear with it.

Case 3: ***Presence of integration and presence of exchange relations in the region.*** This is the case of a large, more or less integrated firm which maintains exchange relations with other actors located in the region (suppliers, customers, sub-contractors, research centres, training centres). In contrast with the previous case, some of its activities are externalised, i.e.: they are executed by other independent firms. The effects in terms of exogenous development depend largely on the nature of the relations which are established between the firm and its partners, in other words, on whether the milieu has an effect.

Two different situations are possible. In the one situation the firm and its partners cooperate and there is complementarity, exchange of knowledge, of know-how and of technology. As a result of the generated synergies interdependencies grow between the various partners of the large firm. In actual fact, the partners of the large firm do not only depend on it. In this context, the large firm is territorialised, it is an important stimulus for the development of the region. It represents a significant development and it takes part in the renewal of the territory's specific resources. In the other situation, on the contrary, the firm's exchange relations are with dependent partners. The relations which evolve are of the kind between principal and sub-contractor. The exchanges are strictly of a trading nature. At best, there is some transfer of knowledge or of technology to the sub-contractors bound to the firm, but without any of the resulting synergy mentioned above because the sub-contractors have no other partners than the firm and merely execute orders.

Even though the firm cultivates relations with the territory, the threat of relocation away from the region remains. In this case, the consequences may be serious for the territorial production system: the territorial production system may become unhinged. Everything will depend on the partners' degree of dependence on the firm. Inasmuch as the firm is their customer and their supplier, the move away from the region may signify the end of their activities. This is why the greater the effect of the milieu is, the costlier it is for the firm to leave a territorial production system because it will lose the advantages supplied by the latter (trusted partners, synergies, specific resources, etc.).

Case 4: ***Absence of integration and presence of exchange relations in the region.*** In this case we are dealing with a territorial production system made up of numerous small specialised and independent firms belonging to a delimited or a partially delimited area of production. The rule in such a system is interaction between the various actors. Mechanisms of co-operation/competition have been established to ensure the system's co-ordination and coherence. The characteristic of these systems is the fact that co-ordination between the various stages of production is not organised according to the hierarchic model of the large firm, but results from a complex set of relations and rules which ensure the coherence and flexibility of the whole (Maillat, 1998). In fact, the organising principle is the milieu which acts like a cognitive element on the way the system functions. The industrial districts function according to this principle and are illustrations of this case. Because of the permanent interaction between the actors, there is no appropriation of specific resources in such systems, and the system only functions effectively if the actors are able to keep up co-operation.

Compared with the previous case, this type of territorial production system is less risky for the region since the territorial production system's functioning does not depend on a single firm. In actual fact, the disappearance of a firm does not affect the existence of the others. The development potential, however, of such a system, resting exclusively on small firms, is obviously weaker inasmuch as it does not possess the mobilising effects a large firm can produce.

Naturally, the four cases described above are not mutually exclusive. In reality the territorial production systems overlap. Furthermore, there is no reason to view this typology as being static. Quite on the contrary, it helps illustrate the transition from one to the other. Let us examine some examples.

The firms of the first case are evidently not territorialised since they have no relations with the other firms in the region. Nonetheless, the circumstances could change and certain large firms seeking new skills or new synergies might be led to arrange relations of a horizontal type with other firms established in the region. So one moves from case 1 to case 3. In the same way, a large firm corresponding to case 2 may find it interesting to externalise certain of its activities and to thus establish relations with other firms in the region, or even to favour spin-offs. It may be the seed for the genesis of a territorial production system. A milieu must evolve, however, for the partners of the large firm to be able to escape from its domination. Thus, one moves from case 2 to case 3. If the large firm, of case 3 with partial, vertical integration, becomes more integrated either by taking over its competitors or by absorbing its sub-contractors or because the sub-contractors disappear, we move to case 2. The effect of the milieu may vanish and the region may be back in a situation of dependence on a large firm which could, from one day to the next, relocate. It could also happen that the large firm with partial, vertical integration of case 3 might disappear or leave the region. In this case, everything will depend on the effect of the milieu. If it is sufficiently strong, one could move from case 2 to case 1. Of course, all these evolutions do not occur automatically and it is possible, in some case, to steer them. Everything depends on the capacity of the actors in the region to act and the type of learning they have undergone or are able to acquire.

12.4.3 Learning Region and Steering Territorial Production Systems

The various cases described above are not exhaustive, of course, but they do help illustrate the dynamics that govern territorial production systems. Territorial logic can, in fact, only be maintained if the actors have understood the necessity of fostering mobility and the reproduction of specific resources and a conventional system of trading and non-trading relations based on trust, on non-contract, in the co-ordination of their productive activities. This entails that the principle of co-ordination/competition and of renewal of specific skills must be managed in the long term. As the theory of the innovating milieu has demonstrated, this implies the simultaneous action of interaction logic and of learning dynamics (Maillat,1998). The approach through the innovating milieux thus makes it possible to understand how the innovating processes are generated and set up in a territorial production system. The concept of the learning region makes it possible to explain the evolution from one territorial system to another. In fact, it emphasises the processes which are required by a territorial production system to acquire a territorial logic, i.e.: the capacity to become sufficiently autonomous in order to take decisions (at the territorial level) and to garner long term competitive advantages. This requires the setting in motion of those processes which enable the territorialisation of innovation and of firms and of various learning processes (interactive learning to ensure co-operation, organisational learning to ensure flexibility, institutional learning to ensure the institutional framework's adaptation and learning by learning to speed up the effects of experience accumulated).

Take as an example the evolution from case 2 to case 3 (see 12.3). For the large firm to be able to deverticalise itself, it has to become involved in the process of interactive learning (learn to co-operate), of organisational learning (modify its structure) and of institutional learning (change its routines, habits, standards, values). At the level of the region the necessary partners must also exist or emerge (research centres, training centres, interface organisations, firms) and the institutional framework must be or become favourable. This is only possible if all the actors of the region are committed to or are motivated to commit themselves to the learning processes which will allow them to adjust to the changes in the environment and which will favour the territorialisation of firms.

Thus, the learning concept makes it possible to appreciate the nature of the various types of learning that must be set in motion so that a territorial production system can transform itself to such an extent as to become an innovating milieu which is no less than the ultimate form of a learning region.

12.5 The Learning Region and Regional Development Policy

The concept of the learning region is very fecund from the point of view of regional policy, for it implies that the development of knowledge is the key to regional development. If this is true the concept contributes nothing new to the simple idea that training and education policy is the key to success. In actual fact, the concept of the learning region is misleading inasmuch as it leads us to believe that the region learns. Of course it is the actors who learn, who are the vehicle for the transmission of knowledge, who take the decisions. Moreover, the learning region cannot be discussed in isolation from the territorial production system which circumscribes it. Indeed, it is the nature of the latter which defines the nature of the relations which the actors, be they private or public, have spun and maintain. Thus, it is essential that the actors of a territorial production system be sufficiently autonomous or become sufficiently so for the interactions to become evident and therefore, they should not be encased in a functional logic, but rather set in a territorial one. For us, the learning region is a region in which the actors use interactive, organisational and institutional learning in order to create an innovative milieu and to keep it dynamic. The concept of learning helps indicate the ways, but also the constraints, to achieve this.

From the point of view of the local authorities, which pursue or wish to pursue a regional development policy, the concept of the learning region - in that it demonstrates the processes - has the advantage of providing a reading and analysis aid that can be applied to all the possible cases. In fact, by starting out with the learning processes, one can determine the weak and the strong points of the various territorial production systems in any type of region. In this case, the role of politicians and other leaders is to guide the learning processes with a view to bringing about interactions, to territorialising firms, to developing new knowledge or even to modifying institutions.

The concept may be attractive and seductive, but it is still very young. There yet are only very few empirical studies of the learning regions. Furthermore, the learning mechanisms are very complex and are usually connected to the region's history. They need time to grow. Becoming a learning region cannot be decreed and does not come about in a day. A regional policy with long-term aims must be put into place. Nonetheless, the learning region approach, despite the difficulties which characterise it, is a pertinent, operational method.

References

Asheim, B., 1995, *Industrial Districts as Learning Regions. A Condition for Prosperity,* Studies in Technology, Innovation and Economic Policy, University of Oslo, Oslo.

Cook S.D.N. and D. Yanow, 1996, "Culture Organizational Learning", in M. D. Cohen and L.S. Sproull, (eds.), Organisational Learning, Sage Publications, Thousand Oaks.

Crevoisier, O. and D. Maillat, 1989, "Milieu, organisation et système de production territorial: vers une nouvelle théorie du développement spatial", Dossier de l'IRER 24, IRER, Université de Neuchâtel, Neuchâtel.

Florida, R., 1995, "Toward the Learning Region", *Futures,* Vol. 27, 5:527-536.

Gelsing, L., 1992, "Innovation and the Development of Industrial Networks", in B. Lundvall (ed.), *National Systems of Innovation,* Pinter Publisher, London.

Gregersen, B. and B. Johnson, 1997, "Learning Economies, Innovation Systems and European Integration", *Regional Studies,* Vol. 31, 5:479-490.

IRES – Instituto Ricerche Economiche e Sociali, 1998 " Learning Region ", internet document, Rome.

Johnson, B., 1992, "Towards a New Approach to National Systems of Innovation", in B.-Å. Lundvall (ed.), *National Systems of Innovation: Towards a Theory of Innovation and Interactive Learning,* London.

Journé, B., 1996, "L'entreprise créatrice de savoir", Analyses de la SEDEIS, no 111:13-16.

Lazonick, W., 1993, "Industry Cluster versus Global Webs: Organizational Capabilities in the American Economy", *American Economy: Environment and Planning D, Society and Space,* pp. 263-280.

Le Bas, C. and E. Zuscovitch, 1993, "Apprentissage technologique et organisation", Economies et Sociétés, Série Dynamique technologique et organisation, Vol. 1, 5:153-195.

Lorenzen, M., 1996, "Communicating Trust in Industrial District", Paper presented in the Erasmus Intensive Seminar, Turin.

Lundvall, B.A., 1988, "Innovation as an Interactive Process – From User-Producer Interaction to the National System of Innovation", in G. Dosi et al. (eds.), *Technical Change and Economic Theory,* Pinter Publisher, London.

Lundvall, B.A. and B. Johnson, 1994, "The Learning Economy", *Journal of Industry Studies*, Vol. I, 2:23-42.

Maillat, D. and N. Grosjean, 1999, "Globalisation and Territorial Production Systems", in M. Fischer and L. Suarez-Villa (eds.), *Innovation, Networks and Localities,* Springer-Verlag, Berlin, forthcoming.

Maillat, D., 1998, "From the Industrial District to the Innovative Milieu: Contribution to an Analysis of Territorialised Productive Organisations", Recherches Economiques de Louvain, vol. 64, no 1, Département des sciences économiques, Université catholique de Louvain, Louvain-la-Neuve, pp. 111-129.

Maillat, D., O. Crevoisier and B. Lecoq, 1994, "Innovation Networks and Territorial Dynamics: A Tentative Typology", in B. Johansson, C. Karlsson and L. Westin (eds.), *Patterns of a Network Economy,* Springer Verlag, Berlin.

Maskell, P. and A. Malmberg, 1995, "Localised Learning and Industrial Competitiveness", Brie Working Paper 80, Berkley.

Midler, C., 1994, "Evolution des règles de gestion et processus d'apprentissage», in A., Orléan (ed.), *Analyse économique des conventions,* PUF, Paris.

Morgan, K., 1995, *The Learning Region: Institutions, Innovation and Regional Renewal,* University of Wales, Cardiff.

Németi, F. and M. Pfister, 1994, *Aspects de la compétitivité de l'industrie microtechnique suisse,* EDES, IRER, Université de Neuchâtel, Neuchâtel.

Paquet, G., 1994, Technonationalism and Meso Innovation Systems", Draft of a discussion paper prepared for the Trinational Institute on innovation, Competitiveness and Sustainability, organised by the Center for Policy Research on

Science and Technology of Simon Fraser University at Whistler B.C. on August 14-21 1994.

Pedler, M., et al., 1991, *The Learning Company,* McGraw-Hill, London.

Perrat, J., 1992, "Stratégies territoriales des entreprises transnationales et autonomie du développement régional et local", *Revue d'économie régionale et urbaine*, 5:795-814.

Perrin, J.C., 1995, "Apprentissage collectif, territoire, milieu innovateur, un nouveau paradigme pour le développement des régions en crise", in J. Ferrão (ed.), Políticas de inovação e desenvolvimento regional e local, actas do encontro realizado em d'Evora, 23 november 1995, Lisboa.

Pratt, A., 1997, "The Emerging Shape and Form of Innovation Networks and Institutions", in J. Simmie (ed.), *Innovation, Networks and Learning Regions?,* Jessica Kinsley, London.

Schmitz, H., 1997, "Collective Efficiency and Increasing Returns", Working Paper no 50, Institute of Development Studies, University of Sussex.

Simmie, J., (ed.), 1997, *Innovation, Networks and Learning Regions?,* Jessica Kinsley Publisher, London.

Thuderoz, C., 1997, Sociologie des entreprises, Editions La découverte, Paris.

Walther, R., 1998, "European Experience with Local Training Partnerships for Global Competition", Internet document.

Weinstein, O., 1992, "High Technology and Flexibility", in P. Cooke et al. (eds.), *Towards Global Localisation,* UCL Press, London.

Wolfe, D.A., 1997, "The Emergence of the Region State", Papier préparé pour le Bell Canada Papers 5, The Nation State in a Global Information Era: Policy Challenges, John Deutsch Institute for the Study of Economic Policy, Queen's University, Kingston, Ontario.

13 Clustering and Economic Change: New Policy Orientations – The Case of Styria

Michael Steiner
Institute of Technology and Regional Policy – InTeReg – Joanneum Research, Elisabethstraße 20, A-8010 Graz, Austria

13.1 Changing Strengths and Weaknesses

According to an (if not true, then well found) anecdote the world's champion in heavy weight boxing walking along a nice sunny beach wants to take a swim. Taking off his clothes he thinks to protect them from getting stolen by putting a note on them: "These clothes belong to the world's heavy weight champion in boxing." Returning after a nice swim he finds his clothes gone, the note still there with the words written on the other side: "Thank you. Sincerely yours, world champion in long distance running."

What are strengths, what are weaknesses of regions? According to recent developments in economic theory locational advantages have turned from comparative (being relatively cheaper) to competitive advantage relying on more qualitative elements. This new way of thinking about locational strength is founded on a special profile – what a region is able to do with specific lines of production. Without a coherent set of special things and abilities or so called core competences, regions are unable to be attractive for cooperation in the long run. Otherwise it will remain in a constantly repeating short run situation where it has to compete with other (mostly cheaper) suppliers similarly weak relative to other places in the world. In other words: without recognizable competences and strengths based on highly qualified linkages of intraregional cooperation regions get caught in the so-called 'globalization trap'.

Clusters, as a special form of regional strength, are based on such qualified links of cooperation: interlinked activities of complementary firms (in production and service sectors) and their cooperation with public, semi-public and private research and development institutions that are envisioned to create synergies thereby increasing producitivity that creates economic advantages. Hence, regions should develop such linkage strengths and policy should be oriented to create, develop and support clusters.

How strong are clusters? The existence of old industrial areas – now an infamous type of region representing a previous and now stagnant form of cluster – points to potential weaknesses of cluster oriented strategies. The literature describing such cases which are based on regional theories of the product cycle (see e.g. Norton/Rees, 1979; Markusen, 1985; Steiner, 1987; Tichy, 1987) brings ample evidence of the imminent dangers of a cluster oriented policy. Namely, regions become closed systems, the stagnation of the cluster structures impedes innovation and subsequently leads to an inability to adjust to new situations thus creating a "viscious cycle" rather than a system with positive feedback and increasing returns (Arthur, 1987).

Below, the strengths and weaknesses of a cluster orientated regional policy are discussed based on recent economic developments in Styria, Austria. First the regional economy of Styria, with her old and new cluster orientation is briefly described. This is followed by an examination of alternative methodological approaches for analysing clusters within this region. Several policy strategies and instruments are then discussed and evaluated. The general implications for a cluster orientation of regional policy are considered in the concluding part of the paper.

13.2 A Short History of Structural Change in Styria

Styria is one of the nine provinces of the Federal Republic of Austria (see map) with presently a population of about 1,2 million, an area of 16,ooo km2 (about 19% of Austria), a share in total employment of 14%, a share in the Austria's GNP of around 12%. Accordingly Styria is ranked 4th, 2nd, 4th and 4th respectively compared with the other eight provinces of Austria. Its GRP per head is markedly below the Austrian average which puts it at the 8th place in a ranking of the provinces.

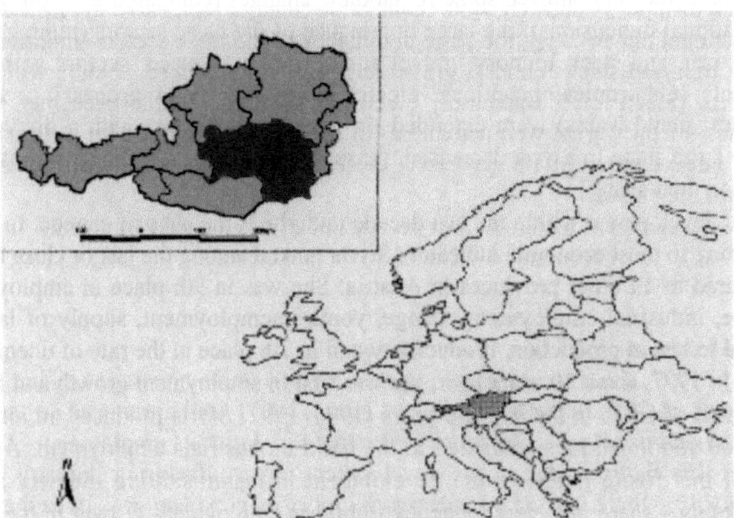

Map 13.1 The location of the province of Styria

As a regional economy Styria's present situation is formed by a long history. She was one of the world's leading iron producers in the 16th and 17th century. At that time the northern parts of the province had a sophisticated value chain based on ore and charcoal, but consisting also of a network of production, distribution and marketing firms. At that time Upper Styria was a very succesful cluster but whose European importance declined steadily after England learned to substitute charcoal for coal in the 17th century. Locally ore and charcoal remained as dominant industries, but the cost disadvantage of not having coal and more advanced products enforced concentration.

Styria's other economically prominent parts – the area around its capital Graz and the periphery in the South – were also formed by history. These parts of the region were throughout the Middle Ages a frontier exposed to attacks from invading forces coming from the South and East. The border character of the area was enforced in this century by the fact that both after World War I and II economic linkages with the neighbouring regions were reduced to a minimum. Especially after 1940, Styria was a region at the outskirts of a market oriented system situated at its dead end.

Not surprisingly therefore Styria, at the beginning of the 1950's, was a region with a huge agrarian sector (43.3% of total employment), an industrial employment share of 35.9% (one third of it in iron production), a small service sector with a share of 20.3%. Yet even on this rough scale a tremendous change within the last 40 to 50 years took place: in 1991 employment in primary sector declined to 8.6%, in industry it rose slightly to 37.7%, in services to 53.7% (the figures for the total of Austria are 5.8%, 35.6% and 58.7% respectively).

Behind these figures which indicate the vast dimensions of structural change in Styria are essential changes on both a meso and a micro level. The sectorial composition of industry showed some remarkable changes (compared to Austrian and international dimensions): the large dominance of the basic sectors (mining, glas, wood, iron and steel, foundry, gravel and soil) was reduced, sectors with final products (electronics, machines, electrical supply, wood processing, motor vehicles, metal wares) were extended. In general the sectors with a disproportionate large share in Styria decreased, those sectors which were underrepresented enlarged their share.

The development within the last decade underlines the pace of change. In 1988 according to most economic indicators Styria ranked among the last or close to last compared to all other provinces in Austria: She was in 9th place in employment change, industrial employment change, youth unemployment, supply of labour, second to last in production, productivity and in 7th place in the rate of unemployment. In 1997, about 10 years later, she was first in employment growth and fourth in growth of GRP. In the last two years (1996, 1997) Styria produced an increase of 6 600 additional jobs compared to the trend in Austriaís employment. Almost half of this figure is due to the development in manufacturing industry and – remarkably – there is a clear tendency to higher employment growth in the technologically advanced sectors.

These figures are an indication of a remarkable structural change: within the last decade Styria underwent a process of development from the province with the most aggravating problems (especially in terms of labour market indicators) to the one with the highest level of job creation.

13.3 Methodological Approaches for Cluster Identification in Styria

This recent succes of the Styrian economy was preceded by numerous studies and projects trying to identify the region's strengths and weaknesses and to find a basis for a new policy orientation. Among others this led to pronounced innovation and cluster oriented activities. The idea of regional networks and clusters has many roots and diverse theoretical foundations (for a recent survey see Feser, 1998). Not surprisingly there are different empirical approaches for identifying clusters within a region such as Styria stemming both from different cluster concepts and from different data bases. Below, five different methodological approaches to cluster identification and assessment used in the Styrian case are described and evaluated.

a) Input-Output Linkages and Spatial Association

One of the first approaches for measuring agglomerative tendencies and to verifying the existence of spatial clusters compared material linkages between sectors and the spatial concentration of these sectors. This approach raises the question: "do input-output linkages lead to spatial concentration"? Beginning with Florence (1944) and later by Streit (1969), Richter (1969), Czamanski and Czamanski (1977) and Harrigan (1982), this approach starts with the assumption that 'industrial agglomerations are not due or not only due to the common attraction or agglomeration in urban centres, but to interaction among the various industries' (Czamanski and Czamanski, 1977). Industrial complexes accordingly are defined by the linkage of industries by input and output flows and their locational proximity.

Using an input-output table for Austria (dating from 1983) and sector employment data at the district level (the lowest regional level with a consistant database) Kubin and Steiner (1987) showed different degrees of intensity of regional industrial (and also service-oriented) concentration. By means of correlation coefficients, cluster analysis and graph-theoretical methods, different intensities of 'functional' linkages (i.e. input-output linkages) and 'spatial association' (i.e. a similar distribution of sector employment across all regional units) were compared.

Figure 13.1 shows the graph of functional linkages of 31 sectors with different intensities of input-output relations (measured in terms of input-output coefficients which are weighted by means of graph-theoretical concepts; for more details see Kubin and Steiner 1987).

Figure 13.2 shows groups of sectors that are spatially associated. As mentioned above a similar distribution of employment between the respective sectors across all regions, i.e. districts, is regarded to indicate spatial proximity; as indicator the correlation coefficients between all percentages of employment of the specific sectors across all regions are taken (Streit, 1969). The numbers in the figure represent these correlation coefficients indicating the degree of similar distribution of sector employment and thereby the degree of proximity. In an additional step homogenous groups of spatially associated sectors are extracted by means of cluster analysis.

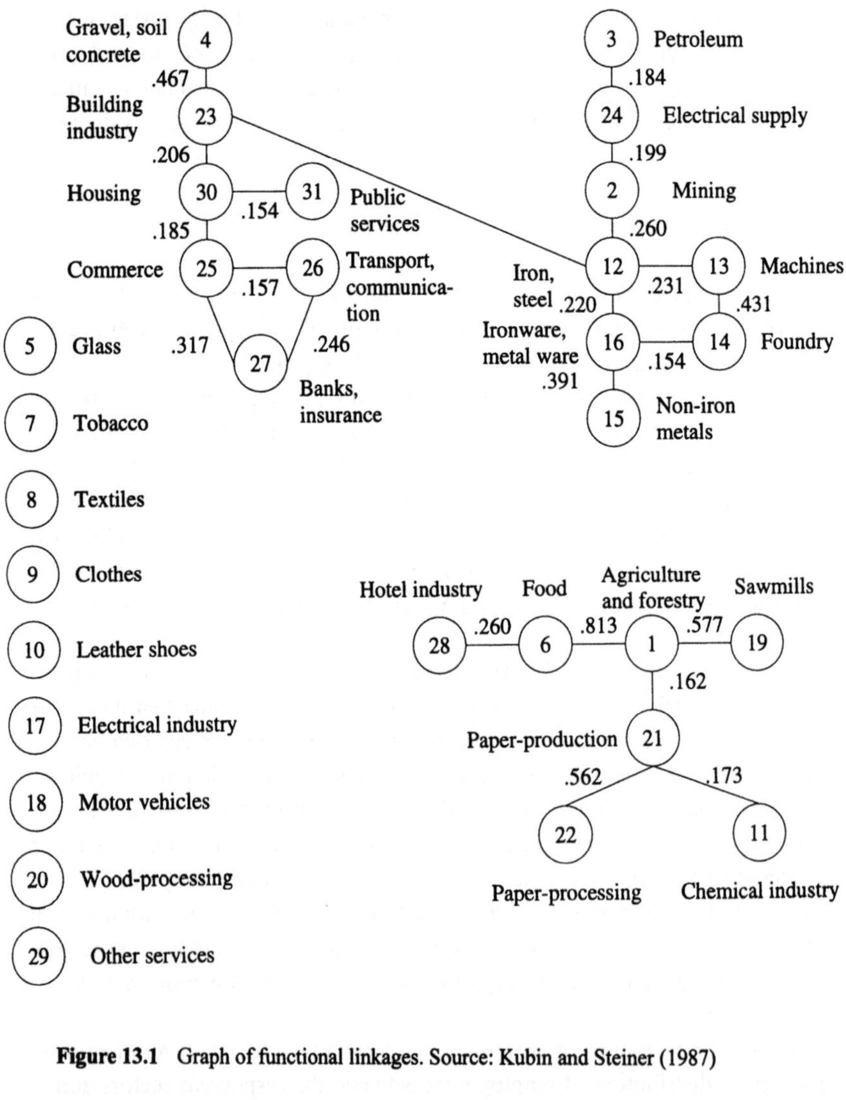

Figure 13.1 Graph of functional linkages. Source: Kubin and Steiner (1987)

Figure 13.3 indicates the 'coincidence' of functional linkages between sectors and their spatial concentration. The graphical symbols such as circles, triangles etc.

stand for spatial concentration of the respective sectors, the functional relations being the same as in Figure 13.1. Especially group 2 (mining, glass, iron and steel, machines) and group 3 (ironware, foundry, non-iron metals) show both functional linkages and are spatially associated (as can be seen by the coincidence of having strong functional linkages and of belonging to the same spatially homogenous group).

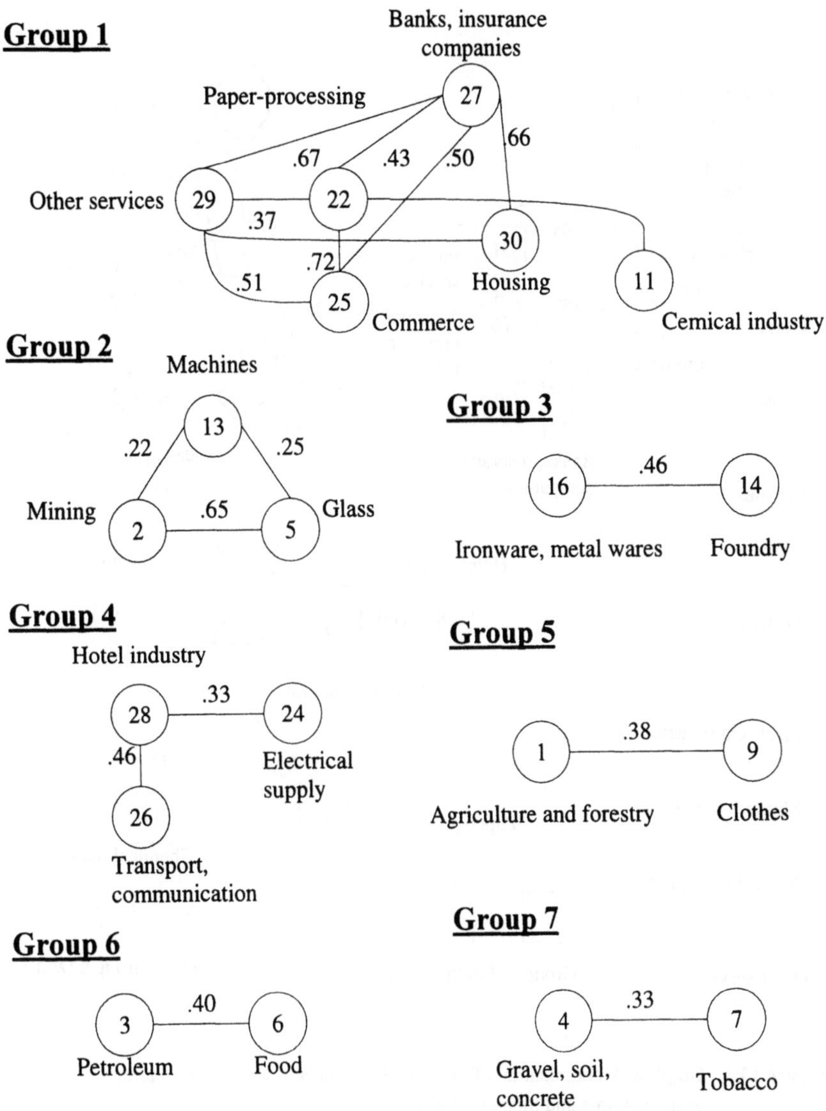

Figure 13.2 Group of industries with spatial association. Source: Kubin and Steiner (1987)

Yet only a small part of all sectors has linkages that lead to spatial concentration. This refutes somewhat the classical hypothesis of location theory that regional industrial concentrations are caused by material linkages – this is the case only for a small number of sectors. Yet it shows, especially for Styria with a still strong concentration on iron, steel, foundry and mining, that the strongest combination of functional linkages and regional concentration occurs in these sectors. The input-output based approach therefore shows the manifestation of the existence of an 'old' cluster within this region in the 1980s.

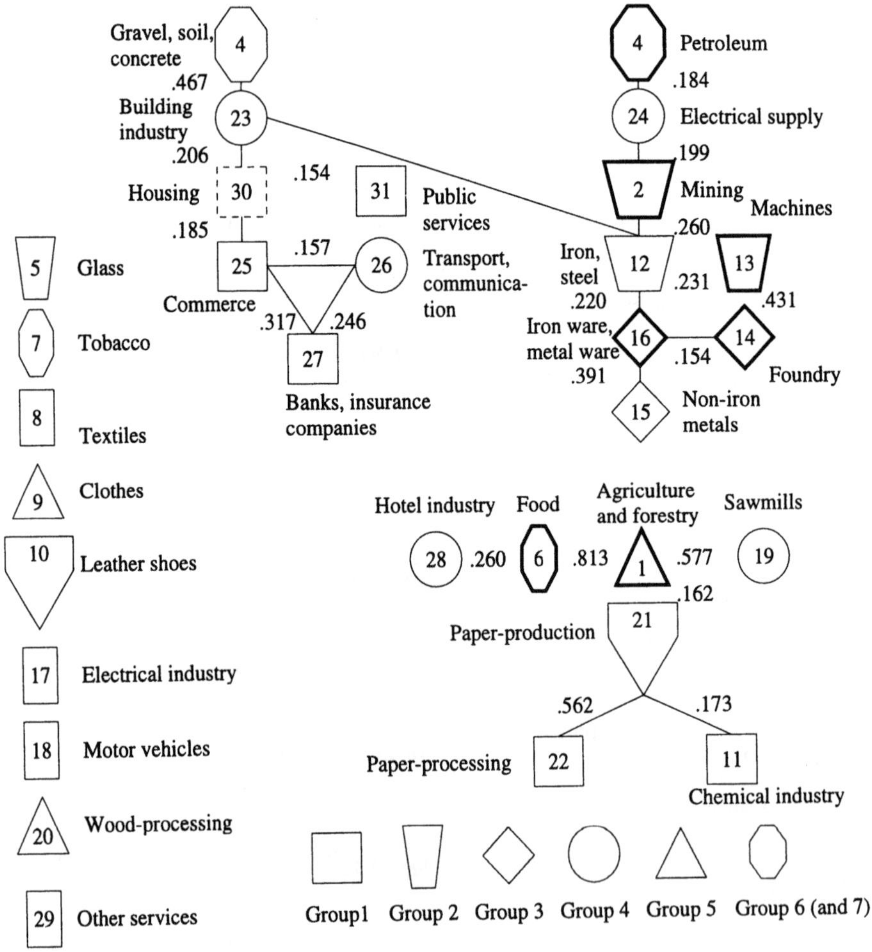

Figure 13.3 Graph with combination of functional linkage and spatial association
Source: Kubin and Steiner (1987)

b) Competitive Clusters

In search of other dimensions of regional clusters a modified Porter approach was used. Porter (1990) uses industries' success in international markets as the primary barometer of the competitive advantage and strength of a nation. The statistical basis for its regional application for Styria was Peneder (1994) who analysed the competitiveness of Austria's industry by means of diverse internal and external factors such as international market shares, foreign trade specialisation, the degree of international division of labour, export distances and also measures of productivity and profitability. The aim of this study was to define sectors and product groups in terms of different levels of competitiveness.

Using this basic information on the whole of Austria's industry the data were transferred to the industrial structure of Styria to find out the extent to which competitive clusters exist in this region (Steiner et al., 1996). The results are shown in Figure 13.4.

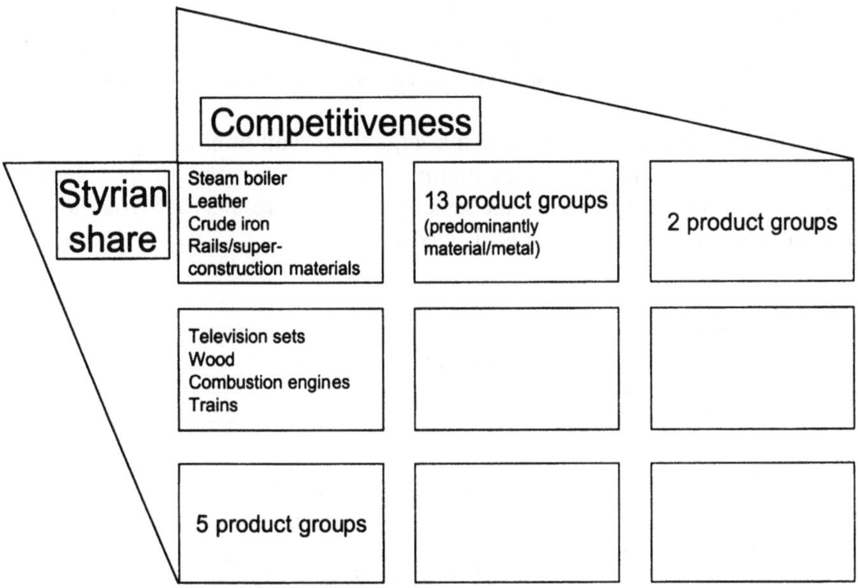

Figure 13.4 Competitive cluster
Source: Peneder (1994), Steiner et al. (1996)

In this matrix, clusters with different degrees of competitiveness (as identified for the whole of Austria) are aligned with different intensities of representation for these industries in Styria. A high degree of competitiveness with, simultaneously, a strong representation in Styria can be found in the production of steam-boilers, leather, iron and rails. In general, the results showed that Styria has a considerable portion of Austria's high competitive sectors. This can be interpreted as the presence of a regional concentration and clustering of these industries. Yet these sectors represent only a small part of the whole of Styria's industry, and are to be regarded rather as innovative niches of successful firms within a total mix of rather traditional industries.

c) Technological Clusters

Technological clusters are based on the idea that firms having a similar patent behaviour and structure are also technologically similar and therefore form such 'technological clusters' (Jaffe, 1989; Jaffe et al., 1993). Firms within such clusters are able – according to this thesis – to use technological spillovers. In contrast, firms which concentrate their production on single technologies – within their regional environment – have little chance of benefiting from external effects.
Such technological clusters were, based on Jaffe's work, identified for Austria (Hutschenreiter, 1994). Firms with similar patent activities were structured and grouped into clusters. In this way it is possible to identify potential 'horizontal spill-overs' or immaterial linkages between firms with similar technological activities: informal contacts, exchange of information, a common pool of qualified workers and a common educational and research infrastructure are examples of such horizontal linkages.

For Austria as a whole there were five large fields with firms having a similar patent structure: electro/electronics, transport, construction, sport articles, pharmaceuticals/chemicals. When using the same approach, the results for Styria showed the following pattern(s) (Steiner et al., 1996):

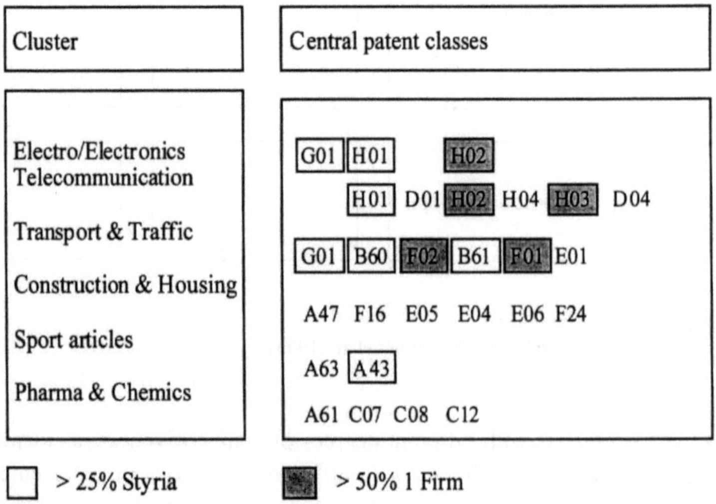

Figure 13.5 Technological cluster

In Figure 13.5 those patent classes are marked which have a strong representation in Styrian firms (i.e. more than 25% of the patenting firms are from Styria); in the two clusters of construction & housing and pharmaceuticals & chemicals Styria is only weakly represented, but the electro/electronics and transport clusters are dominated by Styrian firms (the numbers represent specific patent classes such as F02=internal combustion engine or H03=electronic circuits). This again can be interpreted in support of the existence of regional clusters: technologically similar firms are regionally concentrated and form the basis for technological spill-overs and positive external effects of knowledge diffusion.

Yet these basically positive results (from a Styrian perspective) have to be qualified: rather small technological niches within these fields play a dominant role; it is mostly small and medium-sized firms which concentrate on a specific field of technology within the patent class and, more important, the number of firms is rather small, i.e. many patents are performed by a limited group of innovative firms (here: more than 50% of the patents are from one single firm) so that the potential for spill-overs is reduced.

d) Learning within the Materials and Metal Cluster in Upper Styria

This approach for identifying a special aspect of cluster formation focussed on firms within the material and metal processing sector in Upper Styria (Steiner and Hartmann, 1998).

Out of a total 186 firms related to raw materials and metal in the region 38 firms were identified as being embedded in local regional networks and inter-firm linkage structures. Qualitative interviews and a baseline-survey were employed to identify different forms of firm-specific knowledge acquisition, technology spill-overs and cluster-specific inter-firm collaboration and learning.

Furthermore different concepts and approaches to 'organizational learning' were used (e.g. lower-higher level learning, deutero learning – for these concepts see e.g. Argyris and Schon, 1978; Fiol and Lyles, 1985; Dodgson, 1993).

At the firm level learning by doing could be identified in most cluster corporations in the qualitative investigations as well as in the 'survey' carried out. In addition to these findings spillover effects that are present at the cluster level could be detected. The most important sources of spillovers identified were 'waterholes' (informal meeting places), training sessions and seminars, and the incorporation of new members into the firms were identified. The above results may be viewed as supporting the presence of agglomeration economies in the cluster.

At the cluster level it was possible to identify different kinds of inter-firm links. Five upstream, three downstream and five horizontal links were found. In addition it was possible to localize two SME-networks within the cluster: one building and construction network consisting of 17 member firms, and one network offering goods and services for the metal processing industry consisting of 10 member firms.

Through these inter-firm links (upstream, downstream or horizontal) the firms were able to carry out lower-level learning and to a certain extent also higher-level learning. Technological learning through joint R&D efforts dominated these inter-firm activities.

In the networks the member firms were all able to perform in particular higher-level learning activities with such contents as technological learning through intense joint R&D efforts, management learning through the continuous improvement of the routines and procedures carried out together and marketing learning through the development of new products together with the clients.

Clusters may be characterized as learning organizations when learning at the inter-firm (cluster) level is present. The findings of this case study presume indeed that clusters are acting as learning organizations albeit loosly structured organisations in most cases. In almost all cases of inter-firm activities within this specific cluster learning activities such as lower-level and higher-level learning could be detected. As Dodgson (1996) already points out inter-firm links may provide opportunities for 'higher level' learning.

But the findings of this case-study suggest as well that the specific kinds of links between the firms may have an influence whether higher-level or only lower-level learning activities will be carried out. While the horizontal links examined brought only forth such lower-level learning efforts, the up- and downstream links seemed to foster higher-level learning as well. Thus we may conclude that client-supplier relations and supplier-client links have, in this particular cluster, obviously a higher potential to enable the partners to carry out higher-level learning than horizontal co-operations.

Networks seem to have in this cluster an even bigger ability to bring about higher-level learning activities. In this case study all learning efforts carried out in the examined networks could be classified as higher-level learning. This greater ability of networking to foster higher-level learning activities may lead to the suggestion that clusters may be better learning organizations the more they have real network structures.

A final interpretative remark: The cluster of this case study consists of slowly emerging links and cooperation between SMEs in a region long dominated by large and often nationalized companies. The decline of this old industrial area in the 70s and 80s was to a large degree caused by an inability to learn new forms of behaviour and a preservation of old ones (Geldner, 1998). The reported attempts of developing a culture of learning among SMEs may be also interpreted as a regional process of 'unlearning' old habits (something that is harder than learning new ones). The SME-cluster creates for the region as a whole, therefore, positive spill-over and demonstration effects as to the importance of joint learning.

e) Cooperative Behaviour and the Strategies of Styrian Firms

To complement this regional cluster-oriented analysis of input-output linkages, competitiveness, technological cooperation linkages and forms of learning, a survey of Styrian firms was conducted to learn about their strategies, their market situation and related competitive pressure (Sturn and Steiner, 1995). The survey comprised all manufacturing firms with more than 20 employees (altogether 730 firms with a response of 195). Special emphasis was put on strategic and cooperative behaviour to obtain information on the tendency and the willingness to form networks, to make use of research and transfer institutions and to integrate producer oriented services. Concentrating on the relative importance of cooperation as an entrepreneurial strategy for the future, the firms were asked what their strategic priorities for the next decade were.

The strategies pursued can be largely divided into three groups: very important strategies (more than 80% of the responding firms consider them as decisive), important ones (more than 40% of the firms regarded them as decisive), less important ones (less than 40%). (See Table 13.1).

The following, for example, were regarded as very important strategies: orientation towards higher qualification, quality, looking for market niches; orientation towards clients and the opening markets in Eastern Europe. Altogether, Styrian firms pursue an offensive, market-oriented strategy. Yet, cooperative strategies are considered to be of minor importance. Only 27% of the firms intend to cooperate with other firms, 34% with research institutions, and 26% with production-oriented services. The strongest form of cooperation is considered to be informal contacts (41% of the responding firms). This points to a low readiness for networking and (strategic) alliances. In combination with the other results concerning cooperation (as presented above) where a low degree of cooperation is obtained, this additional piece of information reveals that firms obviously have little interest and intention to do so in the future. Forming clusters by means of (formal and informal) cooperation and alliances is not a decisive strategy.

Table 13.1 Stategies of firms grouped according to their degree of importance

Strategies of firms
Very important for more than 80% of all firms
– qualification, quality, techjnology
– non-cooperative strategies for Eastern Europé
– non-cooperative strategies for EU
– client-specific strategies
Very important for 80-40% of all firms
– cooperative strategies for the European Union
– defensive strategies
Very important for less than 40% of all firms
– cooperative strategies for the Eastern Europé
– access to external services and research
– inter-firm cooperation

13.4 Implications for Technology Policy in the Regional Context

13.4.1 Elements of a Cluster Oriented Policy in Styria

Summarizing these empirical approaches for an analysis of strengths and weaknesses from a cluster oriented perspective, the following points seem essential:

- Styria had up to the 1990s an old cluster formed by strong input-output linkages in basic industries such as coal, ore, iron. Yet there are slowly emerging forms of joint learning processes and cooperation among small firms in this area showing the potential for new clusters.
- A considerable portion of Austria's highly competitive industries is regionally concentrated in Styria, albeit forming but a small part of the overall activities of the firms.
- Technological clusters – identified by similar and regionally close patent activities – do exist and are focussed on transport/traffic and electro/electronics/ telecommunication industries.
- Cooperation and network organization is not (yet) considered as an essential strategy by firms in the region.

Also on the basis of these cluster oriented approaches therefore there are signs of structural change. Policy intervention to support this change was consciously and actively undertaken in the last two decades. In the early 1980s, attempts were made to achieve stabilization by traditional policy tools (especially loss compensation). In the middle of the 1980s this approach began to be reconsidered and the initial structural reorganization schemes were introduced. At the beginning of the 1990s, the main future-oriented concepts and organizational structures were developed culminating in an integrated policy proposal entitled 'Technology Policy Concept Styria' (Steiner et al., 1996). In a highly simplified presentation these steps may be summarized as follows:

- Restructure nationalised industry to make it more flexible and adaptable; enforce its division into smaller enterprises and concentrate management within the plant itself (instead of headquarters in the remote capital) – i.e. reorganize the old cluster.
- Improve technology transfer in all of its possible forms to generate innovative impulses for the regional economy. Innovation and collaboration between productive entities in Styria should be promoted (firms, research, transfer, advisory institutions etc.) and the emergence of new clusters should be supported.
- Create an infrastructure to support diffusion and absorption of new knowledge and to improve the qualification of the workforce.

On the basis of these strategies technology policy measures were elaborated which are on one hand operative and on the other hand coordinative. They were based on international experiences of technology policy in a regional context (see e.g. Cooke and Morgan, 1991; Balthasar and Knöpfel, 1993; Staatsministerium Baden-Württemberg, 1994; Batt, 1994; Harrison 1995; for an overview see Jud and Sturn, 1996) and transformed and reinterpreted for the Styrian context. The operative measures are the tip of the active cluster orientated technology policy in Styria. They should weaken the barriers to innovation, especially for small and medium-sized enterprises due to the acceleration and distribution of new technologies to initiate cooperation and to support the building-up of a competitive cluster structure. The co-ordinative measures serve to consolidate the existing range of instruments and to combine the new ones with an effective goal-directed technology policy.

Among the instruments proposed the following are especially focussed on the support of cluster formation:

a) Instruments to Improve Cooperative Ability

Pilot projects. The development plan for a pilot project for Styria consists of two modules:

- organization of cooperation events for selected clusters (e.g. automobile, wood/ construction);
- financial support of cooperative innovation projects for the respective clusters.

The first module is focused on the active initiation of cooperation, the second on the reduction of innovation- and cooperation risk.

The organization of a soft cluster promotion scheme. Through this program the internal cluster relations should be intensified and focused on the integration of existing clusters: information about existing cooperation projects should be passed on, the potentials of firms wishing and having the need for cooperation should be coordinated and made public. The focus of these activities rests on the removal of information barriers, cooperation risks and organisational hindrances.

Foundation of special research institutes in Styria. Through these institutes (comparable to centres of expertise, additional research facilities at existing university institutes) the active technology and knowledge transfer from the universities to the economy should be improved. These institutes focus on increasing the know-how-base of complex innovation projects, reducing the risk for the included enterprises. These institutes should create research facilities with a tight connection to the universities. They are required to do research work which is strongly related to practical work and allows for continous problem concentration, something which is not possible in universities with their orientation towards basic research and teaching. The work of these institutes must suit the needs of the enterprises. For this reason the enterprises should play a decisive part in the selection of the research focus and should actively be involved in cofinancing these institutions (for more details see Sturn, 1997).

b) Measures to Increase Absorption Ability and Diffusion

Co-ordinated information initiative. For the implementation of an information offensive for the Styrian economy a central institution has to be created, which should make it easier for the enterprises to acquire a service of research and technology facilities, promotion agencies, educative institutions to get an overview over the services of important suppliers on regional, national and international levels. Through such a measure the existing information barriers would be minimized and the diffusion of new technology accelerated.

Demonstration centres. Demonstration centres focus on the elimination of the lack of technology-specific information and qualification especially for the SMEs. These are institutions, which demonstrate the practical uses of new technology and which offer start-up consulting and further education events. So they are characterized by the integrated supply of business-oriented consulting, technological consulting, demonstration of new technique and trainee programmes. To be more effective, they should be arranged above all in technological cross-sections (e.g. mechanical engineering, information and communication technique) and should serve as instruments for cluster promotion (for example, material technology, environmental technology). They can also be used thus encompassing the pure industrial domaine. Demonstration centres should employ specialists with appropriate experience, who test together with interested enterprises the possibility of the use of the new technologies in the firms and who elaborate concepts for implementation.

c) Quality and Qualification

Cluster-specific qualification initiative in accordance with labour-market policy. These activities have to fulfill cluster-specific qualifications and to open up synergies resulting from the teamwork between technological, educational and labour-market policy. Specific measures comprise e.g. the establishment of so-called "Fachhochschulen" (comparable to technical colleges) which concentrate on the skills appropriate for clusters (one example is a Fachhochschule for automotive engineering). Another measure comprises special educational programmes organized by labour market institutions: instead of connecting qualification programmes to person-specific conditions (as for example age or qualification) they should be oriented towards technological diffusion and important clusters. The plans of new further education modules should involve enterprises, workforce, education, consultation and research institutions.

Innovation assistants for SMEs. The promotion of innovation assistants aims at abolishing the existing qualification barriers to alleviate the innovation, cooperation and absorption processes. Especially to overcome such inhibition levels by the employment of universities and technical colleges the innovation assistant represents an excellent instrument for a stronger integration of SMEs in the Styrian clusters and to intensivate their national and international co-operation relations. Besides its technology policy effects this measure has also a labour-market policy effect, making it easier for the university graduates to enter a professional career (similar activities have been coordinated on an international scale by the so-called t3net, see Adametz, 1998).

All these elements of a cluster oriented approach represent a "re-engineering" of regional policy and changed its character: it is confronted with a larger degree of uncertainty as to the means – end relation, it involves more actors on different levels and from diversified institutions, and it incorporates more and more knowledge elements. Information exchange, enforcement and support of communication processes and of cooperation become therefore important elements and necessitate a more experimental approach. A cluster oriented policy also can be understood as a fitness training for the partners involved, as a program to develop its own strengths. From this perspective regional policy design becomes more and more a concept in progress, a process of "learning in developing the results".

13.5 Concluding Remarks: The Pros and Cons of Cluster Orientation

Yet what are the possible drawbacks of such an approach, what are its potential risks? Should regions specialize on specific strengths and thereby accept all potential dangers resulting from specialisation?

The first answer is straightforward: Yes, clusters are more indispensable than ever (Tichy, 1998). This straightforward response of course is based on all of our explicit and implicit understandings of the competitive advantages of regions. If a region's special image is linked to a set of specific competences, if the innovative ability and, consequently, the competitiveness of its firms depends on being embedded in interlinked activities with other firms and institutions (i.e. depends on being part of a cluster), there is no longer room for the dominance of traditional instruments of regional policy. Cluster orientation and specialisation then has to be understood as the active response to the potential challenges of globalization to regional competitiveness.

The directness of the answer is of course subject to qualifications.

The first counter-argument points to the risks of specialisation (Fritz et al., 1998): specialisation increases efficiency but is also associated with risk and makes the specialised region more vulnerable. Regional policy makers therefore are confronted with a risk-return trade-off.

These risks are of two types: in analogy to the regional portfolio concept both 'structural' and 'cyclical' risks can be differentiated. The first kind of risk manifests itself in the well known case of 'old industrial areas' where a permanent decline in specific sectors leads to the decline of whole regions that were specialised in these industries.

The second type of risk refers to economic stability: unnecessary fluctuations cause inefficieny and welfare losses. Since the effects or impacts of business cycles are unevenly distributed among industries, the region's industrial structure and its optimal mix is a goal per se, yet one that is in competition to the gains of regional specialisation. The more a regional economy is dominated by a cluster the greater the risk associated with the regional portfolio.

Another point is the spatial scale and the geographical dimension of cluster orientation – how regional are clusters? Geographical proximity is but one aspect of cluster formation. Of equal – if not higher importance – is organisational proximity: firms of equal size and knowledge orientation, geographically dispersed branch plants belonging to the same legal firm generally show stronger forms of cooperation than do accidentally spatially close enterprises.

This gradual extension at the interregional level of tight relationships of co-makership leads to new tasks of policy support (Cappellin, 1997): Investments in the development of modern logistics and distribution services become a critical factor for a wider interregionalization and internationalization of the involved firms. There is a need for major decisions to undertake large projects on infrastructure network developments and on other modern collective services. Just because of an enlargement of the spatial dimension of cooperation the organization and quality of the territory will become a key factor in international competitiveness.

One of the recent questions within the cluster debate is the emphasis on changed business behaviour and the importance of learning and knowledge exchange both at the level of the firm as well as between firms. Yet recent research shows that learning within clusters, i.e. organisational learning, is no universal phenomenon among the different forms of clusters (DE.L.O.S., 1998). Training and logistical support policies and initiatives, especially for SMEs, need to be carefully targeted rather than generic. Policy instruments need to reflect the different configurations of clusters and their respective learning organisational behaviours. From this experience a number of policy implications for the support of training and joint learning in favour of SMEs can be made (Cullen, 1998):

- In relation to training and support policies three main constituent components of "organisational learning" need to be targeted: information gathering, knowledge acquisition, competence consolidation and development implying different training and logistical support capabilities. They should incorporate both provision of "formal" services, together with actions designed to enhance informal networking arrangements.
- SMEs are relatively active in knowledge acquisition activities, but not in lower-level market intelligence gathering or higher level competence development. This underlines the need for awareness-raising campaigns – SMEs need to be made aware of the need to balance these three different components in their human resource development planning and management.
- The entrepreneur turns out to be the pivotal figure in decision-making within SMEs, yet the evidence suggested that a large proportion of SMEs are in "crisis management" rather than pro-active learning organisations.There is therefore a need to encourage SMEs to adopt a more participative style of collective learning.
- Microentreprises and new start-ups are particularly prone to "crisis-management", and the lack of coherent organisational learning strategies. Since this situation is almost certainly associated with lack of resources, it would suggest the need for support services that can provide pooled resources for SMEs.

Another problem of policy induced clustering is that clusters presuppose a mentality of partnership, the willingness to cooperate, the insight of its necessity, a supportive incentive structure and, especially, the existence of trust. This implies above all that no kind of policy can substitute for the dynamism and social organisational skills that must exist on a local and regional level for cluster building policies to succed. This also implies that without decentralisation, without a larger political autonomy of local and regional institutions, without federalist reforms network creation becomes difficult.

After all these revisionist and limiting qualifications, is there any legitimization left for the straightforward support for a strategy of regional specialisation without again referring to its inevitability? Maybe we could recur to a metaphor used in theological circles where a second level of naivity – being aware of all doubts – allows for reflective kind of belief. Such a belief in the usefulness of a cluster orientation then is founded on the necessity of "seeing regional economies whole" (Bergman, 1998).

This enlighted view would point to different complementing levels of cluster analysis (macro – meso – micro) and to different strategies and instruments for different segments of the resulting clusters having specific needs. Their combined needs then require a portfolio of suitable policies. These policies then would have to adopt strategies and instruments suited to a continually changing mix of industries.

This perspective also requires to view cluster policy as a whole. A cluster approach may then serve as a structuring principle of different domaines of industrial policy and lead to an institutional renewal of economic policy in general.

Acknowledgement

I would like to thank the participants of the Uddevalla workshop for discussion of an earlier version of this paper and an anonymous referee for detailed comments and suggestions.

References

Adametz, C., 1998, "Innovators for SMES", Paper presented at the T3Net-Conference, March 1998, Vienna.

Argyris, C. and D. Schon, 1978, *Organizational Learning,* Addison-Wesley, Reading.

Arthur, B., 1987, "Self-Reinforcing Mechanism in Economics", Mimeo, Stanford University.

Balthasar, A. and C. Knöpfel, 1993, Die Technologiepolitik europäischer Staaten und Japan. Ein Überblick über Deutschland, Frankreich, England, Schweden, Holland und Japan, Studie im Auftrag des Schweizerischen Wissenschaftsrates, Bern.

Batt, H.-L., 1994, Kooperative regionale Industriepolitik, Frankfurt am Main.

Bergman, E.M., 1998, "Industrial Trade Clusters in Action: Seeing Regional Economic Whole", in M. Steiner (ed.), *Clusters and Regional Specialisation,* European Research in Regional Science, vol. 8.

Cappellin, R., 1997, "Regional Policy and Federalism in the Process of International Integration", in K. Peschel (ed.), *Regional Growth and Regional Policy within the Framework of European Integration,* Springer-Physika Verlag, Heidelberg.

Cooke P. and K. Morgan, 1991, *The Network Paradigm,* Regional Industrial Research Report 8, Cardiff University.

Cullen, J., 1998, "Promoting Competitiveness for Small Business Clusters through Collaborative Learning: Policy Consequences from a European Perspective", in M. Steiner (ed.), *Clusters and Regional Specialisation*, European Research in Regional Science, vol. 8.

Czamanski, D. and St. Czamanski, 1977, "Industrial Complexes: Their Typology, Structure and Relation to Economic Development", *Papers of the Regional Science Association,* 38:93-111.

DE.L.O.S., 1998, *Developing Learning Organization Models in SME Clusters,* TSER project (participating partners: Istituto G. Tagliacarne/Italy, Tavistock Institute/U.K., Formit/Italy, ECWS/Netherlands, CCI/France, Infyde/Spain), Project report available at the Istituto G. Tagliacarne/Rome.

Dodgson, M., 1993, "Organizational Learning: The Review of Some Literatures", *Organizational Studies,* Vol. 14, No. 3.

Dodgson, M., 1996, "Learning, Trust and Inter-Firm Technological Linkages: Some Theoretical Associations", in R. Coombs, et al., *Technological Collaboration – The Dynamics of Cooperation in Industrial Innovation,* Cheltenham.

Feser, E.J., 1998, "The Old and New Theory of Industry Clusters", in M. Steiner (ed.), *Clusters and Regional Specialisation,* European Research in Regional Science, Vol. 8.

Fiol, C. and M. Lyles, 1995, "Organizational Learning", *Academy of Management Review,* Vol. 10, No. 4.

Florence, S. 1944, "The Selection of Industries Suitable for Dispersion into Rural Areas", *Journal of the Royal Statistical Society,* 16:93-116.

Fritz, O., H. Mahringer and M. Valderrama, 1998, "A Risk-Oriented Analysis of Regional Clusters", in M. Steiner (ed.), *Clusters and Regional Specialisation*, European Research in Regional Science, Vol. 8.

Geldner, N., 1998, "Erfolgreicher Strukturwandel in der Steiermark", in Wifo-Monatsberichte 3, Österreichisches Wirtschaftsforschungsinstitut, Wien.

Harrigan, F., 1982, "The Relationship between Industrial and Geographical Linkages: A Case Study of the United Kingdom", *Journal of Regional Science,* 22:19-31.

Harrison, R.T., 1995, "Development in the Promotion of Informal Venture Capital", in Fraunhofer Institut für Systemtechnik und Innovationsforschung, Hrsg., EIMS Innovation Policy Workshop: "Innovation Financing: Private Investors, Banks & Technology Appraisal", Luxembourg.

Hutschenreiter, G., 1994, Cluster Innovativer Aktivitäten in der Österreichischen Industrie, tip, Wien.

Jaffe, A.B., 1989, "Charachterizing the "Technological Position" of Firms, with Application to Quantifying Technological Opportunity and Research Spillovers", *Research Policy,* Vol. 18, 2:87-97.

Jaffe, A.B., et. al., 1993, "Geographic Localization of Knowledge Spillovers as Evidences by Patent Citations", *Quarterly Journal of Economics,* Vol. 108, 3:577-598.

Jud, T. and D. Sturn, 1996, "Wie gestalten andere europäische Länder ihre Technologiepolitik?", *Wirtschaftspolitische Blätter,* Vol. 43, 1:35-43.

Kubin, I. and M. Steiner, 1987, "Muster räumlicher und funktionaler Verbindung", Mimeo, Department. of Economics, University of Graz.

Markusen, A., 1985, Profit Cycles, Oligopoly and Regional Development, MIT Press, Cambridge, Mass.

Norton, R. and J. Rees, 1979, "The Product Cycle and the Spatial Decentralisation of American Manufacturing", *Regional Studies,* 13, 141-151.

Peneder, M., 1994, *Clusteranalyse und sektorale Wettbewerbsfähigkeit der Österreichischen Industrie,* Studie im Auftrag des Bundesministeriums für Öffentliche Wirtschaft und Verkehr sowie für Wissenschaft und Forschung, t.i.p., Wien.

Porter, M, 1990, *The Competitive Advantage of Nations,* Harvard University Press, Cambridge, Mass.

Richter, Ch., 1969, "The Impact of Industrial Linkages on Geographic Association", *Journal of Regional Science,* 9:19-27.

Staatsministerium Baden-Würtemberg, Hrsg. 1993, *Bericht der Zukunfts-kommission Wirtschaft 2000,* Stuttgart.

Steiner, M., 1987, "Contrasts in Regional Potentials: Some Aspects of Regional Economic Development", *Papers of the Regional Science Association,* 61:79-92.

Steiner, M., Th. Jud, A. Pöschl and D. Sturn, 1996, Technologiepolitisches Konzept Steiermark, Leykam, Graz.

Steiner, M., (ed.), 1998, *Clusters and Regional Specialisation,* European Research in Regional Science, Vol. 8.

Steiner, M. and Ch. Hartmann, 1998, "Learning with Clusters: A Case Study from Upper Styria", in M. Steiner (ed.), *Clusters and Regional Specialisation,* European Research in Regional Science, Vol. 8.

Streit, M., 1969, "Spatial Associations and Economic Linkages between Industries", *Journal of Regional Science,* 9:177-188.

Sturn, D. and M. Steiner, 1995, "Cooperative Strategies of Firms. Chances and Drawbacks for Regional Technology Policy", in J. Padjen et. al., (ed.), *Industrial Restructuring and its Impact on Regional Development,* Zagreb.

Sturn, D., 1997, *Einrichtung von Kompetenzzentren, Erstellung eines Konzeptes zur Auswahl, Einrichtung und Finanzierung von Kompetenzzentren in Österreich,* Joanneum Research, Wien/Graz.

Tichy, G., 1987, "A Sketch of a Probabilistic Modification of the Product Cycle Hypothesis to Explain the Problems of Old Industrial Areas", in H. Muegge et al., *International Economic Restructuring and the Regional Community,* Avebury, Aldershot.

Tichy, G., 1998, "Clusters: Less Dispensable and More Risky than Ever", in M. Steiner (ed.), *Clusters and Regional Specialisation*, European Research in Regional Science, Vol. 8.

14 The Italian Smallness Anomaly: Coexistence and Turbulence in the Market Structure

Dino Martellato
Università Ca' Foscari, Dipartimento di Scienze Economiche
San Giobbe 873, I-30121 Venezia, Italy

14.1 Introduction

The Italian economy has always been characterized by a relatively large presence of medium and small size firms, a feature that has received a range of definitions and has been the subject of a large number of studies even in the international literature. The continued and sometimes large relative presence of small plants and firms has long been considered a kind of anomaly. However, the alleged inefficiency and unsustainability of small scale productions is in patent contrast with the durability of the phenomenon, with the good economic performance of those regions where small scale production dominates, with the often high innovation performance of small firms, with their export performance and with the high degree of internationalisation of many small and medium size firms. Small wonder then that in the last twenty years there has been a complete revision of the interpretation of the small firm's role in economic theory, mainly along the lines of evolutionary economic theory (Section 14.2).

The permanence of firms of limited size and a skewed-size distribution is clearly an obvious feature of market structure and not an anomaly, but there are other modalities by which market structure evolves. Coexistence of firms varying in terms of behaviour, innovation, organization and performance, not only in size, is a more than evident feature. Another often reported fact is market turbulence, i.e. high firm entry and exit flows. As is obvious, such features deserve an explanation, but seem to defy what Nelson and Winter called appreciative theory, which is rather common in the field. Moreover, products have a life cycle which changes the composition of market transactions. The market structure is always in a state of flux and the resulting dynamics in terms of firms and products, not static equilibrium, is its dominant feature (Section 14.3).

Section 14.4 examines one concept of equilibrium, namely the stationary regime of a dynamic process, which is not at all common in economics. Concepts of product space and firm space in order to give a somewhat formal treatment of the issue are then introduced. The simple dynamic model presented, which is borrowed from biology, is aimed at showing the conditions under which firms that are heterogeneous in their performance, rather than in their size, coexist. Market turbulence and product differentiation exist even in equilibrium.

14.2 The Italian Smallness Anomaly

A quite well-documented feature of Italian industrial organization is the relative abundance of small and medium size firms – a feature that in Italy has always been more evident than in many other European economies (see Table 14.1, top). The relative abundance of medium and small scale business firms looks like the increasingly dominant modality by which the Italian manufacturing industry extends its territorial, and product coverage as the average size of the firm is tending to decrease. (Table 14.1, bottom).

Table 14.1 Italy, manufacturing industries, 1984 and 1992[1]

	Firms 1984	Firms 1992	Employees 1984	Employees 1992
< 20	296,895	287,116	1,310	1,421
20 – 99	27,289	31,438	1,092	1,186
100 – 499	4,740	4,494	913	855
> 499	697	598	1,214	939
Total	330,161	342,382	4,529	4,401
<20	.899	.839	.289	.322
20 – 99	.083	.092	.241	.269
100 – 499	.014	.013	.201	.194
> 499	.002	.002	.268	.213
Total	1	1	1	1
<20			4.4	4.9
20 – 99			40.0	37.7
100 – 499			192.6	190.2
> 499			1741.7	1570.2
Total			13.7	12.8

Considering the alleged inefficiencies of small business firms together with the obvious structural deficiencies of the Italian economy, one might be surprised to see that Italy, on average, is neither poor nor fares worse than the rest of Europe. Indeed, small and medium size business firms have been the most vital component in the Italian economy since the Italian lira entered the European Monetary System[2].

The Italian industrial structure may nevertheless seem rather exotic[3], but if one compares a typical Italian industrial district with a standard corporation, one realizes that the former is something like the latter – a virtual corporation which produces a few specialized tradables, with separate cost centres and fragmented but

1 See: Tattara-Occari (1997).

2 I.e. after 1979; Table 11.1 (middle) shows the increase in the small and medium-size firms' share of employment in manufacturing industries observed from 1984 to 1992.

3 Still exotic or more and more exotic, according to one's point of view.

direct control over the distinct phases and functions of the production process. In some way if the Coase-Williamson firm is a device for keeping out the market, but is in the market, Italian network firms and networks of firms bring the market inside the network itself, but do not raise funds in the stock market.

Economic theory considers small business in a mixed way. Alfred Marshall himself gave a dual interpretation of scale economies in that while scale economies at firm level yield a firm's bigness, scale economies at the level of sector or industrial district imply sustainability of a firm's smallness. Steindl (1947, pp.59-60), however, considered the continued existence of small firms an anomaly which can be explained only by the existence of some special factors. (1) Bigger firms are inevitably bound to grow and smaller ones to disappear, but this is a gradual process which takes time; (2) there is imperfect competition caused by transport costs, product differentiation and limited consumer information and irrationality which all favour small firms; (3) there is no sufficient incentive for oligopolistic firms to fight smaller firms; (4) small entrepreneurs accept unusually high risks at very low remunerations and are thus gamblers.

Ten years later, Wellitz (1957), in explicitly considering the Italian mechanical industry, observes that the survival of so many and so small enterprises may be explained not so much by the low intensity of competition between large and small firms and by the segmentation of the market, but rather by the existence of a real symbiosis between them, where subcontracting helps both small firms in surviving and big ones in not growing too much. The contribution of Wellitz is interesting in that it incorporates a precise psychological factor (p.123). Large companies do not want grow too much in view of the riskiness of being big when labour is a quasi-fixed input of production. Twenty years ago Lucas (1978) explained the enduring existence of small and suboptimal scale enterprises with the existence of a distribution of persons by managerial talent. While successful entrepreneurs manage their firms in growing and attaining optimal size, less successful entrepreneurs cannot do the same. They simply turn over in the market.

According to the increasing body of relevant literature, however, small firms and entrepreneurship are playing a much more important role than scholars have previously been inclined to acknowledge[4]. One of the most distinctive facts concerning the whole process seems to be the fundamental shift in the size distribution of firms which has occurred in the last twenty years all over the industrialized world. This fact has had an important impact on the growth performance of regions where small firms prevail.

There is indeed now a considerable distance between the old theory which saw in the small firm almost an anomaly and the current premise that small firms, far from being burdened by any inherent size disadvantage, are able to offer distinctive and valuable contributions to the economy. According to Acs-Audretsch (1993, p.4), small firms: (1) have a higher entrepreneurial content, which fosters technological change and innovation; (2) contribute to market turbulence and thus provide regeneration; (3) provide a mechanism for regeneration through newly

[4] In the last twenty years, however, there has been a complete revision of the interpretation of the potentiality and role of small enterprises (in particular: Piore and Sabel, 1983 and Acs-Audretsch, 1990).

created niches; (4) offer a larger share of newly created jobs. These contributions are clearly very positive for any region trying to develop its economy and it is obvious that local and urban development cannot take off without the birth and the subsequent consolidation of locally controlled activities. Acs-Audretsch (1990, p.4) enumerate some of the factors which would seem to explain the new role of small firms and the changes in the firm-size distribution. However, it is questionable whether the dynamics itself and general economic growth can be understood by studying firm-size distribution alone. Firm size is neither a good proxy for competitiveness nor a good predictor of growth. Neither large firms nor medium and small firms are, necessarily, superior. Many scholars are clearly aware of this. In a recent paper (Dosi et al., 1995), it has been argued that industrial dynamics defies received economic theory which in one way or another adopts an equilibrium concept. They believe that it is helpful and perhaps necessary to have some other representation of the system when it is in motion and they have accordingly formulated a model in which innovation is the driving force of change. But unfortunately they stick to firm-size and firm-age distributions as indicators.

It is surprising that despite the intricacy of the subject of market structure, the general approach adopted in the analysis of the organization of industry[5] appears to be rather informal. It is as if old and new ideas regarding the process[6] by which firm behaviour and market structure are jointly determined over time cannot receive a formal treatment. The larger part of modern and less modern studies on the issue of the organisation of Italian industry is made up of examples[7] of what Nelson and Winter (1982, p.16) have called appreciative theory[8], which is the theory aimed at interpreting the facts in an informal way. Giving some insights and judging verbally the value of the phenomenon or understanding in what way it is new and relevant, is really good, but it may not be enough.

14.3 Intra-Industry Dynamics and the Size of Firms

The obvious characteristic of the structure of many production sectors is not stability, but rather motion, change and even turbulence. Thus it does not come as a surprise that more and more attention is paid to the description and interpretation of the dynamics of market structure largely along the lines of the evolutionary theory of economic change as described, among others, by Nelson and Winter

---

5 Particularly the Italian industry.

6 In place of static equilibrium derived from the principle of profit maximisation.

7 Among others: Contini e Revelli (1990), Invernizzi e Revelli (1993), Rullani (1997), Tattara (1997), Bonomi (1997).

8 According to Nelson and Winter (1982, p.16) there are two styles – formal and appreciative – of theorizing . To them, both look necessary and each one influences the other. In this paper, however, the relation goes from appreciative to formal theorizing.

(1982) and Audretsch (1995)[9]. As is well known, the firm-size distribution is the most commonly used descriptive instrument of market structure and the law of proportionate change, or Gibrat's law, has played a prominent role in the explanations of the growth of firms[10]. But the evidence against it has been mounting and Gibrat's law is no longer considered a valuable predictive device. The firm-size distribution, always skewed, is a mere static representation of the market structure, which is almost constantly in a state of flux. There is no clear relation between position in the distribution and growth rate even though there is probably one between size and probability of survival. All this means that intra-industry dynamics cannot be understood by looking at size distribution.

In this paper it has been said so far that the crucial distinction to consider is not the one between small and large firms but that between good and less good firms or, respectively, superior and inferior firms. From the point of view of present and future performance, the distinction between superior and inferior firms, albeit difficult to apply ex-ante, looks more convincing. It is obvious that superior firms (it makes no difference whether small or large) survive and prosper[11], while inferior firms stagnate and then exit. Also from the point of view of employment, the distinction between small and large firms could be profitably substituted by that between inferior and superior firms. Not only are small firms generating a large share of total employment, but the share has increased in the recent past. Furthermore, superior firms can offer more chances to their employees than inferior firms. In what follows we will make use of the concept of space of firms. It is understood that it should be defined in terms of some variables, but in this paper it is assumed that it is known whether firms are superior or inferior and what makes them so. The space of firms can then be defined by similarity. If the space of events for casting a dice is six, the space of events for evaluating a firm, in the present case, is two: superior and inferior firms. It is obvious that if one wanted to find a more refined way of defining the distance between firms, the space could be enlarged accordingly.

Alongside the qualitative distinction between inferior and superior firms is the distinction between new and mature products[12]. In the discussions about the endless motion taking place inside the industrial structure products have often been neglected as the focus of the analysis is on companies. One has to recognize, however, that there is a product space, i.e. the environment in which firms operate. If one turns to the concept of the product life cycle one can find at least three possibilities. There are of course new products and mature products and a small firm can be as good or bad at producing a new (old) product as can a large one. It is extremely difficult to prove that large firms are definitely superior to small ones

---

9 There is of course a long tradition in mainstream economics according to which economic actors are perfect maximizers and the equilibrium market structure is the outcome of the careful use of rational decision-rules.

10 *"According to this law, the probability of a given proportionate change in size during a specified period is the same for all firms in a given industry, regardless of their size at the beginning of the period."* (Mansfield, 1962).

11 In some cases this means that they grow at a rate which is above the sector average.

12 Process innovation has to do with competition between firms.

and viceversa, provided that the latter have attained the minimum scale of production, not to mention the possibility that a really small firm – that is a firm operating below the minimum efficient scale – can exploit what Audretsch (1995, Chapt. 5) calls compensating strategies.[13] New firms enter the product space and sucessively move within this space in order to survive and prosper; in some cases they leave it. In what follows, a simple model will be presented, in which product variety and firms' dissimilarity combine into a description of industry dynamics: a process consisting of unending flows of products and firms entering and leaving the market.

14.4 Competition, Biodiversity and Industrial Demography

14.4.1 Preliminaries

When product variety expands as a result of innovation, increased vertical and horizontal differentiation and deverticalisation of production, the number of incumbent firms increases. It is assumed here that each product can be produced by a single firm or, rather, by a single activity, irrespective of the dimension of the specific product market. This implies that the same activity is at least a temporary monopolist. This approach chosen in order to look at the dynamic process of adaptation taking place inside the productive structure of the economy, is to disregard the size of firms[14] and to focus on the competition for the space of products, i.e. for the market.

The dynamic process is characterized by a continuous entry flux of new products and a parallel flux of old abandoned products, which can be the result of changing innovation content, new design, lower price, higher quality or simply fashion and consumer preferences. At the same time, there are a number of new business firms entering the market. New activities compete for the space of products; but at the same time there are a number of incumbent firms which exit from the market. Their exit may be motivated in several ways: it may be caused by profits, it may be the consequence of the abandonment of an old product or it may happen simply because the incumbent firm has been displaced by another firm.

The evolutionary process examined here has been studied by others – notably by Audretsch (1995), who proposes two different models. The first is the revolving door model, which is suited for sectors where new businesses enter, but where there is a high probability of a subsequent exit. This model is tailored for the routinized regime, i.e. for industries where incumbent firms hold an advantage in terms of innovation or know-how and new entrants do not. The other model is the forest or Marshallian model, which is for sectors where incumbent firms are

---

13 Compensating strategies are aimed at offsetting size-related disadvantages by deploying factor inputs and remunerating them differently than they are by the larger enterprises. (Audretsch, 1995, chapt. 5).

14 As a firm can have more than one activity. In the following, the terms "activity" and "firm" will be considered as equivalent.

displaced by new entrants. The latter seems to be more suitable for sectors dominated by the entrepreneurial regime where new entrants have the innovation advantage. To Audretsch, it does appear that whichever selection model is more applicable depends on various factors, namely technological conditions, economies of scale and demand.

In this paper the focus is not on why and how a business enters, survives, stagnates and exits or, depending on the circumstances, why and how it prospers and grows, as it seems pointless to relate the firm-size distribution to the dynamic process described above. It is assumed that size is not the only interesting or most relevant feature of a firm and consequently that firm-size distribution is not a suitable description of the industry structure.

The intention here is to model the dynamic process of entry end exit both of products and establishments to see the relation between the equilibrium eventually reached in the system (a specific sector in a region for instance) and the intensity of industry turbulence. When small firms are abundant, turbulence is high as both the variety of products and the number of firms which enter[15] and exit from the market is high and possibly increasing. In investigating the demography of the industrial sector, the focus will be on categories which are different from size and sectors and are rather more directly tied to their quality and behaviour: new, mature, old products and good and less good firms. In doing this, I look for the characteristics of dynamic equilibrium where equilibrium is not conceived as the state where the incentive to change vanishes but rather a state of the system in which entry and exit flows are always present, but balance.

14.4.2 The Dynamics within the Product Space

Products can be classified in many ways, but here it is assumed that they can be ordered according to the stage in their life cycle. Firms and activities , in their turn, can be classified according to their competitiveness, rather than their size, because the latter cannot be considered a reliable factor of competitiveness. There is then a space of firms formed by less competitive, or inferior, and competitive, or superior, activities. For simplicity's sake, it is assumed that products and firms are only of two species – new or mature[16]. The competitive firm (or type 1 firms) always displaces the un-competitive (type 2). The product may be a new or a mature one, but even though its maturity may be more or less close, the displacement occurs with probability equal to one.

Let us consider first the dynamics within the space of products. This is done by assuming that there are only two possible states for products: new or mature. New products enter into the space of products, after a while they become mature and subsequently they become old and disappear.

In each instant of time there are (P) products, some are new (N) and some are mature (M). After stating the exogenous dynamics within the two sets, the com-

---

[15] Current statistics report a very high rate of entry.

[16] Old products are simply abandoned.

position in the space of products can be easily established. In each period there are some new products (iP), some new products become mature (bN) and some mature products (cM) disappear altogether. This writes as follows:

$$dP = dN + dM$$
$$dN = (iP - bN) \tag{14.1}$$
$$dM = (bN - cM)$$

In order to determine the fraction of new products and the fraction of mature products which will establish in equilbrium:

$$\alpha = N / P, \text{ and } 1 - \alpha = M / P \tag{14.2}$$

If one differentiates in the first equation of (14.2), makes use of (14.1) and then solves the resulting dynamic equation:

$$d\alpha = i - b\alpha - (i - c)\alpha = 0 \tag{14.3}$$

the two shares are obtained:

$$\alpha = \frac{i}{i+b-c}, \text{ and } 1 - \alpha = \frac{b-c}{i+b-c} \tag{14.4}$$

where it seems sensible to assume $b > c$. The product variety changes at a rate equal to:

$$\frac{dP}{P} = i - c\frac{b-c}{i+b-c}, \tag{14.5}$$

which is positive if the weighted exit rate is sufficiently small in comparison to the entry rate: i.e. if $i > \dfrac{c}{1 + i/(b-c)}$. In the space of products there is a globally stable dynamic equilibrium characterized by a pair of constant shares to which the system bends for all possible initial states. As the average life spans of the two classes of products are the reciprocals of the exit rates, the average life span of all potential products in the economy and in equilibrium is simply the average:

$$s = \frac{1}{b} \cdot \frac{i}{i+b-c} + \frac{1}{c} \cdot \frac{b-c}{i+b-c} \tag{14.6}$$

If the specific rates are equal to one, $b = c = 1$, then s=1. If $b = c$, then $s = 1/b$. In general: $1/b < s < 1/c$ for $b > c$ and viceversa.

14.4.3 The Dynamics within the Space of Firms

To model the behaviour of shares in the space of firms, one has to learn from Marshall that "*the Mecca of economics (is) in economic biology rather than economic mechanics*". Then one can borrow directly from the competition model of population biology[17] assuming that there are only two species of firms. There could be many species, but I prefer to stick to the case of two species only as notation is much lighter even though the model retains all the basic properties. The difference is not, as one may think, between more innovative firms and less innovative firms or even between innovating firms and the others. As the space of products grows from itself, i.e. endogenously, firms are classified according to their ability to invade the available space and retain their position.

The model suggested here is a model of competition and coexistence for a subdivided space. It allows coexistence of competitive species even though species do not have the same ability in competition. The model explains why habitat subdivision can make possible the stable coexistence of an unlimited number of competing species. In explaining why almost any industry has firms flowing in and out, and why it has competitive and large (small) firms coexisting with less competitive and small (large) firms it seems worthwhile to borrow the space competition model and fit it into the submodel for products described above.

There are two species only for each one of the two segments of the space of products, which means that there are four sets of firms. Species 1 is formed by activities, i.e. firms, which always outcompete firms of species 2 in a product of the relevant segment. The reverse is not possible, by definition. In what follows, they will always be referred to in terms of their relative frequencies or contingencies .

In this section all parameters concerning entry and exit are assumed as constant[18], but in the next one this assumption will be relaxed by arguing that entry parameters change in relation to the profit perception of entrepreneurs. As can be demonstrated, there is always some space of product left empty for new entries, two points must be considered. First of all, the product-firm space is divided into six parts, as shown in Table 14.2

Table 14.2 Categories of firms and products

	New products (N)	Mature products (M)
Superior firms (1)	p_{1N}	p_{1M}
Inferior firms (2)	P_{2N}	p_{2M}
Empty space	$1 - p_{1N} - p_{2N} = s_N$	$1 - p_{1N} - p_{2M} = s_M$

[17] The model used in the following, called spatial competition model or competition model for spatially structured habitats or patches was proposed by Levins (1969) and refined by others such as: Hastings (1980), Tilman (1995). Bunchgrasses are interacting species which coexist, but compete, with unequal efficiency, for space and thus for nitrogen.

[18] In the previous section the same was done for the product parameters.

In the second instance, the perception by the entrepreneur of the existence of some empty space is a factor that can explain the changes in the entry rate. But this aspect will be considered in the final part, so first the basic equations of the Levins space competition model are stated without any mention of the relevant class of products, then the complete firm-product model is considered and some conclusions drawn.

The dynamics of occupancy for superior firms is determined by the colonization rate and the exit rate, respectively: c, m, and it is completely unaffected by the presence of inferior firms. The equation is as follows:

$$\frac{dp_1}{dt} = c_1 p_1 (1 - p_1) - m_1 p_1 \tag{14.7}$$

The equation states that the fraction of products offered by superior businesses increases at a rate proportional to the products of the two fractions and decreases in proportion to its relative abundance. The reason is that there is an invasion from the occupied space (p_1) into the free space (p_2) and this occurs at a certain rate (c_1). At the same time the occupied space is left empty by ailing firms, which occurs at a rate (m_1). The differential equation is globally stable in the sense that the fraction p_1 reaches its equilibrium value, for all possible initial states and returns to it after any perturbation, except in the case where the fraction becomes zero after the perturbation. By equating to zero the equation above and solving, the proportion of products occupied at equilibrium ($p_1 = P_1$) is immediately obtained:

$$P_1 = 1 - \frac{m_1}{c_1} \tag{14.8}$$

To survive, the mortality rate of the species must obviously be lower than the colonization rate. The movement from the initial share towards the equilibrium one given by the equation above, is obtained by solving the dynamic equation (14.7):

$$p_1(t) = \frac{p_1(0)(c_1 - m_1)}{p_1(0)c_1 + (c_1 - m_1 - p_1(0)c_1)e^{-(c_1 - m_1)(t)}}$$

which means that for any $p_1(0) > P_1$, the fraction decreases exponentially to P_1, and with $p_1(0) < P_1$, the same fraction increases logistically to P_1.

The equilibrium fraction of superior species just determined, is able to influence the fraction of the inferior species, but not viceversa. More interestingly: even at equilibrium, and even if there is coexistence, there is some empty space which encourages the entry of would-be new entrepreneurs. There are always some new entrepreneurs starting new businesses and the present model rather nicely captures this fact, i.e. the start of new firms even though there is equilibrium. To survive when condition (14.8) holds, the inferior species must meet some conditions which imply a kind of vitality in the species itself. To clarify this point let us consider the equilibrium fraction of the inferior species and the related necessary and sufficient conditions of survival. If one therefore solves for the steady state the equation for the inferior species, the equation writes as:

$$\frac{dp_2}{dt} = c_2 p_2 (1 - P_1 - p_2) - m_2 p_2 - c_1 P_1 p_2 = 0 \tag{14.9}$$

where the first term is the entry rate, the second is the exit rate and the last the displacement operated by the superior species. In view of equation (14.8), the equilibrium or steady state solution $(p_2 = P_2)$is:

$$P_2 = \frac{m_1}{c_1} - \frac{c_1 - m_1 + m_2}{c_2} \tag{14.10}$$

The above equation, together with (14.8), clearly shows that any increase of the specific colonization rate c_2 implies an increase in the equilibrium share of the inferior species. The same equation yields the necessary and sufficient condition for the survival of the inferior species, namely:

$$c_2 > c_1 \frac{P_1}{1 - P_1} + m_2 \frac{1}{1 - P_1}. \tag{14.11}$$

Provided that $c_1 > m_1$, the condition above allows the coexistence of the two species. It makes clear that the colonization rate must not only be higher than the mortality rate, as is the case for the superior species, but it must be higher, the higher the fraction of product space occupied by the first species. Thus:

$$c_2 > c_1, \text{ and } c_2 > m_2 \tag{14.12}$$

In equilibrium there is always some space left open, as the sum of fractions (14.8) and (14.10) occupied by incumbent firms comes out as:

$$P_1 + P_2 = 1 - \frac{m_2}{c_2} - \frac{c_1}{c_2} P_1 = 1 - \frac{c_1}{c_2} + \frac{1}{c_2}(m_1 - m_2) \tag{14.13}$$

Even in the special case of uniformity in the mortality rate (i.e. $m_1 = m_2$), equation (14.11) implies that: $P_1 + P_2 < 1$. The above result seems very interesting as turbulence in the industry – which is present even in equilibrium – is directly related to profit possibilities perceived by potential entrants and therefore to what they perceive as empty space. The fraction :

$$1 - P_1 - P_2 = \frac{c_1}{c_2} + \frac{1}{c_2}(m_2 - m_1) \tag{14.14}$$

can be considered as the really empty space fraction, as there is also the occupied, but potentially invadable space.

14.5 The Combined Dynamic Equlibrium

When considering the structural dynamics of market for new and for mature products with the competition model just described, some clarification must be made as to what the differences are in the two segments. Each segment will have its own equilibrium allocation of product space to the activities of the two species. It would seem reasonable to assume that the balance between superior and inferior businesses is not the same. It is presumed that in the mature products segment, competition has already been able to select more superior firms than in the new products one. The selection process should have been able to retain more superior than inferior firms

$$P_{1M} > P_{1N} \tag{14.15}$$

As equation (14.11) above shows, in order to survive, inferior firms must have a colonization rate which increases with the fraction P_1. It follows that the assumption of an expanding P_1, implies:

$$1 < \frac{c_{2N}}{m_{2N}} < \frac{c_{2M}}{m_{2M}} \tag{14.16}$$

Equation (14.13) shows that there is always some empty space. This means that, in equilibrium, the average product life-span is given by the following more complex expression, rather than the one given in equation (14.6):

$$s = \frac{1}{b} \cdot \frac{1}{1 + \frac{b-c}{i} x} + \frac{1}{c} \cdot \frac{1}{1 + \frac{i}{b-c} \frac{1}{x}} \tag{14.17}$$

where x is the ratio between the two total shares, namely:

$$x = \frac{P_{1M} + P_{2M}}{P_{1N} + P_{2N}} \tag{14.18}$$

Equation (14.17) is only approximately equal to (14.6) as $x \neq 1$. According to the above assumption the ratio is indeed greater than one. This is clearer in the case of equal mortality rates, where the last ratio is in its turn approximately equal to:

$$x \cong \frac{1 - \frac{c_{1M}}{c_{2M}}}{1 - \frac{c_{1N}}{c_{2M}}}. \tag{14.19}$$

In order to take into consideration market turbulence – i.e. large and unstable entry and exit flows – one must distinguish the flows implied by the dynamic process which brings the system to equilibrium and keeps it there, from the flows resulting from the changes in the model parameters. To deal with the latter, one must relax the assumption of constant parameters made so far. The model used here actually has three sets of parameters. There are four entry parameters, four exit parameters and three product parameters. The focus here has been only on entry parameters by assuming that profit expectations are directly tied to perceived empty product space (equation 14.14). Self-employed people, small, medium and large firms – which can be new or simply already operating in different products, sectors and countries may also be interested in diversifying their current production – are presumably attracted by the perception of existing really empty space. As the flow related to invasion of already occupied space is already present in the last term of equation (14.9), one can safely limit oneself to really empty space as defined by equation (14.14).

As covered shares (p_1) and (p_2)progressively increase, entry parameters decrease. A reasonable assumption then is that entrepreneurs of species 1 react to the perception of a more abundant empty space by increasing the entry rate and by decreasing the same rate when empty space is scarce. There is, however, a complication. As entry flows are determined by entry coefficients and market shares there are always two separate effects to consider. An increase in market shares has two distinct effects on entry flows. There is one positive effect which is related to the increase itself in market shares, and a negative one related to the adjustment of coefficients which is in order whenever the system is out of equilibrium.

Inspection of equation (14.14) reveals that when $(c_1)$is increasing, empty space becomes more abundant, which implies a positive feedback. The same equation (14.14) shows, however, that an increase in (c_2) implies a contraction of empty space which implies a negative feedback. The maintained assumption according to which entry parameters are positively related to available empty space can then be translated into following simple linear reaction function for c_{ij}, the current entry coefficient of firms of type i, product j:

$$\frac{dc_{ij}}{dt} = \alpha_{ij} - \beta_{ij} c_{ij} \tag{14.20}$$

where : $1 > \beta_{1j} > \alpha_{1j} > 0$, for: $i = 1.$, and $j = N, M$
and $1 < \beta_{1j} > \alpha_{1j} > 0$,for: $i = 2$, and $j = N, M$.

The alpha is the forcing term that can be interpreted as the intrinsic entry rate while the beta is a positive reaction coefficient. In the state of dynamic equilibrium, the following steady state entry coefficient holds:

$$c_{ij} = \frac{\alpha_{ij}}{\beta_{ij}}, \tag{14.21}$$

but out of equilibrium the changing reactions add an element of instability to the entry flows. Under the above condition, equation (14.11) therefore implies:

$$\frac{\alpha_{2j}}{\beta_{2j}} > \frac{\alpha_{1j}}{\beta_{1j}} \cdot \frac{P_{1j}}{1-P_{1j}} + m_2 \frac{1}{1-P_{1j}} \tag{14.22}$$

for: $j = N, M$

The equation shows that, in equilibrium and for given parameters of type 1 firms, the survival of inferior business activities is explained by a high intrinsic entry rate $(\alpha_{2j})$and/or by a low reaction coefficient (β_{2j}). When the system is out of equilibrium, entry and exit flows from any one of the two markets, depend not only on equations (14.7) and (14.9), but also on the current values of the entry parameters which are provided by solving equation (14.20):

$$c_{ij}(t) = \frac{\alpha_{ij}}{\beta_{ij}} + \left[c_{ij}(0) - \frac{\alpha_{ij}}{\beta_{ij}} \right] e^{-\beta_{ij}t} \tag{14.23}$$

for $i = 1.2,\ j = N, M$.

The above equations are not only able to explain the coexistence of differentiated products and firms but also the turbulence which does not disappear even in the state of dynamic equilibrium. One could say that coexistence exists and increases and turbulence decreases, but not completely even in a state of dynamic equilibrium.

As far as exit parameters are concerned, the best assumption could be to relate the exit probability of any firm to such entities as firm size and firm growth rate as it has been proved that larger and initially fast growing firms are less likely to exit. But this task will not pursued as in this model, size and growth are not considered at all. The parameters regarding product space (respectively: i, b, c) can change a lot by adding to this a further motion component to that expressed by the differential equations above and by the very state of dynamic equilibrium. It is difficult to say if any increase in the rate of introduction of new products $(i)$is backed by a parallel increase of the other two transfer rates $(b)$and (c).

Turbulence, which is directly related to the current value taken by these parameters and to the availability of empty space, may increase when the dynamics within the space of products, which has been considered so far as invariant, undergoes a change. This change can take the form of a new rate of change of innovation or a new rate at which new products become mature (and mature ones become old). Such changes too imply a variation in the empty fraction of the space of firms. If any such change does occur, the initial dynamic equilibrium would be displaced and firms would react to the perceived new situation by starting a new dynamic phase during which entry and exit flows could exhibit large changes and eventually reach a new dynamic equilbrium.

14.6 Conclusions

If one looks at the literature regarding the market structure one gets the impression that the firm-size distribution and the degree of competition between firms are the two most relevant features. This paper is not an exception as, starting from a few remarks on the Italian case, it offers some arguments about the just mentioned features. However, the focus here is more on competition rather than on size.

The firm-size, and particularly optimal firm-size, are obviously related to a firm's performance and to its growth. The emphasis placed on this issue probably helps to explain why the continued existence of small – i.e. sub-optimal – firms beside large firms was initially considered by the economic profession as a real anomaly. It is only during the last twenty years, and certainly in view of the increasing role of small firms not only in Italy, that there has been a re-appraisal of the role of small business firms. However it is difficult to find any real advance from the early contribution of Marshall and others, in the explanation of the "why and how" of small firms. The novelties lie in the new interpretation of the economic role of small businesses in creating jobs, in innovating, and in keeping the production sector vital.

The firm-size distribution obviously is an often reported fact which does not obey Gibrat's law and which has little to say about the growth performance of specific firms, as growth performance seems not to depend on size. But the coexistence of firms differentiated by size is here to stay and it is something too big to be completely sidestepped. In the paper, however, one has been more interested in the market structure per se than in its impact on the economy, and in considering the issue it has been chosen to disregard size and focus on competition and particularly the effects of competition on market turnover. It has been also considered product variety by distinguishing new from mature and old products and the related turnover, as competition is strictly tied to product variety. The ensuing arguments regarding market structure are then referred to the dynamics within what it has been called the "space of firms" and the "space of products" (Section 14.3).

With the formal model offered in the final part of the paper an attempt has been made to investigate some remarkable, but different from size, aspects of industry demography. The unending flows of entry, exit of firms and products is indeed an aspect worthy of further investigation. The competition model, which is borrowed from biology, shows the conditions under which there is a dynamic equilibrium between flows when firms are different in their performance. The equilibrium turns out to be characterized by the coexistence of differentiated firms and products, and market turbulence.

The model is only a modest and preliminary tentative in that it is only a first step towards a formal analysis of the market process, but one, however, which seems able to add something to what we can understand from the mere appreciative theory of market processes.

References

Acs, Z. and D. Audretsch, (eds.), 1990, *The Economics of Small Firms, A European Challenge*, Kluwer, Dordrecht.

Acs, Z. and D. Audretsch, 1993, *Small Firms and Entrepreneurship: An East-West Perspective*, CUP, Cambridge.

Audretsch, D., 1995, *Innovation and Industry Evolution*, MIT Press, Cambridge, Mass.

Bonomi, A., 1997, *Il capitalismo Molecolare, La società al lavoro nel Nord Italia*, Einaudi, Torino.

Contini, B. and R. Revelli, 1990, "The Relationship Between Firm Growth and Labour Demand", in Z. Acs and D. Audretsch (eds.), *The Economics of Small Firms, A European Challenge*, Kluwer, Dordrecht.

Dosi, G., O. Marsili, L. Orsenigo and R. Salvatore, 1995, "Learning, Market Selection and the Evolution of Industrial Structures", *Small Business Economics*, 7:411-436.

Hastings, 1980, "Disturbance, Coexistence, History and Competition for Space", *Theoretical Population Biology*, XVIII:363-373.

Invernizzi, B. and R. Revelli, 1993, "Small Firms and the Italian Economy: Structural Changes and Evidence of Turbulence", in Z. Acs and D. Audretsch (eds.), *Small Firms and Entrepreneurship: An East-West Perspective*, CUP, Cambridge.

Levins, R., 1969, "Some Demographic and Genetic Consequences of Environmental Heterogeneity", *Bulletin of the Entomological Society of America*, Vol. XV, 2:237-240.

Lucas, R.J., 1978, "On the Size Distribution of Business Firms", *Bell Journal of Economics*, Vol. IX, 2:508-523.

Mansfield E., 1962, "Entry, Gibrat's Law, Innovation, and the Growth of Firms", *The American Economic Review*, VII:1023-1051.

Marshall, A., 1948, *Principles of Economics*, 5th edition, Macmillan, London.

Nelson, R. and S. Winter, 1982, *An Evolutionary Theory of Economic Change*, Harvard Press, Cambridge, Mass.

Piore, M.J. and C.F. Sabel, 1985, "Italian Small Business Development: Lessons for US Industrial Policy", in J. Zysman, and L. Tyson (eds.), *American Industry in International Competition*, Cornell UP, Ithaca.

Rullani, E., 1997, "Rapporti tra Imprese", in P. Feltrin (ed.), *Quale Società della Piccola Impresa,* Nuova Italia Scientifica, Roma.

Steindl, J., 1947, *Small and Big Business, Economic Problems of the Size of Firms*, Basil Blackwell, Oxford.

Tattara, G. and F. Occari, 1997, Struttura e Organizzazione del sistema produttivo veneto nell'ultimo quindicennio, in P. Feltrin (ed.), Quale Società della Piccola Impresa, Nuova Italia Scientifica, Roma.

Tilman, D., 1994, "Competition and Biodiversity in Spatially Structured Habitats", *Ecology,* Vol. VXXV, 1:2-16.

Wellitz, S.H., 1957, "The Coexistence of Large and Small Firms: A Study of the Italian Mechanical Industries", *Quarterly Journal of Economics*, VXXI:116-131.

15 Knowledge Workers, Communication, and Spatial Diffusion

Niles Hansen
Department of Economics, University of Texas
Austin, TX 78712, USA

15.1 Introduction

Although local innovations may have little effect on the total production system of a nation, some generic technologies involve spillovers from one innovation to another, and from one sector producing the technology to others using it. Some key innovations may be complementary, giving rise to dynamic increasing returns to scale through clustered interdependencies among innovation, investments, information and knowledge. If the continuous upgrading and diversification of goods and services is extrapolated to the national level, innovation becomes endogenous within each country. As a result, the relative growth of internationally open economies is closely related to the ability to innovate and gain market shares, in contrast to neoclassical models that typically assume closed economies with exogenous technological change. Because newer models recognize that there is a constant flow of innovations coming from the competitive process itself, they are dynamic and sequential, and they indicate that no one unique equilibrium growth path usually exists.

The new emphasis on the roles of innovation and positive externalities – benefits for which no payment is made – in economic progress has been accompanied by a new interest in the geography of innovation diffusion, within nations as well as globally. Particular attention has been given to regional clusters of economic activity on the ground that proximity of firms should enhance communication possibilities and thus promote innovation diffusion through dynamic externalities that occur at the level of individual production establishments. These dynamic externalities may arise from such sources as access to research and development (R and D) information through formal or informal personal contacts and firm networks, access to skilled personnel, and cooperative relationships with suppliers and sub-contractors in place of relationships with such firms based simply on competitive price bidding.

The following section presents a number of different theories concerning how the geographic clustering of firms promotes the competitive advantage of the firms concerned and that of the regions in which they are located. Although differing in emphases, these theories all imply that firms in a peripheral location are likely to be disadvantaged by not belonging to an urban agglomeration. In contrast, the second section presents evidence that firms can benefit from participation in national and global innovation networks, whether or not they participate in local or regional innovation networks. In the third section it is argued that whatever the spatial nature of networking processes, much of the knowledge required for innovation is gained from sources outside of a firm or sector, and that innovations are often made as a result of user-driven needs. The fourth section examines the role of information and communications technologies (ICT) as a means for firms to not only communicate their innovations to potential users, but also to communicate their needs for innovation to potential suppliers of innovative activities. Although these technologies have contributed to a greater spatial dispersion of some manufacturing and service activities, they have not eliminated the need for face-to-face communication, especially when learning processes involve complicated ideas. Advanced firms and regions have benefited from the network externalities of ICT, but ICT has not been sufficient to generate economic development in regions that lack knowledge workers and the necessary organizational and institutional capacity. In the concluding section it is argued that while ICT cannot create local innovation networks where none existed before, it can contribute greatly to the capacity of peripheral areas to engage in external networking on a global scale and to innovate, if the appropriate educational and civic environment has been established. Learning and innovation require above all knowledge workers who not only receive information, but who can also recognize the significance of information and knowledge and use them to innovate successfully.

15.2 Competitive Advantage and Geographic Clustering

Alfred Marshall was the first person to specifically recognize the mutual advantages (localization economies) that could be obtained from the geographic clustering of firms in the same industry, especially if they are small and medium-size enterprises (SMEs). In particular, he remarked upon the benefits arising from the coordination of distinct stages among differing segments of the production process, the development of education and skills, and the spread of ideas (knowledge) among firms within the core production system, which he termed the industrial district (Marshall and Marshall, 1879, pp.52-3; Marshall, 1920, p.271).

Renewed interest in the significance of localization economies in industrial districts came about in the 1980s, largely due to the work of Piore and Sabel (1984) on the third Italy, a term used to distinguish it from the old industrial area in northwest Italy and from the less developed area south of Rome. Within the Third Italy there is a pronounced tendency for production to be carried out efficiently by SMEs, and for localities to specialize in the production of a range of related goods. The competitive advantage of SMEs is secured by the use of flexible manufacturing technologies, specialization in niche markets, and collaboration among competing firms, which is facilitated by local social relationships. Similar examples of localization economies in industrial districts have been found in Denmark (Hansen, 1991), Germany (Best, 1990), and elsewhere in Europe. Indeed, there has been an abundant literature in the last decade on local "innovative regional milieux" in Europe (Maillat and Lecoq, 1992; Maillat, 1995; Camagni, 1995), indicating that clustered development has been successful in small and medium-size localities where there is an entrepreneurial culture and civic involvement in promoting competitive advantage.

Porter (1990), in his study of the competitive advantage of nations, argues that many successful industries locate within a single town or region within a nation to such an extent that it is questionable whether the nation is a relevant unit of economic analysis. The reasons given for such clusters are essentially Marshallian, though he proposes that a concentration of rivals stimulates innovation,whereas Marshall emphasizes cooperation rather than rivalry. Porter also finds that the trust and reciprocity needed for collaborative success among regionally clustered firms is less evident in the United States and the United Kingdom than in many areas of continental Europe.

Saxenian (1994) maintains that Silicon Valley is a U.S. counterpart to the innovative industrial districts of Europe, and that it has been more resilient and adaptable than Boston's Route 128 electronics complex. Despite similar origins and technologies, the two regions have evolved different forms of industrial organization. Silicon Valley has an industrial system based on regional networks that promote entrepreneurship, collective learning, flexible adjustment, and experimentation. In contrast, Route 128 is dominated by a small number of relatively vertically-integrated corporations that keep largely to themselves. These experiences indicate that regional economies based on networks are more flexible and technologically dynamic than those in which learning is confined to individual firms. To survive and flourish, these networks need a region's institutions and culture to ensure the repeated interaction that builds mutual trust while also maintaining competition. The clustering of firms in a given area does not by itself create decentralized processes of collective learning and continual innovation.

Krugman (1991) argues that regions that have a head start, typically because of historical accident, will attract industry and growth from regions with less favorable initial conditions. For Krugman, the interaction of external economies of scale with transportation costs is the key to explaining regional concentration and the formation of regional centers and peripheries. Although Krugman invokes certain types of Marshallian externalities, he has preferred to concentrate on externalities that can be modeled rather than on more elusive spillovers of technological knowledge (Martin and Sunley, 1996).

In contrast, Glaeser et al., (1992) and Henderson et al. (1995) have used sectoral analyses in attempts to evaluate the nature and significance of dynamic knowledge externalities. In each case, they examine U.S. cities on the ground that urban concentration provides an environment where ideas flow locally and thus rapidly from persons to persons, resulting in an expansion of knowledge and innovation. As in the case of static externalities, dynamic externalities may be localization economies that involve interactions among local firms in the same industry; or they may be urbanization economies that result from an expansion of knowledge due to the interactions among a variety of geographically proximate industries. For all economic activities taken together, Glaeser et al. conclude that dynamic externalities are all accounted for by urbanization economies. Using a more detailed sectoral breakdown, Henderson et al. find that for mature capital good industries there is evidence for dynamic localization externalities, but not for urbanization externalities. However, for new high-technology industries both types of externalities are found, suggesting that new industries do best in large, diverse agglomerations; but with maturity, production decentralizes to smaller, more specialized cities, according to the concept of spatial-temporal product cycles.

Finally, Audretsch and Feldman (1996) find that in the United States industries in which new economic knowledge plays a relatively important role also tend to have a greater geographic concentration of production. However, even after controlling for the concentration of production their evidence indicates that industries where knowledge spillovers are more prevalent – that is, where industry R and D, university research, and skilled labor are the most important – have a greater propensity for innovative activities to cluster than is the case for industries where knowledge externalities are less important. In an analysis of the three most rapidly-growing industries in Israel, Shefer and Frenkel (1998) found that agglomeration economies significantly increased innovation potential in the high-technology electronics industry, though this was not the case for the plastics and metals sectors, which utilized more traditional technologies.

15.3 Dynamic Externalities and Innovation Diffusion: Is Proximity Really Necessary?

The evidence just presented suggests that firms located in geographically peripheral areas lose competitive advantage to firms benefiting from dynamic externalities arising from urban concentration. But is it the case that "Externalities are highly dependent upon distance," and that "Proximity is thus a strong necessary condition to take advantage of externalities generated by other firms" (Antonelli, 1992, p.20).

Appold (1995) argues that the agglomeration of manufacturing establishments is often erroneously interpreted as apparent evidence for the existence of locally-bounded advantages; and that many researchers, by reporting for the most part on allegedly successful agglomerations, have ignored the possibility that firms with characteristics similar to those in agglomerations, but which are not located in such areas, may be just as productive and innovative. Using a random sample of almost 1,000 U.S. metalworking plants, Appold found that establishment performance was enhanced by successful interfirm collaboration, but that neither performance nor collaboration depended on location within an agglomeration. Cosmopolitan firms that draw on wide areas of resources gain externalities that make them competitive on a world scale. In contrast, firms dependent on local resources may represent a second tier that creates neither good jobs nor technological innovation.

Angel (1995) found that about one-third of a random sample of 495 manufacturers in the chemical, instruments and electronics industries engaged in some form of collaborative technology development activity with customers, suppliers or other firms. Enterprises thus engaged are on average the more innovative firms within an industry. Interfirm collaboration is more prevalent among large firms than small firms, and among establishments serving international markets. However, only a small proportion of collaborative partnerships are with local firms; most are national and international in scope. The local benefits from collaboration therefore come mainly from the strengthened performance of participating local firms, rather than from the emergence of localized networks among manufacturers, customers and suppliers. Hansen and Echeverri-Carroll (1997) obtained responses from 178 high-technology establishments in Texas concerning the importance of various types of formal and informal interfirm relations for their business performance. The most important relations were with non-local firms engaged in the same activity, followed by relations with local firms in the same activity and local firms in other activities. Non-local relations were particularly important for manufacturing firms, independent establishments, large establishments, and establishments that exported. Kaufman et al. (1994) found that New Hampshire's leading industries benefit from collaborative relationships with sophisticated customers in such high-technology sectors as aerospace, scientific and industrial instrumentation and information technology. Contrary to Porter's proposition concerning the importance of geographic concentration and firm rivalry, New Hampshire companies do not compete with one another; their competitors are widely distributed throughout the United States and the world. Geographic location has little effect on firm rivalry and product innovation. Moreover, 52 percent of New Hampshire manufacturers reported that geographic proximity to customers was of little or no importance, and another 32 percent stated that is was only moderately important. Communication and transport technologies made proximity unimportant. Upton (1995) found that SMEs in the U.S. petroleum distribution sector have gained competitive advantages through collaborative networks sustained by trust, understanding and continuity. However, these networks are composed of company presidents located in widely separated areas.

Suarez-Villa and Walrod (1997) analyzed establishment-level data from seven R and D-intensive electronics sectors in Los Angeles, Orange and Ventura counties (the Los Angeles basin) to determine whether spatial clusters are more supportive of R and D than non-clustered locations. The establishments located in non-clustered areas tended to be both smaller and more R and D-intensive than those located in clusters. R and D-intensive firms may have sought locations away from the larger firms found in clusters in order to maintain their independence and creativity and to secure greater secrecy for their R and D activities. The operational strategies considered, outsourcing and just-in-time deliveries, were found to promote R and D more strongly in the non-clustered plants. Moreover, in a comparison with Silicon Valley establishments, the non-clustered Los Angeles basin plants had stronger operational support for R and D.

Denmark, where entrepreneurial product-specialized industrial districts are common, has one of the most export-oriented economies in the western world. Nevertheless, there is hardly any district in the country where manufacturers are able to organize an entire value-added chain for a product that is eventually exported. Instead each individual in a location relates to a complex network of regional, national and international firms for knowledge concerning inputs, markets and technologies. Even if the producer for the final market subcontracts to other regional firms, they in turn obtain their inputs from regional, national and international sources. By using its specialized assets flexibly for different customers, the individual SME participates through indirect linkages in a geographically wide network of value-added chains (Kristensen, 1994).

Although it has a small domestic market, high costs, a high degree of regulation and a scarcity of venture capital, Sweden has developed an important export-oriented, technologically-advanced electronics sector. The geographic division of labor in the industry contradicts assumptions concerning the need for agglomeration of high technology industries (Suarez-Villa and Karlsson, 1996). There has been a progressive dispersion of firms owing to the creation of new enterprises in hinterland areas. No county, except Stockholm, has more than 10 percent of the establishments or employment in electronics. Many indicators show that hinterland location actually has advantages for research-intensive production. The development of infrastructural access as well as human resources infrastructure has played a key role in this process. Local initiatives with respect to labor training, managerial support and facility construction have also been important. The development of trust-based strategic alliances and other forms of mutually advantageous interfirm cooperative arrangements for subcontracting and research has been a particular characteristic of the Swedish experience.

Among all counties in the United Kingdom, Hertfordshire has the highest relative concentration of high-technology employment. Towns in Hertfordshire are also the locations for some of the highest relative and absolute concentrations of R and D employment. On the basis of survey research involving innovative firms in this county, Simmie (1997) found that innovation is a much more chaotic activity than suggested by prevalent theories. Numerous seemingly unrelated innovations have emerged from a complex knowledge base that is embodied in the highly-educated, professional work force that has chosen to live in and around Hertfordshire. The availability of venture and long-term capital from sources in nearby London is also a major contributory factor to innovations in the area. Successful innovating firms compete in global markets from their base in Hertfordshire, but there is little evidence of either systematic networking or high-level linkages contributing to award-winning innovations. In particular, the importance of any local networks in contributing to innovation is rated as "low" or "none at all" by local firms. The few local linkages found mostly appear to involve low-level support services such as office suppliers and cleaners. Thus, while demand pull from global markets is a key stimulus to innovation, local interfirm relations do not appear to contribute much to knowledge and innovation diffusion.

Wiig and Wood (1997) examined the innovation activities of manufacturing firms in the county of Møre and Romsdal in Central Norway. They selected this county because it is widely recognized as a region for innovation activities in traditional industries, especially furniture, fabricated metal products (primarily shipbuilding), and fish products. It has a history of entrepreneurial skills and in both 1982 and 1992 it received a high number of patents in relation to other counties. An earlier tradition of collective entrepreneurship through cooperation gave rise to an economic vitality that in turn paved the way for initiatives taken mainly by individual entrepreneurs. The latter are conscious that they are self-sufficient in undertaking innovation; most have little internal research but rather engage in development and trial work. Large firms, however, look more to the external environment for support of their innovation activity. Entrepreneurs see the presence of related firms as of little importance to their activities, but, especially for small firms, the lack of cooperation possibilities is not seen as an obstacle to innovation. Thus, while the evidence indicates that many firms do in fact undertake innovation in products and processes, it also suggests that "individual firm strategies and networks actually work against the formation of a visibly integrated regional innovation system" (p.95).

Whatever the spatial nature of interfirm networking, it is necessary to understand what drives innovation processes that stimulate regional development and to consider the extent to which modern information and communications technologies can stimulate and enhance such processes. These issues are addressed in the next two sections.

15.4 User-Driven Innovation and Regional Development

Scientific discoveries clearly broaden the possibilities for useful innovation, but Schmookler's classic study of nearly 1000 innovations in the railway, petroleum refining, paper making, and farming sectors indicated that in no instance was the stimulus for the invention a particularly scientific discovery. Rather, in nearly every case "the stimulus was the recognition of a costly problem to be solved or a potentially profitable opportunity to be seized, in short, a technical problem or opportunity evaluated in economic terms" (Schmookler, 1966, p.199). Scott (1989) has similarly argued that inventions and investment are both motivated by their expected profitability. Moreover, evidence suggests that expected profitability is likely to be higher in industries other than that which originated the invention. Scherer (1982) observed that U.S. research and development outlays allocated to industries of use were generally a more important determinant of productivity growth than such expenditures allocated by industry of origin. Geroski's (1991) analysis of sectoral sources of innovations used in the United Kingdom similarly found that most of the innovations adopted in any particular sector were produced in some other sector. Thus productivity growth in any particular sector is likely to depend as much on flows of knowledge throughout the economy as it does on the generation of new knowledge in the particular sector alone. The policy implication he draws is that efforts to stimulate the diffusion of existing knowledge are more likely to enhance competitiveness than are efforts to stimulate the generation of new knowledge. Howell's (1990) study of the development of global research networks indicates that newly emerging technologies and industries do not arise smoothly from existing firms and industries, but rather rely upon Schumpeterian creative destruction, whereby much of the knowledge required for innovation must be gained from outside the firm and sector. The important ramifications of the biotechnology sector for the pharmaceutical, chemical and energy sectors is a case in point.

A problem in the relationships between producers and users of innovations arises when producers dominate because of their superior financial resources or technical competence. In this situation there is an inherent tendency to develop costly innovations that may not be well suited to the needs of users (Lundvall, 1993). In contrast, innovation diffusion is more efficient when producers coordinate their innovative capabilities with the requirements of potential users.

If firms are to remain competitive and able to move into new markets they must be able to maintain close relationships with their existing and potential customers, because the sources of new ideas for new products and innovations increasingly are user firms and industries (Howells, 1990). Moreover, it is increasingly evident that sophisticated producer services play a strategic role in the process of creating dynamic information and knowledge externalities (Hansen, 1990). Cornish (1997) argues that information about markets, or market intelligence (MI), is a particularly vital service with respect to product innovation. In her empirical study of the Canadian software product sector, she determines the geographic sources of the various components of MI and addresses the implications of distance between producers and markets for the acquisition of information such as MI. The findings indicate that while MI is a crucial input to product innovation, proximity between producers and markets plays only a limited role in effective product innovation. Continuous feedback from distant markets is a key factor in continuous innovation, and firms that acquire more market intelligence have higher sales, faster growth, and greater profitability. In New Hampshire, collaborative relations with customers have helped the leading industries to improve their products and production processes. These industries work over long distances with sophisticated customers to understand their specific needs and use this information to design and make products that meet these requirements (Kaufman et al., 1994).

Given the importance of health care to the quality of life, and indeed often to life itself, one might suppose that easy communication possibilities would make medical practice fairly uniform in a country such as the United States. Although there is no evidence that Americans who have more medical care or who pay more for it are healthier, regional variations in medical costs and in methods of treatment are so pronounced that they would not have happened by chance (Kolata, 1996). Moreover, areas with the highest supply of medical services have the highest costs, indicating that supply is the driver of demand, and supply in turn is driven by "professional norms," which are essentially fads or fashions that vary by region. There is an analogy in manufacturing. Empirical analyses by Rigby and Essletzbichler (1997) show that within U.S. manufacturing there are substantial long-run variations in regional techniques of production that cannot be explained by industry mix or business cycles. Technology tends to move along different trajectories conditioned by local learning processes.

What are the policy implications? Regional agencies rarely recognize that innovation is so often a response to the needs of users. From the perspective of innovation diffusion, a systematic effort should be made on a larger geographic scale to communicate specific problems that regional firms need to have solved to firms outside of the region, so that they can realize specific profitable opportunities from innovation diffusion to firms within the region. Given that firms usually know more about their problems than do medical patients, it would be particularly advantageous to have the opportunity to explore the different innovative technological possibilities potentially available from different regions. This type of information system would in effect be regional policy in the same sense as an efficient national labor market information system, which has frequently been advocated as a means to facilitate regional economic adaptation.

15.5 Information and Communications Technologies (ICT) and Innovation Diffusion

ICT is a term often used to designate the combined use of electronics, telecommunications, software, decentralized computer workstations, and the integration of information media (voice, text, data and graphics). Manufacturing has in the past been treated as a separate function from research, development, or design, but now it is increasingly being integrated with these operations within the context of flexible production. Moreover, new ICT allows firms to overcome geographical constraints and to restructure business relationships by bypassing intermediaries while networking with desired organizations.

Computer networks have long been part of manufacturing firms and other businesses, but only recently have separate firms started to tie their networks together to take advantage of data interchange as a form of cooperation and integration. In a survey of U.S. manufacturers carried out in 1996, only 13 percent of the respondents indicated that the Internet affected their operations in any way. Most viewed the World Wide Web primarily as a marketing tool and as a way to be perceived as a technological leader; very few regarded the Internet as a technology integral to their manufacturing efforts. In any case, nearly two-thirds of the respondents expected to have a Web site by the end of 1997 (Chapman, 1997).

As the Internet becomes more efficient, through greater band width provided by telecommunications carriers, manufacturing will increase in sophistication. Designers will work together simultaneously in many parts of the world. Managers will have remote, real-time access to video overviews of production lines. And engineers in widely dispersed locations will be able to control machine tools and robotic production machinery anywhere in the world and see the results of their interventions in real time. Eventually there will be a world-wide Internet-accessible data base of information about production needs, distribution channels, sales, and marketing. This in turn will accelerate "agile" manufacturing, job-sharing, toolsharing, more efficient use of inputs, and reduced inventories.

What does the new ICT technology imply for the geography of innovation and production? With respect to manufacturing, it could be a reversal of decentralization on a global scale, which has been driven by a search for inexpensive labor for manufacturing at the tail end of the product cycle, where standardized outputs are produced in routine production runs. To the extent that labor is being replaced by automated machines, labor would not be a major determinant of location. Moreover, because the speed of technological change keeps introducing new versions of a product, product cycles often do not reach the inexpensive-labor phase. These phenomena suggest that production would be located close to large industrialized-country markets.

However, there is an alternative scenario. In recent years, for example, General Motors operations in Brazil have accounted for 25 percent of its operating profit (Blumenstein, 1997). As demand for autos grows in developing-country markets, GM is shifting the proportion of its total production outside of North America from 20 percent to 50 percent. In keeping with this strategy, GM is essentially building the same plant in Argentina, Poland, China, and Thailand so that if there is a problem in one plant it can be solved by calling one of the others. The plants are designed to receive a large array of already assembled parts, which cannot be done in the United States because of labor union resistance, and they will also utilize flexible manufacturing practices precluded by U.S. unions. Technological innovation, once confined to U.S. engineers, will also be managed globally, often by less-expensive but highly-competent Mexican engineers.

A division of General Electric has created a Trading Process Network (TPN) that already provides some twenty big industrial firms with lists of suppliers from which to shop. General Electric itself already buys $1 billion of goods through TPN each year and plans to purchase all of its industrial supplies electronically by 2000. By selling the use of the system, it hopes to realize a vast electronic clearing-house in which hundreds of thousands of firms can exchange trillions of dollars of industrial inputs (The Economist, 1998).

Service industries are the biggest users of ICT, and the most rapidly-growing subsector, business services, is particularly oriented toward communicating information. Just as many blue-collar jobs have migrated overseas in the past two decades, today there is a movement of data processing and other "back office" jobs offshore, to places where wages are lower, people speak English, and state-of-the-art telecommunications facilities allow instantaneous links with the U.S. companies' host computers. Nor are these just simple data entry tasks. For example, India's software industry, which barely existed 10 years ago, had sales of $1.2 billion in 1996 and is growing at over 40 percent a year (The Economist 1996). About half of the industry's revenues come from exports, and many clients are in the United States. Until recently teams of software engineers had to fly back and forth. Now, at the end of the working day, U.S. clients can e-mail their software problems to Bangalore, the industry's center, and have solutions back the next morning.

It has been argued that if telecommunications are a substitute for face-to-face interactions, then the need for the latter will decline as telecommunications improve, and that cities will eventually lose their importance as sites where people meet and exchange information (Naisbitt, 1995; Negroponte, 1995). There is also contrasting evidence suggesting that the new telecommunications infrastructure favors large cities, especially those that are centers for corporate headquarters and financial services (Malecki, 1997, pp.88-90). Moreover, Gaspar and Glaeser (1996) argue that improved telecommunications may increase face-to-face interactions because improved telecommunications make possible more contacts and relationships, some of which will eventually lead to face-to-face meetings. As telecommunications have improved, there has in fact been a sharp rise in business travel. The need for face-to-face communication is especially important when learning processes involve complicated ideas, and few would claim that the world is becoming any simpler. For example, electronic communication in R and D seems to be effective only if it is reinforced regularly by face-to-face contact (Coffey and Bailly, 1996; Malecki, 1997, p.151). Telecommunications may thus be a complement, or at least not a strong substitute, for cities and face-to-face interactions.

Echeverri-Carroll (1996) argues that while interfirm electronic network systems are the technologies that enable greater supplier-buyer coordination over great distances, these technologies also impose higher investments for interfirm linkages and more stringent requirements with respect to labor skills and flexibility, both of which constrain the location of industry. Salomon (1996) similarly maintains that the effects of telecommunications on urban and regional systems depend on the abilities of individual agents to seize opportunities that fit their particular situations. Drawing on European experiences, Capello (1994) and Rodrigues et al. (1997) have shown that while advanced firms and regions gain from the network externalities of advanced telecommunications technologies, backward regions seem to be unable to achieve economic advantage from these technologies because of their low level of organizational and managerial know-how. The introduction of advanced telecommunications services in less-developed regions is not in itself sufficient to create economic development in the absence of policies that effectively promote a relatively highly-skilled work force and adequate regional organizational and institutional capacity.

15.6 Conclusion

In recent years increasing attention has been devoted to knowledge and innovation as the fountainheads of economic development. Alfred Marshall long ago recognized that the clustering of firms in industrial districts could enhance innovation diffusion and generate positive externalities for the businesses concerned. After many years Marshall's insights have been revived and extended along a number of lines. Empirical evidence from Europe and the United States indicates that mutually beneficial externalities can be realized when firms are geographically concentrated, especially if firms participate in networks that promote entrepreneurship, collaboration as well as competition, and collective learning. On the other hand, there is also evidence that proximity is not necessary for the creation and continuation of collaborative networks that are rich in dynamic externalities arising from the diffusion of information and knowledge.

It has often been maintained that rapid improvements in ICT will result in national and global decentralization of manufacturing and business services. But other observers suggest that there will be greater concentration in cities, because ICT infrastructure is oriented toward cities that have many major corporate headquarters and financial activities, because ICT usage increases the total level of business interactions, including those that require face-to-face communication, and because R and D and other complex learning activities need a great deal of face-to-face communication.

What do these often seemingly contradictory findings imply for the development of economically peripheral regions? First, a preliminary note of caution that the fallacy of misplaced concreteness must be avoided in discussions of regions, networks and ICT, which in themselves can neither learn nor innovate. Learning and innovation require above all highly qualified professional workers who not only receive information, but who can recognize the significance of information and knowledge and use them to innovate successfully. This in turn requires high standards of education and a civic environment that encourages the innovation potential of a region.

Cities have a major role to play in this process, especially if they have the locally socially-embedded collaborative networks hypothesized in some theories of networks and industrial districts. However, one also finds innovative geographic clusters of firms with a relative abundance of skilled workers in contexts where the presence of related firms is of little importance. New technologies such as ICT cannot create local innovation networks where none existed before. However, even in economically peripheral areas these technologies can contribute greatly to the regional capacity to innovate – given that the appropriate educational and civic environment has been established – by significantly expanding opportunities for external networking on a global scale. And it is primarily SMEs that have the most to gain by local efforts to link regions and cities to global networks. Moreover, in view of the fact that innovation is so often demand-driven, the Internet should be used not only as an advertising tool, but as a means to pose problems faced by local firms that can potentially find innovative solutions elsewhere. In these respects, as in others discussed in this paper, it should be emphasized that the "goal of network economics is to increase our understanding of flow patterns in a modern differentiated society, where new forms of barriers, mobility, accessibility, complexity and self-organization caused by interaction through links not always directly connected with geographical distance or territorial adjacencies dominate the development" (Karlsson and Westin, 1994, p.3).

References

Angel, D.P., 1995, *Interfirm Collaboration in Technology Development,* Economic Development Administration, U.S. Department of Commerce, Washington, D.C.

Antonelli, C., 1992, "The Economic Theory of Information Networks", in C. Antonelli (ed.), *The Economics of Information Networks*, Elsevier Science Publishers, Amsterdam.

Appold, S.J., 1995, "Agglomeration, Interorganizational Networks, and Competitive Performance in the U.S. Metalworking Sector", *Economic Geography* Vol. 71, 1:27-54.

Audretsch, D.B. and M.P. Feldman, 1996, "R and D Spillovers and the Geography of Innovation and Production", *American Economic Review,* Vol. 86, 3:630-640.

Best, M.H., 1990, *The New Competition*, Harvard University Press, Cambridge.

Blumenstein, R., 1997, "GM is Building Plants in Developing Nations to Woo New Markets", *Wall Street Journal*, August 7, p.1.

Camagni, R.P., 1995, "The Concept of Innovative Milieu and its Relevance for Public Policies in European Lagging Regions", *Papers in Regional Science,* Vol. 74, 4:317-340.

Capello, R., 1994, *Spatial Economic Analysis of Telecommunications Network Externalities,* Avebury, Aldershot.

Carlsson, B. and S. Jacobsson, 1993, "Technological Systems and Economic Performance: The Diffusion of Factory Automation in Sweden", in D. Foray and C. Freeman (eds.), *Technology and the Wealth of Nations*, St. Martin's Press, New York.

Chapman, G., 1997, "The Internet and Manufacturing", *Texas Business Review*, October pp.1-3.

Coffey, W. and A. Bailly, 1996, "Economic Restructuring: A Conceptual Framework", in W. Lever and A. Bailly (eds.), *The Spatial Impact of Economic Changes in Europe*, Avebury, Aldershot.

Cornish, S.L., 1997, "Product Innovation and the Spatial Dynamics of Market Intelligence: Does Proximity to Markets Matter?", *Economic Geography,* Vol. 73, 2:143-165.

Echeverri-Carroll, E., 1996, "Flexible Production, Electronic Linkages and Large Firms: Evidence from the Automobile Industry", *Annals of Regional Science,* 30:135-152.

The Economist, 1998, January 17, 62.

Gaspar, J. and E.L. Glaeser, 1996, "Information Technology and the Future of Cities", National Bureau of Economic Research Working Paper No. 5562, Cambridge, Mass.

Glaeser, E., H.D. Kallal, J.A. Sheinkman and A. Schleifer, 1992, "Growth in Cities", *Journal of Political Economy,* Vol. 100, 6:1126-1152.

Geroski, P.A., 1991, "Innovation and the Sectoral Sources of UK Productivity Growth", *Economic Journal,* 101:1438-1451.

Hansen, N., 1990, "Do Producer Services Induce Regional Economic Development?", *Journal of Regional Science,* Vol. 30, 4:465-476.

Hansen, N., 1991, "Factories in Danish Fields: How High-Wage, Flexible Production has Succeeded in Peripheral Jutland", *International Regional Science Review,* Vol. 14, 2:109-132.

Hansen, N. and E. Echeverri-Carroll, 1997, "The Nature and Significance of Network Interactions for Business Performance and Exporting to Mexico: An Analysis of High Technology Firms in Texas", *Review of Regional Studies,* Vol. 27, 1:85-99.

Henderson, V., A. Kuncoro and M. Turner, 1995, "Industrial Development in Cities", *Journal of Political Economy,* Vol. 103, 5:1067-1090.

Howells, J., 1990, "The Internationalization of R and D and the Development of Global Research Networks", *Regional Studies,* Vol. 24, 6:495–512.

Karlsson, C. and L. Westin, 1994, "Patterns of a Network Economy: An Introduction", in B. Johansson, C. Karlsson and L. Westin, (eds.), *Patterns of a Network Economy*, Springer Verlag, Berlin.

Kaufman, A., R. Gittell, M. Merenda, W. Naumes and C. Wood, 1994, "Porter's Model for Geographic Competitive Advantage: The Case of New Hampshire", *Economic Development Quarterly,* Vol. 8, 1:43-66.

Kolata, G., 1996, "Regional Incongruity Found in Medical Costs", *New York Times*, January 30, p. B9.

Kristensen, P.H., 1994, "Spectator Communities and Entrepreneurial Districts", *Entrepreneurship and Regional Development,* Vol. 6, 1:177–198.

Krugman, P., 1991, *Geography and Trade,* MIT Press, Cambridge

Lundvall, B., 1993, "User-Producer Relationships, National Systems of Innovation and Internationalization", in D. Foray and C. Freeman (eds.), *Technology and the Wealth of Nations*, St. Martin's Press, New York.

Maillat, D., 1995, "Territorial Dynamic, Innovative Milieus and Regional Policy", *Entrepreneurship and Regional Development,* Vol. 7, 1:157–165.

Maillat, D. and B. Lecoq, 1992, "New Technologies and Transformation of Regional Structures in Europe: The Role of the Milieu", *Entrepreneurship and Regional Development,* Vol. 4, 1:1–20.

Malecki, E.J., 1997, *Technology and Economic Development*, 2nd edition, Longman, Essex.

Marshall, A., 1920, *Principles of Economics*, 8th edition, Macmillan, London.

Marshall, A. and M.P. Marshall, 1879, *The Economics of Industry*, Macmillan, London.

Martin, R. and P. Sunley. 1996, "Paul Krugman's Geographical Economics and its Implications for Regional Development Theory: A Critical Assessment", *Economic Geography,* Vol. 73, 3:259-292.

Naisbitt, R., 1995. *The Global Paradox*, Avon Books, New York.

Negroponte, N., 1995, *Being Digital*, Vintage Books, New York.

Piore, M.J. and C.F. Sabel, 1984, *The Second Industrial Divide*, Basic Books, New York.

Porter, M.E., 1990, *The Competitive Advantage of Nations*, Free Press, New York.

Rigby, D.L. and J. Essletzbichler, 1997, "Evolution, Process Variety, and Regional Trajectories of Technological Change in U.S. Manufacturing", *Economic Geography,* Vol. 73, 2:269–284.

Rodrigues, C., C. Jensen-Butler, E. de Castro and J. Millard, 1997, "Advanced Telecommunication Services in Peripheral Regions", Paper presented at the

European Regional Science Association Summer Course, Meraker, Norway and Are, Sweden, June 1997.
Salomon, I., 1996, "Telecommunications, Cities and Technological Opportunism", *Annals of Regional Science,* 30:75-90.
Saxenian, A., 1994, *Regional Advantage: Culture and Competition in Silicon Valley and Route 128,* Harvard University Press, Cambridge.
Scherer, F.M., 1982, "Inter-Industry Flows and Productivity Growth", *Review of Economics and Statistics,* Vol. 64, 3:627–634.
Schmookler, J., 1966, *Invention and Economic Growth,* Harvard University Press, Cambridge.
Scott, M., 1989, *A New View of Economic Growth,* Oxford University Press, New York.
Shefer, D. and A. Frenkel, 1998, "Local Milieu and Innovations: Some Empirical Results", *Annals of Regional Science,* 32:18-200.
Simmie, J., 1997, "The Origins and Characteristics of Innovation in Highly Innovative Areas: The Case of Hertfordshire", in J. Simmie (ed.), *Innovation Networks and Learning Regions?,* Jessica Kingsley Publishers, London.
Suarez-Villa, L. and C. Karlsson, 1996, "The Development of Sweden's R and D Intensive Electronics Industries: Exports, Outsourcing and Territorial Distribution", *Environment and Planning A,* Vol. 28, 5:783–817.
Suarez-Villa, L. and W. Walrod. 1997, "Operational Strategy, R and D and Intrametropolitan Clustering in a Polycentric Structure: The Advanced Electronics Industries of the Los Angeles Basin", *Urban Studies,* Vol. 34, 9:1343–1380.
The Economist. 1996, March 23, p. 67.
The Economist. 1998, January 17, p. 62.
Upton, H., 1995, "Peerless Advice from Small-Business Peers", *Wall Street Journal,* May 8, p. A14.
Wiig, H. and M. Wood, 1997, "What Comprises a Regional Innovation System? Theoretical Base and Indicators", in J. Simmie (ed.), *Innovation Networks and Learning Regions?,* Jessica Kingsley Publishers, London.

16 Regional Growth Theories and Local Economic Development: Some Case Studies

Bernard L. Weinstein
Center for Economic Development and Research, University of North Texas, P.O. Box 310469, Denton, Texas, 76203 USA

16.1 Introduction

From June 14 to 16, 1998, an international workshop on "Theories of regional development – lessons for regional economic renewal and growth" was convened in Uddevalla, Sweden. During the workshop, presentations and discussions focused on a wide range of theories and models that help explain patterns of regional development and market structure. These "theories" included the following:

(i) ***endogenous versus exogenous growth***: Endogenous growth refers to a system where the long-run growth rate is determined by the working of the system itself. This can be contrasted with exogenous growth, where the long-run growth rate is primarily determined by factors outside the system or region, by population increase, or by an exogenously given rate of technical progress.

(ii) ***changes in financial markets and institutions***: The size, structure and flexibility of institutions that accumulate and redistribute savings for purposes of investment can influence the pace of regional economic development.

(iii) ***size distribution of firms***: Typically, the distribution of firms within a given region tends to be skewed, with many small firms and relatively few large ones. It has been suggested that regions with a lower degree of "skewness" tend to show higher rates of business formation and technological innovation which, in turn, lead to higher growth rates than those found in regions with greater "skewness."

(iv) ***agglomeration and scale economies***: Agglomeration economies are those available to individuals or firms in large concentrations of population and economic activity. They arise from the wider choice and greater range of specialized services that usually occur within conurbations. Economies of scale refers to the factor making it possible for larger organization or regions to produce goods or services more cheaply than smaller ones.

(v) ***human capital, networks and universities***: Most simply, human capital can be considered the present discounted value of the additional productivity associated with higher levels of education and training. Universities often perform the dual roles of upgrading human capital and assisting in the dissemination of knowledge through formal and informal networks.

(vi) ***learning and forgetting regions***: Some regions adapt more quickly and easily to market and technological changes than others. Regions most successful at industrial transformation are those able to "forget" their traditional functions and "learn" new ones.

(vii) ***leadership, entrepreneurship, chance and spontaneity***: Regional economic change cannot always be predicted or anticipated. Entrepreneurial spontaneity frequently alters the economic orientation of a region, as in the case with Microsoft in Seattle, Washington.

(viii) ***core/periphery economic relations***: The core is the central region of an economy typified by good communications and high population density. Outlying, or "peripheral," regions are sparsely populated with poor communications, though this is changing with the revolution in telecommunications. Regional economists have traditionally argued that disparities between core and periphery tend to increase as migrants and investment are attracted to the core.

(ix) ***globalization and trade liberalization***: The rapid growth of international trade and investment has allegedly made far-flung regions more economically interdependent. Theory holds that more open regions can use international or interregional trade as a means of enhancing the rate of economic development. This is also referred to as "export-led" growth.

(x) ***regulatory and public policy influences***: The governmental, legislative and regulatory climate can strongly influence the pace of regional economic development. Some policies may be growth-enhancing and others growth-inhibiting.

(xi) ***technological change***: Over time, invention and innovation can change the economic prospects of a region. As old industries disappear, and new ones emerge, some regions suffer – at least temporarily – while others prosper.

The Center for Economic Development and Research at the University of North Texas has provided research, technical assistance and policy advice to a large number of communities – mainly in the southern United States – desiring to expand or diversify their economic bases. Two case studies are presented below that help to illustrate the diverse nature of regional economic development while at the same time affording an opportunity to validate some of the theories advanced at the Uddevalla conference.

16.2 Case 1: The Dallas-Fort Worth Metropolitan Area: Agglomeration in Information Technology

Since 1980, Dallas-Fort Worth, Texas has been the fastest-growing major metropolitan area in the United States. Currently, the region boasts a population of 4.5 million, an employed workforce of 2.6 million, and an unemployment rate of just over three percent – the lowest for any large urbanized region.

16.2.1 The Size and Scope of the Telecommunications Industry in the Dallas-Fort Worth Metropolitan Region

The telecommunications industry is one of the largest and fastest-growing in the Dallas-Fort Worth Metroplex. Comprised of both the manufacture of equipment and components, as well as the provision of communications services, the industry recorded about 125,000 Metroplex workers in 1995, the latest year for which published Census data are available (Richardson Chamber, 1998b). In view of expansions announced by major telecom employers in the area since 1995, a reasonable estimate of the current level of employment is probably about 140,000.

Most of the region's communications employment is concentrated in the "Telecom Corridor," encompassing Far North Dallas, Richardson and Plano (see Figure 16.1). Indeed, Telecom Corridor's 518 companies constitute the largest telecommunications hub in the United States. The Corridor's 95,000 workers account for 70 percent of Texas' manufacturing jobs in communications equipment and more than 50 percent of the state's semiconductor manufacturing jobs (Richardson Chamber, 1998a).

The future bodes well for the telecommunications industry in the DFW area, both because of anticipated technological changes and the presence of some of the industry's major players – e.g., AT&T, Southwestern Bell, Texas Instruments, MCI, GTE, EDS and a number of large foreign-owned telecom companies. Three of the nation's five biggest paging services companies are located in the Dallas area, and the region will benefit from anticipated growth in cellular communications, PCS equipment and on-line services in the years ahead. Projections by the North Central Texas Council of Governments indicate that the Corridor alone will house an additional 40,000 telecom jobs by the year 2010 (NCTCOG, 1996).

16.2.2 Characteristics of the Telecommunications Industry

Probably the most striking characteristic of the telecommunications industry is its rapid pace of technological change. For example, digital switching and transmission, the internet, and satellite-based personal communications services (PCS) were virtually nonexistent a decade ago yet are commonplace today. What's more, the pace of diffusion of technology and innovation in the telecommunications industry is accelerating. It no longer makes sense to speak of American, European or Asian technology because the knowledge base is increasingly universal and global rather than proprietary.

Secondly, the telecommunications business is highly competitive – not only among companies but between regions as well. Competitive pressures, driven in part by the rapid pace of technical change, have resulted in a high degree of industry agglomeration, which is why we find competitors and suppliers typically clustered in a limited number of locations such as the Telecom Corridor. For instance, when PrimeCo, a major developer in the PCS market, decided to locate its national headquarters in the DFW area several years ago, it cited the desire to be close to other "major players" in the wireless industry as the primary reason for choosing the area.

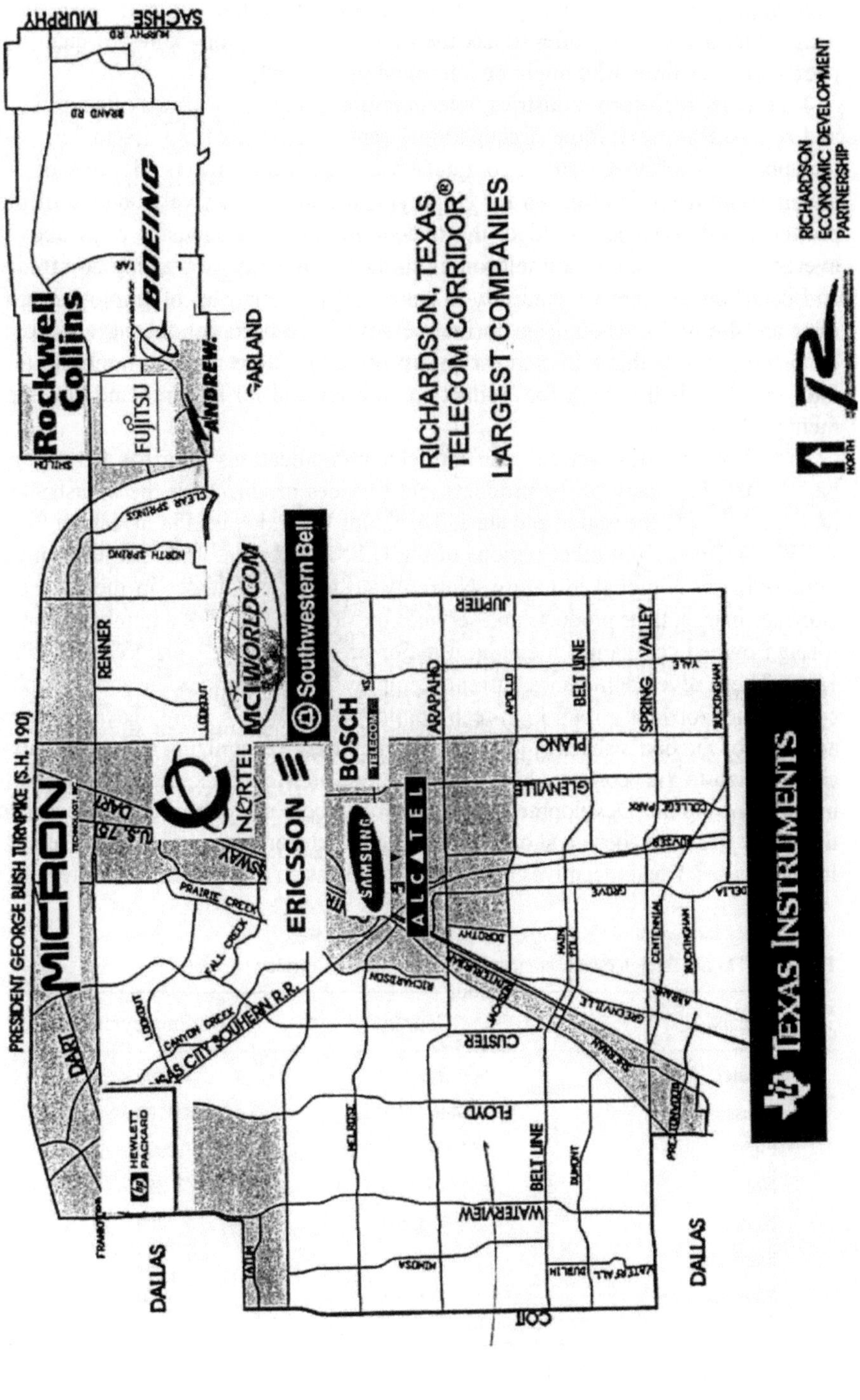

Figure 16.1 The Telecom Corridor

Being in close proximity to each other allows telecom companies to share ideas and develop formal and informal networks. It also provides them with access to a large and diverse local talent pool. These networks are often utilized in recruiting new employees. For example, it is common practice in telecom and other high-tech industries for companies to ask their own workers if they know of qualified people in other firms who might be interested in changing jobs.

Like other high-tech industries, telecommunications requires a skilled, trained and re-trainable work force. Occupational categories range from assemblers and technicians to software writers, computer scientists and engineers. Because of the advanced skill levels required for employment and the high value-added of the products and services provided, the telecommunications industry pays above-average wages. Because the telecommunications industry is growing so rapidly and qualified workers are currently in short supply, high rates of employee turn-over are the norm. Though no formal surveys have been conducted, anecdotal evidence suggests that a 20 percent annual turnover rate is not uncommon in this industry. This is true both for production workers and for research and development professionals.

Significantly, most activities in the telecommunications industry are export-based, meaning many of the products and services produced by the industry are marketed outside the region and state. This helps bring dollars and investment into DFW and Texas from other regions of the U.S. and abroad. The telecommunications industry is global in nature. Not only do many businesses in the Telecom Corridor market their products and services outside the U.S., but a number of large foreign-owned companies maintain a major presence in the Corridor (see Table 16.1). These seven companies currently employ 17,050 workers, or about 18 percent of the Corridor's total high-tech employment. What's more, Texas citizens account for the lion's share of jobs at these facilities. Recognizing their substantial contribution to the economic health of the state and Metroplex, the Texas Department of Economic Development and local chambers of commerce have worked diligently to attract these and other international telecom companies to the region. In some cases, fiscal incentives have been offered as an inducement to locate here.

Table 16.1 Major foreign owned companies in the Telecom Corridor

Company	Country	1998 Employment
Alcatel	France	2800
Ericsson	Sweden	3700
Fujitsu	Japan	2000
Nokia	Finland	250
Nortel	Canada	7700
Samsung	Korea	300
Siemens	Germany	300
	TOTAL	17,050

Source: Richardson (Texas) Chamber of Commerce

After eight years of an expanding national economy and a virtual explosion in telecommunications employment, the job market for high-tech workers is extremely tight. This is true not only in the Dallas-Fort Worth area but in other high-tech centers such as Austin, San Jose, Boston, and the Research Triangle. According to the U.S. Department of Commerce (1998b), the U.S. is experiencing a shortage of approximately 190,000 information technology workers, including programmers, engineers, and hardware/software specialists. The Information Technology Association of America has estimated that one in 10 computer-related positions goes unfilled, which amounts to 346,000 job vacancies (TechWeb, 1997). The current unemployment rate for engineers is about one percent, while that for computer scientists and programmers is even lower.

Against this backdrop, it should come as no surprise that salaries have been rising rapidly while job-hopping has become more prevalent in telecommunications and other high-tech industries. Indeed, Fortune 500 companies are reporting turnover rates of 25 to 35 percent annually for information technology workers (U.S. Department of Commerce, 1998a).

16.2.3 What can we Learn from the Dallas-Fort Worth Experience?

The success of the telecommunications industry in the Dallas-Fort Worth area can be explained by several of the regional development "theories" listed above. Certainly, economies and networks associated with agglomeration have been important. As Figure 16.1 illustrates, a significant clustering of information technology firms is evident in north Dallas and the north Dallas suburbs. Deregulation of the telecommunications industry, which began in the early 1980s, has been another important factor in that it changed a tightly controlled industry with relatively few players into a highly competitive one, as reflected by the formation of new firms and the inflow of foreign investment. Globalization and standardization in the industry have also contributed to the attractiveness of north Texas as a situs for telecom manufacturers and service companies. And certainly the rapid pace of technological change in the communications industry has created new products and markets for Dallas-Fort Worth based manufacturers and service providers.

Significantly, the telecommunications industry was not targeted for industrial development by local officials until it had already achieved a critical mass. The formation of the Telecom Corridor Technology Business Council a decade ago was almost an afterthought, once the importance of the industry to the local economy was recognized. What's more, area universities had little to do with the initial growth of the industry, though they have just recently started offering graduate and continuing education programs targeted at software engineers and other technical professions within the information technology sector.

The Dallas-Fort Worth area has successfully "forgotten" its economic past. As recently as 30 years ago, the dominant industries were energy, agriculture and finance. Today, these sectors account for a relatively small share of employment and regional output while telecommunications and other high-tech industries have flourished.

16.3 Case 2: Dalton, Georgia: Diversification from Carpet-Making

Dalton, Georgia is the town made famous by Paul Krugman in "Geography and Trade" (1991). Krugman uses Dalton as an example of why industries within the United States tend to be localized and can often trace their genesis to some trivial historical accident. In the case of Dalton, the historical accident was a bedspread made in 1895 by a teenager named Catherine Evans. Ms. Evans applied the craft of tufting, which had been common in the 18th and early 19th centuries but had fallen into disuse by 1895. In the 1940s, a machine was developed that could produce tufted carpets, which proved cheaper to manufacture than woven carpets. Dalton eventually emerged as the preeminent carpet manufacturing center of the United States.

16.3.1 Overview of the Dalton Economy

With a population of 86,000, Dalton-Whitfield County Georgia is the largest urbanized region between Atlanta and Chattanooga. Though not yet a statistically-defined metropolitan area according to the Bureau of the Census, recent growth trends suggest Dalton will cross the required thresholds for this designation within the next several years.

Dalton-Whitfield County is also a comparatively prosperous community. Indeed, the area's per capita income exceeds that of the state of Georgia and is just slightly less than the U.S. average. What's more, poverty rates in Whitfield County are lower than those of the state and the nation. The prosperity of the region can be explained partly by its strategic location along Interstate 75 and partly by the unusually high proportion of employment found in the manufacturing sector. Manufacturing accounts for 48 percent of total employment compared to the state percentage of 15 percent and a national share of only 13 percent. Textile mill products account for more than three-fourths of Whitfield County's manufacturing employment.

For decades, carpet and rug production has dominated the economy of Whitfield County and much of north Georgia, and this industry will likely remain the principal employer for the foreseeable future. But the industry currently faces serious competitive challenges. For instance, carpeting's share of overall spending on floor covering has dropped from 80% to 58% since the mid-1980s as homeowners have shifted to wood, ceramics and laminates in recent years. In fact, carpet sales nationally are growing at just two percent per year. Thus, technological innovation, foreign competition, and changing consumer tastes are challenging the long-term viability of tufted carpet-making with obvious implications for Dalton.

Dalton-Whitfield County faces two major challenges: (1) how to keep the carpet industry healthy and competitive, and (2) how to diversify the economy to lessen the region's overall dependence on the carpet market. Against this backdrop, the Center for Economic Development was retained in 1997 to help Dalton's business leadership identify opportunities for employment and income growth beyond the carpet industry, especially in non-manufacturing sectors (UNT, 1998d).

16.3.2 Business Deficiency Analysis

In order to identify non-manufacturing opportunities for business development efforts in Dalton-Whitfield County, we first examined the most recent data on existing employment and total payroll across 2-digit Standard Industrial Classification (SIC) categories. These data were retrieved from *1995 County Business Patterns*, which is prepared by the U.S. Department of Commerce. We defined the local market area as a five-county region including Whitfield, Catoosa, Murray, Gordon and Walker counties.

In 1995, northwest Georgia's wage and salary employment totaled about 108,000 in the specified non-manufacturing industries shown in Table 16.2. Among the area's service industries, retail trade is the largest in terms of employment with 18,050 workers. Business and personal services rank second in employment but first in terms of payroll. Wholesale trade also looms large in northwest Georgia, probably because of its ties to textile manufacturing. The fourth largest non-manufacturing industry in the region is health services with employment of 6,431 and an annual payroll of almost $174 million.

To ascertain potential target industries for the Dalton area, we first identified those which are underrepresented in the local economy. The methodology used to determine the relative level of representation for a particular industry is the "location quotient," which compares the employment share for a given SIC in northwest Georgia with that of a "reference area." The reference area is selected to closely resemble the Dalton region in terms of economic geography, population demographics (in percentage terms) and comparable infrastructure development. To minimize the impact of reference area peculiarities, it is common to "create" a hypothetical reference area by combining the characteristics of more than one county. For the purpose of this analysis, we have used "greater Tupelo" as the reference area, which includes the six-counties of Lee, Pontotoc, Itawamba, Union, Monroe and Chicasaw, Mississippi.

Twenty non-manufacturing industries in the Dalton area show location quotients less than one, indicating a potential for additional jobs and income in each (see Table 16.3). If all the underrepresented industries realized their potential, up to 8,151 new jobs would be supported in northwest Georgia, which would constitute an employment gain of about 7.5 percent in the service industries. What's more, these new jobs would increase area payroll by some $173 million annually.

A number of the industries arrayed in Table 16.2 show location quotients greater than one, indicating an overrepresentation relative to the reference area. However, this overrepresentation does not imply that employment losses will occur over time. Instead, it suggests that the Dalton area possesses strong comparative advantages in those industries due to location, access to infrastructure, labor costs or other factors. This is particularly true for wholesale trade and certain retail categories such as home furnishings.

To summarize, it is clear the Dalton area has substantial untapped potential for growth in service sector employment and payroll. The addition of jobs and income would help to diversify the area's economic base and substitute local services for imported services. At the same time, growing and attracting new establishments in industries where the area already possesses comparative advantages – i.e., location quotients larger than one – can give an additional boost to the regional economy.

Table 16.2 Location quotients for Northwest Georgia

SIC Code	Description	Lee, MS area Total employm.	Lee, MS area Total employm.
----	TOTAL	85,715	$1,734,597
40--	Transportation and public utilities	2,835	$81,946
4100	Local and interurban passenger transit	67	$56
4200	Trucking and warehousing	1,678	$47,901
4500	Transportation by air	72	$1,532
4700	Transportation services	99	$412
4800	Communication	581	$12,558
4900	Electricity, gas, and sanitary services	419	$10,445
50--	Wholesale trade	3,819	$96,021
5000	Durable goods	2,249	$57,084
5100	Nondurable goods	1,625	$36,167
52--	Retail trade	14,319	$169,736
5200	Building materials & garden supplies	653	$9,507
5300	General merchandise stores	2,797	$20,123
5400	Food stores	2,167	$21,130
5500	Auto dealers & service stations	1,470	$30,022
5600	Apparel & accessory stores	892	$7,221
5700	Furniture & home furnishing stores	552	$6,967
5800	Eating & drinking places	4,205	$27,972
5900	Misc retail	1,431	$19,457
60--	Finance, insurance and real estate	2,566	$61,886
6000	Depository institutions	1,514	$37,112
6100	Nondepository institutions	309	$6,152
6200	Security and commodity brokers	70	n.a.
6300	Insurance carriers	269	$6,469
6400	Insurance agents, brokers, and service	227	$5,097
6500	Real estate	251	$3,404
6700	Holding and other investment offices	20	n.a.
70--	Services	16,085	$335,590
7000	Hotels & other lodging places	403	$2,716
7200	Personal services	656	$7,868
7300	Business services	2,589	$22,419
7310	Advertising	60	$0
7320	Credit reporting & collection	188	$1,719
7330	Mailing, reproduction, stenographic	42	$390
7340	Services to buildings	294	$2,933
7350	Misc rental & leasing equipment	127	$2,140
7360	Personnel supply services	1,435	$7,439
7370	Computer & data processing services	195	$0
7380	Misc business services	479	$1,876
7500	Automobile repair, service, & parking	519	$9,310
7600	Misc repair services	238	$3,738
7800	Motion pictures	214	$155
7900	Amusement & recreation services	267	$2,734
8000	Health services	7,551	$221,758
8100	Legal services	392	$10,390
8200	Educational services	611	$911
8300	Social services	1,037	$10,878
8400	Museum, botanical, zoological gardens	10	n.a.
8600	Membership organizations	999	$9,234
8700	Engineering & management services	760	$23,428

Notes: Annual pay is reported in thousands. The Lee, MS area is the reference group for the location quotient. Source: 1995 County Business Patterns, U.S. Department of Commerce

Table 16.2 Cont.

SIC Code	Description	Whitfield, GA area Total employm.	Whitfield, GA area Total annual pay	Location Quotient
----	TOTAL	107,921	$2,428,319	
40--	Transportation and public utilities	4,467	$93,077	1.2515
4100	Local and interurban passenger transit	113	$1,678	1.3395
4200	Trucking and warehousing	2,965	$53,290	1.4034
4500	Transportation by air	80	n.a.	0.8825
4700	Transportation services	267	$4,305	2.1420
4800	Communication	903	$5,130	1.2344
4900	Electricity, gas, and sanitary services	389	$6,080	0.7374
50--	Wholesale trade	8,512	$230,280	1.7702
5000	Durable goods	6,371	$166,301	2.2499
5100	Nondurable goods	2,286	$10,762	1.1173
52--	Retail trade	18,050	$238,781	1.0012
5200	Building materials & garden supplies	656	$11,705	0.7979
5300	General merchandise stores	2,223	$26,558	0.6312
5400	Food stores	3,528	$38,326	1.2931
5500	Auto dealers & service stations	1,880	$40,903	1.0158
5600	Apparel & accessory stores	1,172	$3,228	1.0436
5700	Furniture & home furnishing stores	1,094	$20,680	1.5741
5800	Eating & drinking places	5,053	$45,258	0.9544
5900	Misc retail	2,014	$33,707	1.1178
60--	Finance, insurance and real estate	2,357	$58,891	0.7295
6000	Depository institutions	1,210	$30,013	0.6348
6100	Nondepository institutions	252	$4,379	0.6477
6200	Security and commodity brokers	40	$1,449	0.4539
6300	Insurance carriers	149	$3,154	0.4399
6400	Insurance agents, brokers, and service	317	$8,777	1.1091
6500	Real estate	434	$5,636	1.3733
6700	Holding and other investment offices	51	$1,528	2.0253
70--	Services	16,840	$343,393	0.8315
7000	Hotels & other lodging places	565	$5,323	1.1135
7200	Personal services	817	$11,999	0.9892
7300	Business services	2,872	$47,848	0.8811
7310	Advertising	80	n.a.	1.0590
7320	Credit reporting & collection	82	$764	0.3464
7330	Mailing, reproduction, stenographic	108	$448	2.0423
7340	Services to buildings	202	$2,860	0.5457
7350	Misc rental & leasing equipment	117	$2,410	0.7317
7360	Personnel supply services	2,136	$30,610	1.1822
7370	Computer & data processing services	99	$3,732	0.4032
7380	Misc business services	283	$3,572	0.4692
7500	Automobile repair, service, & parking	583	$11,141	0.8922
7600	Misc repair services	375	$7,052	1.2514
7800	Motion pictures	287	$2,292	1.0652
7900	Amusement & recreation services	578	$8,471	1.7194
8000	Health services	6,431	$173,687	0.6764
8100	Legal services	303	$9,922	0.6139
8200	Educational services	1,195	n.a.	1.5534
8300	Social services	908	$12,184	0.6954
8400	Museum, botanical, zoological gardens	10	n.a.	0.7942
8600	Membership organizations	1,304	$12,583	1.0367
8700	Engineering & management services	679	$19,383	0.7096

These gains would provide an enhanced tax base that could be used to improve the region's infrastructure and upgrade its human capital. This, in turn, would provide the Dalton area with competitive advantages for attracting and developing further economic opportunities.

16.3.3 What can we Learn from Dalton about Regional Development?

Referring again to the list above, we can characterize Dalton as a region that initially grew endogenously as a result of historical accident and technological change. But it is also a region unable to "forget" or "unlearn," which explains why its economy remains so specialized today. In addition, Dalton has lacked the leadership and entrepreneurial talent that might have led to industrial diversification. Its conservative power structure has been focused more on hoarding the workforce for employment in the low-skill carpet industry than targeting resources toward upgrading the region's human capital. Though Dalton is home to a two-year community college that emphasizes vocational education, only now are plans being made to establish a four-year university in the Dalton-Whitfield area.

Table 16.3 Dalton area. Potential employment and payroll gains for selected industries

SIC	Industry	Potential employment gain	Potential payroll gain (000)
4500	Transportation by air	11	$ 227
4900	Electricity, gas & sanitary services	139	2,166
5200	Building materials and garden supplies	156	2,965
5300	General merchandise stores	1,298	15,517
5800	Eating and drinking places	241	2,162
6000	Depository institutions	696	17,269
6100	Nondepository institutions	137	ß
6200	Security & commodity brokers	48	1,744
6300	Insurance carriers	190	4,015
7300	Business services	388	6,457
7320	Credit reporting and collection	155	1,442
7340	Services to buildings	168	2,381
7350	Miscellaneous rental and leasing equipment	43	884
7370	Computer and data processing services	147	5,524
7380	Miscellaneous business services	320	4,041
7500	Automobile repair, service, and parking	70	1,346
8000	Health services	3,077	83,095
8100	Legal services	191	6,240
8300	Social services	398	5,337
8700	Engineering and management services	278	7,932
	TOTAL POTENTIAL GAIN	8,151	$173,126

Dalton, like most small communities interested in industrial diversification, has no alternative but to embrace a new endogenous growth strategy. By targeting specific underrepresented industries, capitalizing on potential comparative advantages, and upgrading the skills of the resident workforce, Dalton-Whitfield County may eventually broaden and deepen its economic base.

16.4 Conclusion

This paper has examined two U.S. regions with very different economic histories and industrial structures. A wide range of economic growth theories was considered to help explain the different developmental paths taken by these two regions.

Dallas-Fort Worth (Texas) has successfully transformed itself from a natural resource based economy to one of the nation's most dynamic high-technology regions. Historically, the economic health of the region was determined almost exclusively by exogenous forces, such as the price of petroleum and cotton. Today, growth is driven by both endogenous and exogenous factors. For example, sizeable investments have been made in the region's human capital while local entrepreneurs and engineers have invented new products and manufacturing processes. The size distribution of firms has changed over time, with many new ventures spinning off from existing firms and capturing market share. Theory tells us that technological change and high rates of business formation are more likely to occur in regions with a low degree of "skewness", and that certainly has been the case in Dallas-Fort Worth.

As in the past, DFW's economy is still export-oriented. A huge percentage of the region's information technology output is sold to buyers outside of the area, both domestic and international. Globalization and trade liberalization have also been of benefit to most of the area's high tech manufacturing and service companies.

By contrast, Dalton (Georgia) has seen very little change in its economic base over the past few decades. Its economy was driven by endogenous factors a century ago, and that is still the case today. Dalton is a region unable to "forget," and institutional and political factors have constrained diversification. With a narrow industrial base and a high degree of "skewness" in the size distribution of firms – i.e., a handful of carpet manufacturers engaged in mergers and acquisitions – the pace of technological innovation has been quite slow. Indeed, in order to keep its carpet mills operating, Dalton has to "import" workers from outside the U.S. The pay is low, and the working conditions are unpleasant.

As a result of historic growth, as well as industry consolidation, Dalton's carpet manufacturers have realized sizeable economies of scale and scope. But the demand for carpet is growing only slowly, as homeowners and businesses opt for other types of floor coverings. Thus, the local economy must diversify if it is to survive in the long term. Only by upgrading the skills of the resident workforce and then making a serious effort at business recruitment will Dalton be able to diversify its economic base. The absence of local capital and financial institutions willing to make loans to new businesses must also be addressed if Dalton's economy is to grow.

References

Brimer, D.R., 1998, "Where Global Companies Change the World", *Richardson Magazine,* Telecom Corridor Issue, 1997-98.

Greater Dallas Chamber, 1998, *Technology Guide Dallas*.

Krugman, P., 1991a, *Geography and Trade,* MIT Press, Cambridge.

Krugman, P., 1998b, "Space: The Final Frontier", *Journal of Economic Perspectives,* Vol. 12, No. 2.

North Central Texas Council of Governments (NCTCOG), 1996, *Economic and Demographic Forecasts to 2020.*

Piore M.J. and C.F. Sabel, 1984, *The Second Industrial Divide*, Basic Books, New York.

Richardson Chamber of Commerce 1998a, *1998 Key Executive Report.*

Richardson Chamber of Commerce 1997b, *1998 High Tech Industry Report. TechWeb News*, August 1.

University of North Texas, 1996a, "The Communications/Information Industry in Dallas/Fort Worth", Center for Economic Development and Research.

University of North Texas, 1997b, "A Labor Force Survey and Assessment of Future Occupational Demand by High Technology Industries in the Metroplex", Center for Economic Development and Research.

University of North Texas, 1998c, "The Telecommunications Industry in the Metroplex: Trends, Characteristics and Challenges". Center for Economic Development and Research.

University of North Texas, 1998d, "Non-Manufacturing Business Development Opportunities and Strategies for Dalton-Whitfield County, Georgia", Center for Economic Development and Research.

U.S. Department of Commerce, 1998a, "The Emerging Digital Economy", Secretariat on Electronic Commerce. Washington D.C.

U.S. Department of Commerce, 1998b, "America's New Deficit: The Shortage of Information Technology Workers", Office of Technology Policy, Washington D.C.

Part V

Endogenous Regional Economic Policy Analyses

17 Universities and Regional Economic Development: Does Agglomeration Matter?

Attila Varga
Institute for Economic Geography, Regional Development and Environmental Management, University of Vienna, Roßauer Lände 23, A-1010 Vienna, Austria

17.1 Introduction

Since the early eighties, resulting from major structural changes in modern economies, a new wave of regional economic development policies has begun to emerge both in the US and in Europe (Atkinson, 1991; Isserman 1994; and Osborne, 1994). While traditional approaches (i.e., "smokestack chasing" via providing attractive financial conditions and business climate for relocating companies) were suitable tools for boosting localities in the era of mass production, they are no longer appropriate in the age of technology-led economic growth when economic globalization and the preeminence of knowledge and information in production have given rise to a renewed importance of regions (Acs, 1998; Florida, Gleeson and Smith, 1994: and Scott, 1996). This new set of policies, called "self-improvement" (Isserman, 1994), or "high-performance economic development" (Florida, Gleeson and Smith, 1994) aims at advancing a region's technology base and human infrastructure through the implementation of specific, technology related programs. In collaboration with the regional industry, governments support technology development, assist in industrial problem solving, provide start-up assistance, and help local firms finance new technologies (Coburn, 1995).

Motivated by the success stories of Silicon Valley and Route 128, regional technology programs put a significant weight on promoting technology transfers from universities to the local industry. Not only has the direct support for university research increased (in the US academic R&D grew form $7 billion in 1980 to $17 billion in 1993 in 1987 dollar[1]), a major portion of technology related expenditures of regional governments is being spent on programs requiring different forms of university involvement. For example, according to the data in Coburn (1995), 30% of the budget of state cooperative technology development programs in 1994 went directly to universities located in the state. This category of expenditures includes supporting university-industry technology centers, promoting university-industry research partnerships, and involvement in different forms of equipment and facility access programs. Moreover, about 70% of the total budget of state technology programs is, in part, associated with some kind of university participation. University-industry research centers (UIRCs) appear to be the most favored vehicles of government involvement in academia-supported regional

1 National Science Board (1993).

development. In 1990, federal and state governments spent about $1.9 billion on research and related activities at the estimated 1,056 UIRCs of the US (Cohen, Florida and Goe, 1994). More than that, 40 US states maintain technology extension programs, many of them are located on university campuses. Additionally, 20 states support incubators and research parks, most of them assume significant university involvement (Coburn, 1995).

Despite high expectations regarding positive regional economic effects of technology transfers from academia, scholarly evaluations of technology-based economic development programs are still rare in the literature[2]. Empirical economic research on regional university knowledge effects still struggles with data problems at lower levels of geographic and industrial aggregation and the absence of a comprehensive theory of regional innovation systems[3].

Studies carried out within the classical Griliches-Jaffe knowledge production function framework report strong and significant effects of technology transfers from university research laboratories to regional innovation both at the level of US states and metropolitan areas (Jaffe, 1989; Acs, Audretsch and Feldman, 1991; 1994; and Anselin, Varga and Acs, 1997a). However, this effect exhibits notable sectoral variations (Anselin, Varga and Acs, 1997b; and 1998).

Several observations support the hypothesis that the intensity of academic technology transfers is not stable across regions. For example, Acs, Herron and Sapienza (1992) and Feldman (1994b) point to the case of Johns Hopkins University and Baltimore. Despite the fact that Johns Hopkins is the largest recipient of federal research funds, no significant high technology concentration has emerged in the Baltimore area. Similarly, based on data in the early 1980s, while roughly equal in terms of research activity, Cornell University ($110 million in 1982) and Stanford University ($130 million in 1982) were situated in completely different regional innovative complexes: only 2 innovations were recorded for the production sector in Ithaca, versus 374 in the San Jose region. Regarding technology policy, these observations suggest that the same amount of university research support might affect regions differently, depending on the characteristics of their economic activities.

Besides definite differences in the scope and practical applicability of research programs at universities and regional variances in cultural traditions (Saxenian, 1994), it seems a reasonable assumption that agglomeration might play an important role in explaining spatial variations in university knowledge effects. To explain the modest university impact in Baltimore, Feldman (1994b) points to the possible role of the absence of a "critical mass" of high technology enterprises, the lack of producer services, venture capital and entrepreneurial culture.

In Varga (1998a; and 1998b) an explicit modeling approach was developed to study the effect of agglomeration on regional academic technology transfers. At the aggregate level of "high technology industry", these studies demonstrate diverse regional impacts of the same amount of university research funding depending on the

---

2 According to my knowledge, the study by Bania, Eberts, and Fogarty (1992) is the only major scholarly attempt in this area of research.

3 Regarding university knowledge effects, modeling approaches belonging to the tradition of neoclassical growth theory in Anderson (1981) and Anderson et al. (1989) and the regional investment model in Florax (1992) should be referred to here as major achievements of this research field. For a recent survey of the literature see Varga (1997).

level of concentration of economic activities in the geographic area. However, it is still not clear whether the influence of agglomeration on university technology transfers is stable across particular industries. Spatial concentration of economic activities might affect university-based technology policies differently depending on the sectoral structure of regional high technology production. Moreover, the possibility of the absence of such effect for certain industries cannot be completely ruled out either. Given, that US high technology activities exhibit a notable spatial tendency of sectoral specialization (Anselin, Varga and Acs, 1997b), industrial details regarding the agglomeration effect might be of particular policy relevance.

In this paper, the methodology developed by Varga (1998a; and 1998b) is applied to study sectoral differences in the agglomeration effect on local university technology transfers. Applying a unique data set of innovation counts and professional employ-ment in private R&D laboratories, an MSA level analysis is carried out within the Griliches-Jaffe knowledge production framework (Griliches, 1979; and Jaffe, 1989). The targeted two-digit SIC industries are as follows: Chemicals (SIC28), Industrial Machinery (SIC35), Electronics (SIC36), and Instruments (SIC38).

Section 17.2 presents the empirical model. It is followed by a data introduction and a discussion of estimation issues. Section 17.4 reports the regression results, while Section 17.5 demonstrates the agglo-meration effect on academic technology transfers. Concluding remarks follow.

17.2 The Empirical Model

A major obstacle of testing the effect of agglomeration on university technology transfers is the lack of a comprehensive measure of academic knowledge spillovers. Technology transfers from academic institutions might be captured by university patent citations (as was done in Jaffe et al., 1993), by the number of graduates finding jobs in the area, or by counts of local faculty spin-off firms. However, these variables cover local academic knowledge spillovers only partially.

For modeling purposes, an implicit measure of academic technology transfers is proposed in Varga (1998a; and 1998b). This measure is based on the Griliches-Jaffe knowledge production function (Griliches, 1979; and Jaffe, 1989). The knowledge production function has the form of:

$$\log(K) = \alpha_0 + \alpha_1 \log(RD) + \alpha_2 \log(URD) + \varepsilon, \quad (17.1)$$

where K measures new knowledge produced by high technology companies, RD is industrial research and development, URD is university research in the respective fields of engineering and hard sciences and ε is a stochastic error term. According to equation (17.1), production of economically useful new knowledge depends on two local inputs: the high technology industry's own R&D efforts and local university research. Jaffe points out that a positive and significant coefficient of the university research variable indicates university technology transfer effects on industrial knowledge production (Jaffe, 1989, p. 957). As such, the magnitude of α_2 can be considered as a measure of local academic knowledge spillovers: the higher the value of this coefficient, the more intensive the effect of university knowledge transfers on local innovation activities. This measure has a particular feature: it is not tied to any specific manner of technology transfers. It summarizes knowledge spillovers of any form in a single value[4].

The parameter expansion method of Casetti (1997) is applied in this paper to test for the effect of agglomeration on academic knowledge spillovers measured by the coefficient of the university research variable in equation (17.1). Knowledge transfer mechanisms[5] are classified into three categories: information transmission via the local *personal networks* of university and industry professionals (local labor market of graduates, faculty consulting, university seminars, conferences, student internships, local professional associations, continuing education of employees), technology transfers through *formal* business relations (university spin-off companies, technology licensing), and spillovers promoted by university *physical facilities* (libraries, science laboratories, computer facilities).

It is presupposed that the amount of technological information transmitted to the local high technology industry from the available pool of knowledge at academic institutions is controlled to a large extent by agglomeration. *Concentration of high technology production* is assumed to intensify information flows through the personal networks of university and industry professionals (for example, it increases local demand for faculty consulting services and raises the probability that graduates get jobs in the proximity of universities). Professional assistance from local *business services* (e.g., financial, legal, marketing services) enlarges knowledge spillovers by facilitating faculty spin-offs and technology licensing from academic institutions[6]. In general, relative to large companies, small firms are less endowed with research facilities. It is a major reason why small businesses rely more on university knowledge transfers (Albert Link and John Rees, 1990; and Acs, Audretsch and Feldman, 1994). Consequently, it is expected that *small firm concentration* enhances local university technology spillovers.

The following expansion equation models the dependence of academic knowledge transfers on the concentration of economic activities.

$$\alpha_2 = \beta_0 + \beta_1 \log(\text{PROD}) + \beta_2 \log(\text{BUS}) + \beta_3 \log(\text{LARGE}) + \mu. \quad (17.2)$$

[4] Given that the coefficient of the university research variable in equation (7.1) reflects local academic technology transfers implicitly, this is not a perfect measure of knowledge spillovers. The absence of such a correct measure is the reason of its substitution with a "second best" solution applied in this paper.

[5] The various mechanisms of local university knowledge transfers have been widely discussed in the literature (e.g., National Science Board, 1983; Dorfman, 1983; Lynn Johnson, 1984; Rogers and Larsen, 1984; Wicksteed, 1985; Douglas Parker and David Zilberman, 1993; Saxenian, 1994).

[6] Regional technology transfers are being supported by different types of local service companies. Not only patent attorneys or management services but also several engineering services are considerable sources of significant support in technology spillovers. Unfortunately, industry classification does not support such details in data collection. A proxy, a measure of business service activities has been chosen as a rough indication of local service input to technology transfers.

In equation (17.2), the magnitude of university knowledge spillovers, measured by α_2, is expected to be positively influenced by the concentration of high technology production (PROD) and business services (BUS). Technology transfers from academic institutions are supposed to be negatively affected by the relative importance of large firms (LARGE) in the geographical area (as suggested by Albert Link and John Rees, 1990; and Acs, Audretsch and Feldman, 1994).

Knowledge spillovers from industrial research laboratories measured by α_1 in equation (17.1) are also assumed to depend on agglomeration. It is widely recognized in the innovation literature, that local networks of related firms are major sources of new technological information (Giovanni Dosi, 1988; Eric von Hippel, 1988; and Edwin Mansfield and Elizabeth Mansfield, 1993). By enlarging the pool of available technical knowledge, concentration of production intensifies knowledge flows through the local network of firms (Feldman, 1994a). It has been well documented that locally available business services promote technological spillovers via supporting spin-off firm formation (Dorfman, 1983; Rogers and Larsen, 1984; and Saxenian, 1994). Acs, Audretsch and Feldman (1994) found that knowledge spillovers among private R&D laboratories are more significant sources of innovation for large companies than for small firms. Thus, agglomeration effects on technology spillovers among firms are modeled as follows

$$\alpha_1 = \gamma_0 + \gamma_1 \log(\text{PROD}) + \gamma_2 \log(\text{BUS}) + \gamma_3 \log(\text{LARGE}) + \eta, \tag{17.4}$$

with the same notation as above. It is assumed that concentration of production and business services and the relative importance of large firms influence local inter-firm technology transfers positively.

A substitution of equations (17.2) and (17.3) into (17.1) provides the expanded knowledge production function:

$$\begin{aligned} \log(\text{K}) = {} & \alpha_0 + \gamma_0 \log(\text{RD}) + \gamma_1 \log(\text{PROD})^*\log(\text{RD}) + \\ & \gamma_2 \log(\text{BUS})^* \log(\text{RD}) + \gamma_3 \log(\text{LARGE})^*\log(\text{RD}) + \beta_0 \log(\text{URD}) + \\ & \beta_1 \log(\text{PROD})^*\log(\text{URD}) + \beta_2 \log(\text{BUS})^* \log(\text{URD}) + \\ & \beta_3 \log(\text{LARGE})^*\log(\text{URD}) + [\eta \log(\text{RD}) + \mu \log(\text{URD}) + \varepsilon]. \end{aligned} \tag{17.4}$$

Equation (17.4) will be used for estimation. It models the production of economically useful new technological knowledge as being dependent on industrial and university R&D activities interacting with local agglomeration factors: concentration of production, business services and large companies.

17.3 Data and Estimation

Estimation of equation (17.4) will be based on the same unique data set of US metropolitan areas as is in Anselin, Varga and Acs (1997b; and 1998). New technological knowledge (K) is measured by counts of product innovations introduced on the US market in 1982. Innovation counts come from the United States Small Business Administration (SBA) innovation citation database (Keith Edwards and Theodore Gordon, 1984). This data set is a result of an extensive survey of the new product sections of trade and technical journals. County and MSA aggregates of the innovation data are available in two-digit SIC industry details and only for 1982. To date the SBA data are the best available measure of US innovative activity[7]. Private research activities (RD) are proxied by professional R&D employment. The source of this data is the 17th edition of Industrial Research Laboratories of the United States (Jaques Cattell Press, 1982)[8]. Following the common approach, university research expenditures stand for research activity at academic institutions (URD)[9]. The data are collected from the NSF Survey of Scientific and Engineering Expenditures at Universities and Colleges (National Science Foundation, 1982). Data measuring the concentration of high technology production (PROD), business services (BUS) and the relative presence of large firms (LARGE) come from County Business Patterns (Bureau of the Census, 1983). Concentration of high technology activities is accounted for by the location quotient of sectoral employment in the metropolitan area[10]. Business services activities are measured by employment in SIC 73. The percentage of high technology firms with employment exceeding 500 accounts for the relative importance of large companies in the MSA high technology economy. Given that MSA innovation data are available only at the level of two-digit SIC sectors, the following high technology industries provide the basis for the analysis: Chemicals (SIC28), Industrial Machinery (SIC35), Electronics (SIC36), and Instruments (SIC38). These categories include most of the three and four digit industries that are commonly considered "high technology" sectors

---

7 For a detailed description of the data set and its advantages over the traditionally used patent data see Acs and Audretsch, 1990 and Feldman, 1994a.

8 Although it is a reasonable approach to account for a four or five-year lag between innovations and research (as was done in Acs and Audretsch, 1990; Acs, Audretsch and Feldman, 1991; and in Feldman, 1994), this approach is not followed here. The technical reason is that 1982 is the first year that the Classification Index of the Directory allows for appropriate industry level aggregations. Besides this technical impediment, the validity of the choice of the year 1982 is supported by the trends in R&D lab location. As reported in Malecki (1979, 1980a, 1980b), location patterns of R&D laboratories tend to be stable for a relatively long period of time. This observation suggests that a regression model on lagged research variables would not provide significantly different outcomes from those reported in this study.

9 Since graduates represent one of the most significant knowledge transmitters from universities, it can be argued that number of graduate students in the area's universities would be a more appropriate measure of local academic knowledge resources than research expenditures at academic institutions. The fact that correlations, between the two variables are high (i.e., 0.74, 0.91, 0.91, and 0.78 for SICs 28, 35, 36, and 38, respectively) suggests that probably a similar estimation outcome would be yielded with the variable measuring the number of graduate students. However, to remain consistent with the classical Griliches-Jaffe knowledge production framework, it is decided to follow the original idea and measure academic contributions to innovative activities with university research expenditures.

10 A location quotient relates local and national importance of an industry, based on its relative share in the local and in the national economy. Formally: $LQ = (EMPSEC_{MSA}/EMPTOT_{MSA})/(EMPSEC_{NATION}/EMPTOT_{NATION})$, where EMPSEC and EMPTOT stand for employment in the specific sector and total employment, respectively. $LQ > 1$ shows that industry employment is more concentrated in the region than on average in the nation.

(Varga, 1998a)[11]. For a detailed description of the data see Anselin Varga and Acs (1997b; and 1998).

Three potential estimation problems of the expanded knowledge production function need closer attention: the problems of heteroskedasticity, multicollinearity, and spatial dependence. The fact that the error term of equation (17.4) depends on observation-specific private and university research values may cause heteroskedasticity in the estimated model. Repeated occurrence of the same variables in subsequent terms of the knowledge production function could be the source of serious multicollinearity. In the following analysis, the Breusch-Pagan (BP) heteroskedasticity test (Breusch and Pagan, 1979) and the multicollinearity condition number (David Belsley et al., 1980) are applied to test for misspecifications in the forms of heteroskedasticity and multicollinearity.

Potential statistical problems associated with dependence among observations in cross-sectional data are extensively treated in the spatial econometrics literature (e.g., Anselin, 1988; Anselin and Florax, 1995; and Anselin and Bera, 1998). Two forms of spatial dependence may exist in a linear regression context: spatial lag dependence and spatial error autocorrelation. A presence of any kind of spatial dependence can invalidate regression results. In the case of spatial error autocorrelation, OLS parameter estimates are inefficient whereas in the presence of spatial lag dependence, parameters become not only biased but also inconsistent (Anselin, 1988).

The general expression for the spatial lag model is

$$y = \rho Wy + x\beta + \varepsilon, \qquad (17.5)$$

where y is an N by 1 vector of dependent observations, W is a row standardized spatial weight matrix[12], Wy is an N by 1 vector of lagged dependent observations, ρ is a spatial autoregressive parameter, x is an N by K matrix of exogenous explanatory variables, β is a K by 1 vector of respective coefficients, and ε is an N by 1 vector of independent disturbance terms.

Autocorrelation among regression error terms represents an alternative form of spatial dependence. Spatial error autocorrelation is modeled as follows

---

11 Besides data availability constraints, technical necessities also motivated the selection of these four sectors. The Cobb-Douglas form of the applied knowledge production function permits research to include only non-zero values of the variables on the right hand side of the equation. Additionally, since this research considers the effect of universities on regional knowledge creation where innovative activities are already present, MSAs where innovations are non-existent are excluded from the data set. For more about the selection of observations, see Anselin, Varga and Acs (1998).

12 Relative positioning of observations is modeled in spatial weights matrices. The dimension of a spatial weights matrix W is given by the number of observations of the regression. A matrix element $w_{i,j}$ reflects the spatial relation between observations i and j. Depending on the expected structure of spatial dependence, a matrix element $w_{i,j}$ can represent either contiguity relations between observations or it can model the role of distance in dependence. If two observations are contiguous (i.e., they share a common border or are located within a given distance band), the value of $w_{i,j}$ is larger than zero, and zero otherwise. The larger-than zero value is 1 in case of a simple contiguity matrix and it is a number between zero and one if the elements are row-standardized, that is, every element is divided by the respective row sum. If spatial dependence is expected to be determined by distance relations, a matrix element is based on the distance of observations i and j (i.e., their inverse distance or the square of the inverse distance).

$$y = X\beta + \varepsilon \tag{17.6}$$

with

$$\varepsilon = \lambda W\varepsilon + \xi \tag{17.7}$$

where λ is the coefficient of spatially lagged autoregressive errors $W\varepsilon$ and ζ is an N by 1 vector of independent disturbance terms. The other notation is as before[13].

Three spatial weights matrices are applied in the following empirical study. D50 and D75 are distance-based contiguities for 50 and 75 miles, respectively while the third one, IDIS2, is an inverse distance squared weights matrix[14]. The presence of spatial dependence is tested for by Lagrange Multiplier test statistics (Burridge, 1980; and Anselin and Florax, 1995). Empirical regressions are carried out in SpaceStat, an econometric software designed for the analysis of spatial data (Anselin, 1992).

17.4 Regression Results

Tables 17.1-3 report estimation results for four high technology sectors at the level of US metropolitan statistical areas (MSAs) in 1982. Parameter estimates together with the appropriate test statistics for the original knowledge production function of equation (17.1) are reported in Table 17.1. Only the econometrically correct forms of the equations are presented in the table. While OLS is the appropriate estimation method for SIC 38, knowledge production in SIC 35 is associated with spatial lag dependence whereas innovation creation in the electronics industry exhibits a significant spatial error dependence. Heteroskedasticity-robust estimation was carried out for the chemicals sector.

Table 17.1 demonstrates that the effect of university technology transfers on regional innovative activity is not stable across sectors. Production of economically useful new knowledge builds upon local university research spillovers in the electronics and instruments industries, while innovation in chemicals and industrial machinery relies on internal knowledge resources. A possible interpretation of this result is that university-supported economic growth in Silicon Valley and Boston, where the major high technology sectors are closely related to the two-

---

[13] The applied spatial econometric methodology is well suited for modeling the spatial extent of knowledge spillovers. Spatial dependence in the knowledge production function, either in the form of lag or error autocorrelation, is a sign of knowledge transfers among the spatial units of analysis. In any case of spatial dependence, the correctly specified spatial econometric equation accounts for spillovers both within and among the spatial units (Anselin, Varga, and Acs, 1997a, 1997b; Varga, 1998a).

[14] Two MSAs are considered contiguous in D50 if their center counties are located within a 50-mile distance range. The same reasoning applies for D75. These matrices are intended to reflect potential spatial dependencies within commuting distances around an MSA. IDIS2 captures spatial effects that might come from the whole geographic area of the regression.

digit electronics and instruments industries in Table 17.1, might not be replicated with other sectors.

Table 17.1 Industrial regression results (1982)

Model	SIC28 OLS – Robust	SIC35 ML – Spatial Lag (D50)	SIC36 ML – Spatial Error (IDIS)	SIC38 OLS
Constant	-1.0283	-0.240	-0.740	-0.662
	(0.375)	(0.152)	(0.151)	(0.240)
W_Log(INN)		0.185		
		(0.081)		
Log(RD)	0.484	0.264	0.381	0.322
	(0.132)	(0.055)	(0.049)	(0.074)
Log(URD)	-0.010	0.081	0.139	0.179
	(0.035)	(0.043)	(0.042)	(0.058)
Lambda			0.408	
			(0.124)	
Number of observations	48	89	70	63
R^2 - adj	0.352	0.387	0.557	0.403
LIK		-51.641	-20.196	
B-P test for Heteroskedasticity		4.537	1.492	2.321
LM-Err (D50)		0.194		0.682
LM-Lag (D50)				2.033
LM-Lag (IDIS)			1.619	
LR-Err (IDIS)			9.053	
LR-Lag(D50)		5.286		

Notes: estimated standard errors are in parentheses; critical values for the Breusch-Pagan (B-P) statistic for Heterosdedasticity with 2 degrees of freedom is 4.61 (p=0.05) and 5.99 (p=0.01); critical values for LM-Err, LM-Lag, LR-Err and LR-Lag are 3.84 (p=0.05) and 2.71 (p=0.10); the spatial weights matrices are row-standardized: D50 is distance-based contiguity for 50 miles, D75 is distance based contiguity for 75 miles and IDIS is inverse distance-squared.

Parameter expansion results are reported in Table 17.2 and 17.3 for the electronics and instruments industries, respectively. In Table 17.2, the first column lists estimation results for (17.4). The extremely high value of multicollinearity (with condition number of 168) makes it impossible to reasonably evaluate the relative importance of different agglomeration factors in the processes of local knowledge transfers. In the second and third columns, parameters of the two research variables are expanded, separately. The results show that both university and industrial knowledge transfers are significantly positively affected by the concentration of production and business services. Another common result is that the small firm effect is not significant for either form of research effects. However, high multicollinearity (an inherent shortcoming of the applied parameter expansion methodology) is a technical impediment to accounting for all the possible factors of agglomeration. Instead, the strongest effects are examined in the final model. The model in the fourth column exhibits the best properties in terms of regression fit and multicollinearity. Spatial dependence among regression error terms is taken care of by means of maximum likelihood estimation. Business services are the major agglomeration factors explaining technology transfers from universities while knowledge spillovers among research laboratories are dominantly promoted by production concentration.

Table 17.2 Regression results for Log (INN36) (N=70, 1982)

Model	Full model OLS	RD model OLS	URD model OLS	Final model ML-spatial error
Constant	-0.315	-0.141	-0.130	-0.186
	(0.183)	(0.186)	(0.187)	(0.149)
Log(RD)	-0.061	-0.595	0.174	0.139
	(0.409)	(0.201)	(0.061)	(0.053)
Log(URD)	-0.183	0.081	-0.507	-0.424
	(0.292)	(0.042)	(0.140)	(0.116)
Log(RD)Log(PROD)	0.209	0.039		0.043
	(0.053)	(0.011)		(0.009)
Log(RD)Log(BUS)	0.022	0.173		
	(0.095)	(0.038)		
Log(RD)Log(LARGE)	-0.097	0.009		
	(0.079)	(0.031)		
Log(URD)Log(PROD)	-0.127		0.026	
	(0.039)		(0.009)	
Log(URD)Log(BUS)	0.094		0.134	0.123
	(0.069)		(0.029)	(0.024)
Log(URD)Log(LARGE)	0.073		0.004	
	(0.055)		(0.023)	
LAMBDA				0.376
				(0.111)
R^2 - adj	0.712	0.671	0.653	0.700
LIK	-3.194	-9.627	-11.476	-4.095
Multicollinearity	168	44	42	38
B-P for Heteroskedasticity	5.360	4.755	11.652	4.719
LM-Err (D75)	4.239	9.319	11.141	
LM-Lag (D75)	4.755	5.530	6.948	0.275
LR-Error (75)				9.638

Notes: estimated standard errors are in parentheses; critical values for the B-P statistic with respectively 8, 5, and 4 degrees of freedom are 15.51, 11.07, and 9.49 (p=0.05); critical values for LM-Err LM-Lag and LR-Err statistics are 3.84 (p=0.05) and 2.71 (p=0.10); the spatial weights matrixes are row-standardized: D50 is distance-based contiguity for 50 miles and D75 is distance based contiguity for 75 miles.

Table 17.3 exhibits parameter expansion regression results for the instruments industry. As evidenced in the second and third columns, both types of technology transfers are positively affected by employment concentration and business services. However, high multicollinearity permits to account for only the strongest agglomeration effects in the final model. For the instruments industry, as opposed to the electronics industry, employment concentration is the main driving factor behind academic knowledge spillovers while for private technology transfers, business services are the strongest agglomeration attributes.

Table 17.3 Regression results for Log (INN38) (N=63, 1982)

Model	Full model OLS	RD model OLS	URD model OLS	Final model
Constant	0.307	0.324	-0.095	0.294
	(0.228)	(0.220)	(0.244)	(0.225)
Log(RD)	-1.030	-0.854	0.159	-0.890
	(0.346)	(0.194)	(0.073)	(0.193)
Log(URD)	0.135	0.026	-0.143	0.055
	(0.129)	(0.047)	(0.101)	(0.048)
Log(RD)Log(PROD)	0.057	0.154		
	(0.145)	(0.050)		
Log(RD)Log(BUS)	0.283	0.238		0.237
	(0.077)	(0.039)		(0.038)
Log(RD)Log(LARGE)	0.087	-0.001		
	(0.100)	(0.032)		
Log(URD)Log(PROD)	0.054		0.060	0.075
	(0.078)		(0.032)	(0.022)
Log(URD)Log(BUS)	-0.026		0.065	
	(0.029)		(0.018)	
Log(URD)Log(LARGE)	-0.050		0.007	
	(0.050)		(0.019)	
R^2 - adj	0.718	0.707	0.544	0.667
LIK				
Multicollinearity	72	30	23	29
B-P for Heteroskedasticity	5.460	5.278	4.838	4.769
LM-Err (D50)	2.976	2.991	1.193	3.732
LM-Lag (D50)	2.324	1.836	1.546	1.587

Notes: estimated standard errors are in parentheses; critical values for the B-P statistic with respectively 8, 5, and 4 degrees of freedom are 15.51, 11.07, and 9.49 (p=0.05); critical values for LM-Err and LM-Lag statistics are 3.84 (p=0.05) and 2.71 (p=0.10); the spatial weights matrixes are row-standardized: D50 is distance-based contiguity for 50 miles and D75 is distance based contiguity for 75 miles.

17.5 University Effect and Agglomeration: A Demonstration of the Importance of a "Critical Mass" for Successful Technology Transfers

Regression results in Tables 17.2 and 17.3 clearly evidence that the available pool of technological knowledge at academic institutions exerts diverse impacts on the local economy, depending on the level of concentration of economic activities in a metropolitan area. However, the scale of local economic activities that is sufficient enough to yield substantial academic knowledge transfers still remains an important issue for the analysis.

In order to address the "critical mass" of local economic activities problem, MSAs in the samples are categorized into three different "tiers." The categorization is based on the intensity of local academic knowledge transfers measured by the estimated coefficients of the university research variables in the industrial knowledge production functions. Given that knowledge production is formulated in the form of a Cobb-Douglas function, these coefficients measure innovation elasticities with respect to university research spending.

For the Electronics industry, based on the final model in Table 17.2, the intensity of academic technology transfers in location j is calculated as follows:

$$\text{Elasticity [Innovation, University Research]} = \partial \log(K) / \partial \log(URD) = -0.424 + 0.123*\log(BUS). \qquad (17.8)$$

According to the final model in Table 17.3, innovation elasticity for location j is calculated for the Instruments industry based on the following equation:

$$\text{Elasticity [Innovation, University Research]} = \partial \log(K) / \partial \log(URD) = 0.055 + 0.075*\log(PROD). \qquad (17.9)$$

Metropolitan areas are ordered by estimated local academic technology transfer intensities and classified into three groups with equal numbers of observations. Tables in the Appendix list MSAs ordered by the calculated values of innovation elasticities for the two industries. Tables 17.4A and 17.4B present average values of innovations together with certain indicators of local knowledge production and agglomeration for the three tiers of MSAs.

The first columns of Table 17.4A and Table 17.4B list average elasticities of innovation with respect to university research for the two industries in this study. With the exception of average population in innovation performing MSAs in the instruments field, all the variables follow the same decreasing tendency of innovation elasticities. The tables suggest that innovation activities of the Electronics industry are more sensitive to university research than those of the Instruments sector: the change in innovations associated with one percent change in university research is higher for Electronics than for Instruments in all the three tiers. Despite smaller elasticity values, innovation in the Instruments industry seems to be more "university research intensive" than knowledge production in the electronics field: average levels of university research expenditures in SIC 38 exceed average academic R&D expenditures in SIC 36 in all the respective tiers. For industrial employment, the tables show the opposite tendency. With respect to private R&D employment, MSAs exhibit comparable sizes in all but the first tiers of the two industries. Average city size, measured by population is significantly larger in the first tier of the Electronics industry than that of the Instruments field (3.1 and 1.8 millions, respectively). For the rest of the tiers, the two industries exhibit the opposite tendency.

The fourth columns of the two tables list average values of innovation predictions for the two industries. These predictions are based on parameter estimates in Tables 17.2 and 17.3. Compared to the respective average tier values of observed innovations, the estimated models of knowledge production provide good average predictions for the second and third tiers. However, the models consistently underpredict average levels of innovation activities in the first tiers of both industries. This observation suggests that for first tier MSAs actual university technology transfers are probably higher in their intensity than indicated by innovation elasticity predictions.

Table 17.4 Average values of innovation elasticities, innovation, R&D activities, employment and population by innovation elasticity categories

A. Electronics Industry (SIC36)

	EL(I,U)	INN36	INN36PR	RD36	URD36	EMP36	POP
Tier 1	0.164	23	9	2875	16154	38121	3.1
Tier 2	0.091	4	3	307	3950	9625	0.9
Tier 3	0.032	2	2	224	3119		0.5

B. Instruments Industry (SIC38)

	EL(I,U)	INN38	INN38PR	RD38	URD38	EMP38	POP
Tier 1	0.074	24	16	1054	61095	15784	1.8
Tier 2	0.042	14	11	404	51082	4867	1.8
Tier 3	0.003	2	4	251	23026	1552	1.4

Notes: EL stands for elasticity of innovation with respect to university research; INN is the number of innovations in the MSA; INNPR is predicted innovations RD stands for R&D professional employment; URD is university research expenditures in thousands of US dollars; EMP is industry employment; POP is population in millions of inhabitants.

Figures 17.1 and 17.2 demonstrate the effect of agglomeration on local university technology transfers. In essence, these figures simulate the impacts of a *pure* university-based regional economic development policy in metropolitan areas exhibiting different levels of economic activities. In other words, the effects of increased university research expenditures on innovation are presented, while all the other characteristics of the MSAs are assumed to remain the same. The X axis represents university research expenditures, while the Y axis depicts expected innovations for university research spending sizes and for different MSAs in both figures. University research activities depicted on the X axis in the figures reflects the range between the highest and lowest levels of observed university research expenditures for the respective industries. The three curves stand for different expected innovation outcomes associated with the same amounts of university research spending. Expected innovations for each tier were calculated based on the final model in the last columns of Tables 17.2 and 17.3. For each tier, average values of private research and the two research coefficients were held constant while university research spending was the only variable element in the calculations.

Figures 17.1 and 17.2 demonstrate the dramatic differences in the "productivity" of the same amount of university research spending depending on the size of economic activities in a geographic area. It is a common observation for both industries that first tier metropolitan areas possess the "critical mass" of local economic activities, that is, those cities absorb university effects in the most efficient manner. While increased academic research expenditures have basically no effects on innovation activities in second and third tier cities, the impact of academic research in the first tiers is remarkable. (Over the ranges of respective university research expenditures, innovation activity increases from 5 innovations to 12 in the Electronics industry and from 8 to 12 in the Instruments sector.)

For both industries, a pure university-based regional development policy seems not to be effective enough to "upgrade" geographical areas in the second and third tiers to a higher level of innovative activity. With the exception of third tier cities in the Electronics industry (where knowledge production reaches the lowest level of second tier innovations after about 4 millions of university research expenditures), even the maximum amount of university research spending is not high enough to reach the lowest average level of knowledge production in the next tier of metropolitan areas.

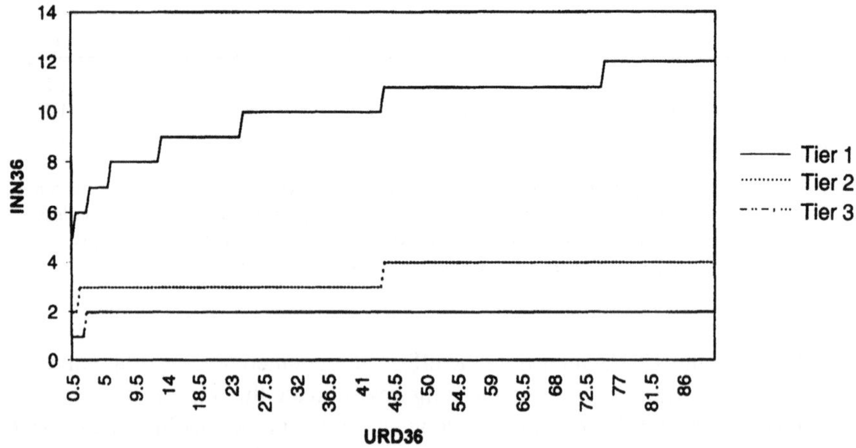

Figure 17.1 Expected innovations: Electronic industry (SIC 36)

Sensitivity of innovative activity to increased university research spending gradually decreases in first tier cities for both industries. While at lower levels of university research activities, boosting local universities seems to be a cost effective way of economic development, this advantage seems to disappear quickly: the larger the amount of university research spending, the higher the cost of each additional innovation.

17.6 Summary and Conclusions

Universities have gained increased attention in modern, technology-based regional economic development policies. Despite high expectations regarding positive economic effects of university support, scholarly evaluations of policies promoting local technology transfers from universities are still scarce in the literature. An important area of research is the effect of spatial concentration of economic activities on university-based regional economic development policies. This paper provided formal empirical evidence of the positive impact of agglomeration on local academic technology transfers for two "high technology" industries, Electronics (SIC 36) and Instruments (SIC 38).

Parameter expansion analyses were carried out within the classical Griliches-Jaffe knowledge production framework. Testing and correcting for spatial effects in regression equations earned a particular attention in the empirical investigations. University technology transfers are most sensitive to the presence of business services in the Electronics sector while concentration of industrial employment is the strongest agglomeration feature affecting local knowledge spillovers for Instruments. For both industries, it was demonstrated that the same amount of university research spending is associated with notable differences in knowledge production depending on the concentration of economic activities in the metropolitan area.

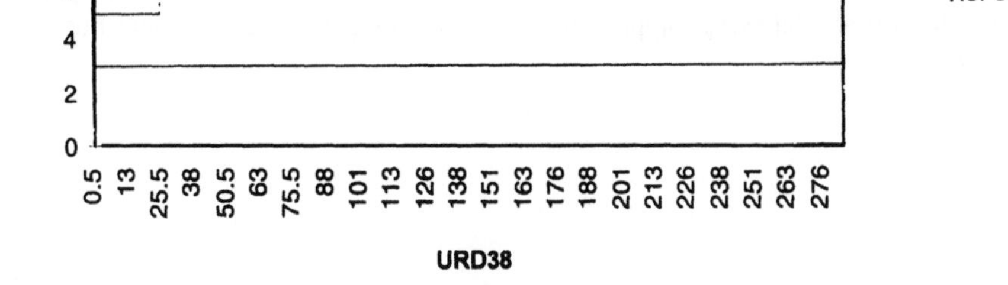

Figure 17.2 Expected innovations: Instruments industry (SIC38)

In addition, it was found that the presence of a "critical mass" of agglomeration in the metropolitan area is required in order to expect substantial local economic effects of academic research. Simulations of university knowledge effects suggest that pure university-based regional economic development policies are not effective enough to "upgrade" localities to a higher tier of innovative activities. Simulation results also suggest that cost-effectiveness of university support is in an indirect relationship with the level of academic research expenditures.

To some extent, the applied data and methodology set the limitations on the interpretations of the results. Since the SBA innovation data are available for one year, only a static analysis is allowed in this study. Consequently, results reflect a "longer term" equilibrium under the assumption that economic variables do not go under significant changes. The innovation data set does not make it possible to differentiate among innovations based on their economic importance. It is possible that some places are over-represented because of their relatively numerous but not necessarily important product developments relative to others where only a few but fundamental innovations were reported. Due to its tendency for quickly increasing multicollinearity, the applied parameter expansion model reflects only the effects of the most important local agglomeration features on academic technology transfers and it cannot be used for a complete modeling approach.

Results of the analyses reflect the general trend of agglomeration effect and should be interpreted this way. Individual cities can (and do) exhibit different combinations of regional economic features while maintaining the same intensity of academic technology transfers. The essence of the results is that individual metropolitan areas cannot be "too far" from the average size in order to preserve tier-specific university effects.

Regarding further research directions, the dynamics of university effects need to be analyzed as soon as time-space data are available for research. It seems to be highly possible that the effect of agglomeration on university technology transfers is endogenous in the system: it determines, and to a possible large extent, is determined by academic knowledge spillovers. Also, a richer time and industrial data can make it possible to carry out direct regional comparisons with respect to the agglomeration effects.

Despite its limitations, the analysis of this paper strongly indicate that university-based economic development policies can be efficient tools for relatively matured high technology agglomerations. For less developed regions the results suggest reducing efforts on university-based regional economic development policies and to concentrate more on the growth of high technology employment (via traditional "chasing" approaches) and widening the base of local business services[15].

---

[15] This result is robust: for the aggregate high technology sector essentially the same consequence was reached in Varga (1998a, 1998b).

Appendix A. Innovation Elasticities, Innovation, Innovation Predictions, R&D Employment, University Research, High Technology Employment and Population by Sample MSAs Electronics Industry (SIC36)

MSA	EL(I,U)	INN36	INN36PR	RD36	URD36	EMP36	POP36
New York	0.259	44	20	1913	22618	37896	8.8
Los Angeles-Long Beach	0.237	39	27	5504	49860	114836	7.5
Chicago	0.221	25	19	4654	20312	108049	7.1
Boston	0.198	60	21	5497	88683	76354	3.7
Houston	0.196	3	7	113	9835	18959	2.9
San Francisco-Oakland	0.193	19	11	2362	16722	18959	3.3
Philadelphia	0.192	10	13	7323	13044	37149	4.7
Washington DC	0.19	6	8	2651	5444	15982	3.1
Dallas-Fort Worth	0.182	24	7	257	5270	53227	3
Detroit	0.164	4	5	2300	1642	8418	4.4
Atlanta	0.162	6	6	125	27318	10571	2
Nassau-Suffolk	0.161	32	8	1378	9125	39648	2.6
Essex county NJ	0.156	19	7	18171	455	26421	2
Anaheim-Santa Ana-Garden Grove	0.155	24	7	729	6457	63000	1.9
San Jose	0.154	151	14	5646	31134	101807	1.3
Minneapolis-St. Paul	0.152	13	8	5333	7793	21227	2.1
Denver-Boulder	0.152	5	7	837	16825	12379	1.6
Cleveland	0.141	14	6	1475	3753	20467	1.9
Baltimore	0.14	1	5	332	3665	29684	2.2
Pittsburgh	0.137	9	4	50	20959	14938	2.3
San Diego	0.135	18	6	825	15698	28107	1.9
Phoenix	0.127	15	5	825	1875	33531	1.5
Tampa	0.122	4	3	284	628	14446	1.6
Seattle-Everett	0.121	13	4	426	8589	8858	1.6
Milwaukee	0.119	6	4	706	1214	28678	1.4
Kansas City	0.117	2	2	188	198	12648	1.3
Cincinnati	0.115	2	3	405	1093	10369	1.4
Bridgeport	0.112	15	2	770	13	18934	0.8
Columbus	0.111	3	4	684	7280	8107	1.1
Hartford	0.105	4	4	1249	4306	13294	1.1
Fort Lauderdale-Hollywood	0.101	3	2	181	9	10360	1
Portland	0.098	6	2	55	1344	10044	1.2
Buffalo	0.095	3	2	40	2551	9202	1.2
Orlando	0.089	2	3	396	521	7587	0.7
Albuquerque	0.088	1	3	141	2294	4522	0.5
Sacramento	0.085	2	2	33	1810	882	1
Raleigh-Durham	0.085	4	3	301	9676	9993	0.5
Riverside-San Bernardino-Ontario	0.085	3	2	7	2391	5476	1.6
Charlotte-Gastonia	0.08	1	1	14	123	5332	0.6
Rochester	0.08	9	3	267	7341	11109	1
New Brunswick-Perth Amboy	0.08	6	3	229	6224	3832	0.6
Dayton	0.078	2	3	230	5901	5278	0.8
Louisville	0.077	2	2	50	263	11306	0.9
Salt Lake	0.077	2	3	778	4552	5057	0.9
Melbourne-Titusville-Cocoa	0.076	6	1	32	21	11381	0.3

MSA	EL(I,U)	INN36	INN36PR	RD36	URD36	EMP36	POP36
Austin	0.073	4	3	176	22496	8126	0.5
New Haven-West Haven	0.073	5	3	121	9231	9851	0.8
Providence-Warwick-Pawtucket	0.071	5	2	94	6315	6658	0.9
Knoxville	0.07	1	2	243	3323	1038	0.5
Trenton	0.069	6	3	625	10660	4309	0.3
Toledo	0.066	2	2	236	417	3477	0.8
Paterson-Clifton-Passaic	0.063	1	2	92	13	9277	0.4
Syracuse	0.056	1	2	97	3534	12494	0.6
Greenville-Spartanburg	0.055	2	1	10	1201	6170	0.6
Jersey City	0.055	4	2	68	422	7095	0.6
Wilmington	0.054	2	2	305	1277	956	0.5
Tucson	0.046	4	2	118	15511	2663	0.5
Springfield-Chicopee-Holyoke	0.04	1	1	12	6058	1860	0.6
Santa Barbara-Santa Maria-Lompoc	0.036	3	3	1330	4885	5129	0.3
Worcester	0.036	3	2	247	500	5642	0.6
Wichita	0.034	1	1	11	281	926	0.4
Ann Arbor	0.031	2	2	105	11618	2379	0.3
El Paso	0.024	3	2	1156	168	3122	0.5
Colorado Springs	0.022	1	1	187	5	4974	0.3
South Bend	0	1	1	3	1529	1391	0.3
New Bedford	-0.011	2	1	65	344	7116	0.5
Binghamton	-0.012	1	1	45	517	25791	0.3
Portsmouth-Dover-Rochester	-0.022	2	1	45	3106	5110	0.4
Lorain-Elyria	-0.057	1	1	20	23	1469	0.3
Janesville-Beloit	-0.069	2	1	36	26	250	0.1

Notes: EL stands for elasticity of innovation with respect to university research; INN is the number of innovations in the MSA; INNPR is predicted innovations RD stands for R&D professional employment; URD is university research expenditures in thousands of US dollars; EMP is industry employment; POP is population in millions of inhabitants.

Appendix B. Innovation Elasticities, Innovation, Innovation Predictions, R&D Employment, University Research, High Technology Employment and Population by Sample MSAs Instruments Industry (SIC38)

TEXT	EL(I,U)	INN38	INN38PR	RD38	URD38	EMP38	POP38
Rochester	0.149	16	19	2060	57081	66435	1
San Jose	0.099	47	19	200	111400	25024	1.3
Bridgeport	0.093	12	5	223	14	11136	0.8
Boston	0.086	114	82	7860	284713	42357	3.7
Worcester	0.083	1	4	164	3302	5545	0.6
New Haven-West Haven	0.081	10	6	51	79432	6225	0.8
Minneapolis-St. Paul	0.075	24	12	107	94204	17947	2.1
New Brunswick-Perth Amboy-Sayreville	0.074	11	6	104	17232	4565	0.6
Anaheim-Santa Ana-Garden Grove	0.074	31	11	86	29037	14506	1.9
Santa Barbara-Santa Maria-Lompoc	0.072	5	4	546	7249	2035	0.3
Nassau-Suffolk	0.067	51	17	560	28344	11967	2.6
Essex county NJ	0.066	46	16	658	16720	11552	2
Denver-Boulder	0.064	5	9	62	57588	10118	1.6
Philadelphia	0.063	51	47	5215	127315	21885	4.7
Salt Lake	0.062	2	5	58	36642	4444	0.9
Wilmington	0.062	3	4	126	3547	2501	0.5
New Bedford	0.059	3	2	9	428	1890	0.5
Chicago	0.059	50	48	1655	131808	31519	7.1
San Diego	0.059	20	11	289	74753	8250	1.9
Madison	0.054	2	4	126	97845	1478	0.3
Ann Arbor	0.053	2	3	300	79533	1229	0.3
Hartford	0.052	9	6	113	47316	4449	1.1
Trenton	0.051	11	4	78	13967	1247	0.3
Cleveland	0.05	15	10	359	39378	6504	1.9
Columbus	0.05	10	8	470	47629	3730	1.1
Paterson-Clifton-Passaic	0.05	5	3	8	213	1407	0.4
Los Angeles-Long Beach	0.049	42	60	2250	176223	26520	7.5
Syracuse	0.047	3	4	68	25559	1792	0.6
Buffalo	0.047	10	7	962	22256	3248	1.2
Pittsburgh	0.046	13	9	279	60246	5944	2.3
Austin	0.045	3	4	149	35458	1852	0.5
Seattle-Everett	0.045	4	8	287	80057	5077	1.6
Fort Lauderdale-Hollywood	0.044	1	3	3	99	2391	1
Lincoln	0.041	2	2	27	9786	563	0.2
Tucson	0.035	1	3	15	51188	1004	0.5
San Francisco-Oakland	0.035	14	16	391	145459	7600	3.3
New York	0.034	65	60	1573	267859	19232	8.8
Springfield-Chicopee-Holyoke	0.034	1	3	4	23461	1119	0.6
Newburgh-Middletown	0.033	1	1	266	125	396	0.3
Milwaukee	0.033	12	7	334	17709	2899	1.4
Tampa	0.033	4	4	27	4256	2677	1.6
Phoenix	0.028	9	8	812	4469	2555	1.5
Dayton	0.028	7	4	89	16933	1314	0.8
Cincinnati	0.025	1	3	1500	3584	862	0.7

TEXT	EL(I,U)	INN38	INN38PR	RD38	URD38	EMP38	POP38
Akron	0.024	1	3	1500	3584	862	0.7
Miami	0.023	1	5	37	32456	2500	1.6
Kansas City	0.022	2	4	39	735	2052	1.3
Toledo	0.021	2	3	55	6248	879	0.8
Houston	0.02	13	7	31	113496	4848	2.9
Charlotte-Gastonia	0.019	1	3	25	526	1043	0.6
Dallas-Fort Worth	0.018	5	12	420	44541	4442	3
Riverside-San Bernardino-Ontario	0.017	6	3	10	8390	1350	1.6
Huntsville	0.016	1	2	500	1729	352	0.3
Atlanta	0.011	9	4	21	59344	2313	2
Detroit	0.01	10	10	730	17955	3395	4.4
Baltimore	0.002	3	7	723	85040	1614	2.2
Jersey City	-0.001	2	2	85	4100	353	0.6
Washington DC	-0.004	5	12	622	32511	2632	3.1
Greenville-Spartanburg	-0.019	3	2	12	7821	284	0.6
Norfolk-Virginia Beach-Portsmouth	-0.02	1	1	5	2554	45	0.8
Davenport-Rock Island-Moline	-0.024	2	2		2	127	0.4
Greensboro-Winston-Salem-High Point	-0.059	4	1	9	9093	110	0.8
Galveston	-0.063	2	1	93	11882	17	0.2

Notes: EL stands for elasticity of innovation with respect to university research; INN is the number of innovations in the MSA; INNPR is predicted innovations RD stands for R&D professional employment; URD is university research expenditures in thousands of US dollars; EMP is industry employment; POP is population in millions of inhabitants.

References

Acs, Z., (ed.), 1999, *Regional Innovation, Knowledge and Global Change*, Pinter, London.

Acs, Z. and D. Audretsch, 1990, *Innovation and Small Firms*, MIT Press Cambridge, Mass.

Acs, Z., D. Audretsch and M. Feldman, 1991, "Real Effects of Academic Research: A Comment", *American Economic Review,* Vol. 81, 1:363-367.

Acs, Z., D. Audretsch and M. Feldman, 1994, "R&D Spillovers and Recipient Firm Size, *The Review of Economics and Statistics,* Vol. 76, 2:336-340.

Acs, Z., L. Herron, and H. Sapienza, 1992, "Financing Maryland Biotechnology", *Economic Development Quarterly*, Vol. 6, 4:373-382.

Anderson A., 1981, "Structural Change and Technological Development", *Regional Science and Urban Economics,* 11:351-361.

Anderson A., D. Batten and C. Karlsson (eds.), 1989, *Knowledge and Industrial Organization*, Springer Verlag, Berlin.

Anselin, L., 1988, *Spatial Econometrics: Methods and Models*, Kluwer Academic, Boston.

Anselin, L., 1992, *SpaceStat Tutorial*, NCGIA, University of California, Santa Barbara,

Anselin, L. and B. Anil, 1998, "Spatial Dependence in Linear Regression Models with an Introduction to Spatial Econometrics", in D. Giles and A. Ullas (eds.), *Handbook of Economics and Statistics*, Marcel Dekker, New York.

Anselin, L. and R. Florax, (eds.), 1995, *New Directions in Spatial Econometrics*, Springer Verlag, Berlin.

Anselin, L., A. Varga and Z. Acs, 1997a, "Local Geographic Spillovers Between University Research and High Technology Innovations", *Journal of Urban Economics*, Vol. 42, 3:422-448.

Anselin, L., A. Varga and Z. Acs, 1997b, "Entrepreneurship, Geographic Spillovers and University Research: A Spatial Econometric Approach", ESRC Centre for Business Research WP 59, University of Cambridge.

Anselin, L., A. Varga and Z. Acs, 1998, "Geographic and Sectoral Characteristics of Academic Knowledge Externalities", Research Paper, Regional Research Institute West Virginia University.

Atkinson R., 1991, "Some States Take the Lead: Explaining the Formation of State Technology Policies", *Economic Development Quarterly*, 5:3-44.

Bania, N., R. Eberts and M. Fogarty, 1993, "Universities and the Startup of New Companies: Can we Generalize from Route 128 and Silicon Valley?", *The Review of Economics and Statistics*, Vol. 75, 4:761-766.

Belsley, D., E. Kuh and R. Welsch, 1980, *Regression Diagnostics, Identifying Influential Data and Sources of Collinearity*, Willey, New York.

Breusch, T. and A. Pagan, 1979, "A Simple Test for Heteroskedasticity and Random Coefficient Variation", *Econometrica*, Vol. 47, 5:1287-1294.

Bureau of the Census, *1982*, *County Business Patterns*, Data obtained from *ICPSR* online data Services.

Burridge P, 1980, "On the Cliff-Ord test for Spatial Correlation", *Journal of the Royal Statistical Society B*, Vol. 42, 1:107-108.

Casetti, E., 1997, "The Expansion Method, Mathematical Modeling and Spatial Econometrics", *International Regional Science Review*, Vol. 20, 1-2:9-32.

Coburn, C., (ed.), 1994, *Partnership: A Compendium of State and Federal Cooperative Technology Programs*, Battelle, Columbus.

Cohen, W., R. Florida and R. Goe, 1994, "University-Industry Research Centers", Carnegie Mellon University.

Dorfman, N., 1983, Route 128: The Development of a Regional High Technology Economy, *Research Policy*, 12:299-316.

Dosi, G., 1988, "Sources, Procedures and Microeconomic Effects of Innovation", *Journal of Economic Literature*, Vol. 26, 3:1120-1171.

Edwards, K. and T. Gordon, 1984, *Characterization of Innovations Introduced on the U.S. Market in 1982*, The Futures Group, U.S. Small Business Administration.

Feldman, M., 1994a, *The Geography of Innovation*, Kluwer Academic Publishers, Boston.

Feldman, M., 1994b, "The University and Economic Development: The Case of Johns Hopkins University and Baltimore", *Economic Development Quarterly*, Vol. 8, 1:67-76.

Florax, R., 1992, *The University: A Regional Booster? Economic Impacts of Academic Knowledge Infrastructure*, Avebury, Aldershot

Florida, R., R. Gleeson and D.F. Smith Jr., 1994, "Benchmarking Economic Development: Regional Strategy in Silicon Valley, Austin, Seattle, Oregon, and Cleveland", H. John Heinz III School of Public Policy Management Working Paper Series, 94-30, Carnegie Mellon University.

Griliches, Z., 1979, "Issues in Assessing the Contribution of Research and Development to Productivity Growth", *Bell Journal of Economics,* Vol. 10, 1:92-116.

Isserman, A., 1994, "State Economic Development Policy and Practice in the United States: A Survey Article", *International Regional Science Review,* 16:49-100.

Jaques Cattell Press, 1982, *Industrial Research Laboratories of the United States,* 17th edition, R. R. Bowker, London.

Jaffe, A., 1989, "Real Effects of Academic Research", *American Economic Review,* Vol. 79, 5:957-970.

Jaffe, A., M. Trajtenberg and R. Henderson, 1993, "Geographic Localization of Knowledge Spillovers as Evidenced by Patent Citations", *Quarterly Journal of Economics*, Vol. 63, 3:577-598.

Johnson, L., 1984, *The High-Technology Connection. Academic/Industrial Co-operation for Economic Growth,* ASHE-Eric Higher Education Research Report, No. 6. Clearinghouse on Higher Education, The George Washington University, Washington D.C.

Link, A. and J. Rees, 1990, "Firm Size, University Based Research, and the Returns to R&D", *Small Business Economics,* Vol. 2, 1:25-32.

Malecki, E., 1979, "Locational Trends in R&D by Large U.S. Corporations, 1965-1977", *Economic Geography*, 55:309-323.

Malecki, E., 1980a, "Corporate Organizations of R&D and the Location of Technological Activities", *Regional Studies,* 14:219-234.

Malecki, E., 1980b, "Dimensions of R&D Location in the United States", *Research Policy,* 9:2-22

Mansfield, E. and E. Mansfield, 1993, *The Economics of Technical Change,* Edward Elgar Publishing Company, Aldershot.

National Science Board, (ed.), 1983, *University-Industry Research Relationships,* National Science Foundation, Washington D.C.

National Science Board, 1993, *Science and Engineering Indicators,* National Science Foundation, Washington D.C.

National Science Foundation, *Academic Science and Engineering: R&D Expenditures, Fiscal Year 1982,* Data obtained from *CASPAR* data files.

Osborne, D., 1990, *Laboratories of Democracy*, Harvard Business School Press, Boston.

Parker, D. and D. Zilberman, 1993, "University Technology Transfers: Impacts on Local and U. S. Economies", *Contemporary Policy Issues,* 11:87-99.

Rogers, E. and J. Larsen, 1984, *Silicon Valley Fever*, Basic Books, New York.

Saxenian, A., 1994, *Regional Advantage: Culture and Competition in Silicon Valley and Route 128*, Harvard University Press, Cambridge.

Scott, A., 1996, "Regional Motors of the Global Economy", *Futures,* 28:391-411.

Varga, A., 1997, "Regional Economic Effects of University Research: A Survey", Research Paper, Regional Research Institute, West Virginia University,

Varga, A., 1998a, *University Research and Regional Innovation: A Spatial Econometric Analysis of Academic Technology Transfers,* Kluwer Academic Publishers, Boston.

Varga, A., 1998b, "Local Academic Knowledge Spillovers and the Concentration of Economic Activity", Research Paper, Regional Research Institute, West Virginia University.

von, Hippel, E., 1988, *The Sources of Innovation,* Oxford University Press, New York.

Wicksteed, S., 1985, *The Cambridge Phenomenon, The Growth of High Technology Industry in a University Town,* London.

18 Change in Manufacturing Productivity in the U.S. South: Implications for Regional Growth Policy

Kingsley E. Haynes and Mustafa Dinc
The Institute of Public Policy, George Mason University, 4400 University Drive, Fairfax VA 22030, USA

18.1 Introduction

For over a century the U.S. South has been considered a lagging region in which its rates of income and employment growth relegated it to second class economic status. This was reinforced by a dominantly rural agrarian economic base. Despite reasonable economic growth rates, the gap between the South and the rest of the country did not seem to close significantly until the late 1960s and early 1970s. As industrialization took hold and urban agglomeration stimulated related supporting production and business service development, economic restructuring began to change the South and in the 1970s, the gap suddenly began to close. Manufacturing seems to have played a central role in closing this employment, income and wealth gap and in 1980s the South became part of growing "sunbelt" which until that time had been dominated only by western states. It is the role of manufacturing that is examined in this study to understand its present role in regional economic growth in the U.S. South.

Over the last three decades, U.S. manufacturing has experienced a relative location shift from the old manufacturing belt in the Mid West of the U.S. to the South and West, the so-called sunbelt, of the United States. Consequently, these new regions have gained substantial employment and capital. This shift has also helped the development of new companies in the South and West, and these regions have emerged as major growth areas of the country (Haynes and Dinc, 1997; Crandall, 1993; Moriarty, 1992; Rigby, 1992; Stough, 1991; Connaughton and Madsen, 1990; Harrison and Blustone, 1988; Richardson and Turek, 1985; Casetti, 1984; Costrell, 1994; Sawers and Tabb, 1984; and Norton and Rees, 1979). Several competing hypotheses have tried to explain the underlying causes of this shift and its outcomes: the filtering down hypothesis of industrial location (Hoover, 1948; Thompson, 1969), the product life cycle hypothesis (Vernon, 1966, 1979; Moriarty, 1983), the labor skill-external economy requirements hypothesis (Erickson, 1976; Erickson and Leinbach, 1979; Moriarty, 1992) and the geographical shift of innovations hypothesis (Norton and Rees, 1979; Norton, 1997). Each of these hypotheses has some merit and explanatory power with respect to this shift.

Although they have not specifically tested any of the above hypotheses, Haynes and Dinc (1997) have investigated employment change in the old Midwest manufacturing belt and the so-called U.S. sunbelt from 1960 to 1992. The findings of that study suggested that (a) there has been a shift in manufacturing from the snowbelt to the sunbelt; (b) the growth in manufacturing in the sunbelt was driven by output growth and non-labor productivity factors (new investments or entries, improvements of infrastructure etc.); (c) recent manufacturing employment gains in the sunbelt were partially due to poorer productivity gains than in the older manufacturing belt. Based on these findings, it was concluded that the improvement in sunbelt manufacturing may seem satisfactory, but in the long run lagging productivity can cause competitiveness problems for the Southern states.

Recently, some scholars have raised similar, if not more pessimistic, concerns about the future of the South. They argue that the southern economy has been declining and the future of the South seems not so bright (Malecki, 1995; Schaffer, 1993). In this paper, we investigate how good or bad the Southern economy is doing in terms of manufacturing sectors.

This paper does not test the above mentioned spatial growth realignment hypotheses directly. Instead, using recent annual data, (from 1987 to 1994) we trace the adjustment of regional manufacturing change in the fifteen Southern states. By relating employment adjustments to changes in output, the role of labor and investment factors, productivity is de-coupled from output expansion. Hence, it is possible to assess the foundations of economic performance and employment change in manufacturing in this important region. To achieve this and to take account of industrial and regional differences in labor and capital (non-labor) productivity and the effects of productivity differences on regional manufacturing employment change we have used the Haynes and Dinc (1997) extension of the shift-share model.

The organization of the paper is as follows. The following section provides an overview of manufacturing in the Southern states. The methodology and data are then discussed in the third section. The fourth major part of the paper includes interpretation and discussion of findings as well as a comparison of the states. Conclusions are presented in the final section.

18.2 Southern Manufacturing

Since the 1960s, manufacturing employment in the south has been increasing. Although the old Midwest manufacturing belt began recovering in the late 1980s (Federal Reserve Bank of Chicago, 1996), a reverse shift in manufacturing employment has not happened. Table 18.1 shows that between 1987 and 1994 manufacturing employment in the South increased by 7.2%. During the same period, total U.S. manufacturing employment increased only 1%. If we exclude the Southern states' share of total U.S. manufacturing employment national manufacturing employment would have declined by 2.1% during the same period. The share of southern manufacturing employment in the U.S. increased from 33% in 1987 to 35.2% in 1994, which means 6.2% growth in seven years. This net growth occurs despite the recession during this period. This is obviously a good sign for the Southern economy in general and manufacturing sectors in particular, although the overall economy has increasingly become more service oriented.

Table 18.1 Manufacturing employment and population in the South (1000 hours)

Year	Total South	Total US	Total US-South	% Share of South	Population of South	Total U.S. population	% Share of South
1987	8,053,800	24,300,600	16,246,800	0.331	83,063	242,307	0.343
1988	8,194,400	24,778,000	16,583,600	0.331	83,890	244,499	0.343
1989	8,225,900	24,662,200	16,436,300	0.334	84,700	246,819	0.343
1990	8,236,800	24,317,500	16,080,700	0.339	85,454	248,718	0.344
1991	7,959,600	23,194,100	15,234,500	0.343	86,928	252,138	0.345
1992	8,198,100	23,581,500	15,383,400	0.348	88,166	255,039	0.346
1993	8,357,200	23,845,100	15,487,900	0.350	89,426	257,800	0.347
1994	8,637,300	24,537,200	15,899,900	0.352	90,712	260,350	0.348
% Change	7.2	1.0	-2.1	6.2	9.2	7.4	1.6

Source: BEA and statistical abstract of the United States, 1996

Table 18.2 shows manufacturing employment change and percentage share of manufacturing employment in total non-farm employment by state. During the study period, in ten states, employment increased at rates well above the nation. Among them Mississippi had the largest increase, 23.1%, followed by Kentucky, 18.9%, Arkansas, 18.1%, Louisiana, 18% and Tennessee, 16.9%. In these states, manufacturing has a relatively high share in total non-farm employment. On the other hand, in four southern states manufacturing employment declined over the same period. Maryland had the largest decline, 16.4%, followed by Florida, Virginia and West Virginia with declines of 4.9, 3 and 0.1% respectively. In South Carolina, manufacturing employment did not change during the study period. In the states with declining manufacturing employment, however, the share of manufacturing in total non-farm employment was relatively low.

18.3 Methodology, Application and Data Methodology

18.3.1 Methodology

In this paper, we have employed the Haynes and Dinc (1997) model of shift-share analysis to measure the impact of output change, productivity gain and other non-labor factors on employment change. In this application, to capture the annual change in regional growth or decline in employment the dynamic shift-share approach of Barff and Knight (1988) is incorporated into the new modification.

Table 18.2 Manufacturing employment change and the share of manufacturing in total non-farm employment

	US	AL	AR	FL	GA	KY	LA	MD	MS
% Change (1987-1994)	1	11.4	18.1	-4.9	2.8	18.9	18	-16.4	23.1
% Share of total empl.	0.16	0.22	0.24	0.08	0.17	0.19	0.11	0.08	0.24
	NC	**OK**	**SC**	**TN**	**TX**	**VA**	**WV**	**Average**	
% Change (1987-1994)	3.1	9.8	0	16,9	15	-3	-0.1	7.1	
% Share of total empl.	0.25	0.13	0.23	0.22	0.13	0.13	0.12		

Source: BEA and statistical abstract of the United States, 1996

Traditional shift-share analysis is a well-known technique for examining regional growth and decline and for making regional comparisons. It is a sectoral decomposition procedure widely used by regional economists, geographers, urban and regional planners, regional scientists and development analysts. It is attractive due to its simple logic, analytic clarity, and relatively modest data requirements. The shift-share method is very practical for assessing the impacts of structural change on regional and local economies and for providing guidance for industrial targeting. For a detailed discussion of shift-share models, see Loveridge and Selting (1997); Dinc, Haynes and Qiangsheng (1998); Knudsen, (1998).

The traditional shift-share model decomposes economic change in a region into three additive components: the reference area component, the industry mix and the differential (regional) shift (Dunn, 1960). The decomposed variable may be income, employment, value added, number of establishments, or a variety of other metric data. The reference component refers to the national economy and is called the national share (for smaller regions such as counties it may refer to the state economy). The national share, (NS) component measures the regional employment change that could have occurred if regional employment had grown at the same rate as the nation. The industrial mix, (IM) component measures the industrial composition of the region and reflects the degree to which the local area specializes in industries that are fast or slow growing nationally. Thus, if a region contains a relatively large share of industries that are slow (fast) growing nationally it will have a negative (positive) industry mix shift. The regional share, (RS) measures the change in a particular industry in the region due to the difference between the industry's regional growth (decline) rate and the industry's reference area growth rate. It may result from natural endowments, other comparative advantages or disadvantages, the entrepreneurial ability of the region, and/or the effects of regional policy. The total shift, (TS) measures the region's changing economic position relative to the reference area, which is the sum of the three components. These components are formulated as:

$$NS \equiv E_{ir}\, g_n \tag{18.1}$$

$$IM \equiv E_{ir}\, (g_{in} - g_n) \tag{18.2}$$

$$RS \equiv E_{ir} \ (g_{ir} - g_{in}) \tag{18.3}$$

$$TS \equiv NS + IM + RS \tag{18.4}$$

where the subscript i indexes the industrial sector in region r. E_{ir}, is employment in sector i of region, r. The growth or decline in total employment in sector i of region r is g_{ir}. The growth or decline in industry i in the reference area n is g_{in}.

Despite its widespread use, shift-share analysis has some limitations. It shows which sectors are growing (declining) in the region but it does not tell why these sectors are growing (declining), and others are not, why the region should target a given industry, or what advantages or disadvantages this region has over other regions. These are fundamental questions to be answered, though some argue that the shift-share analysis is not intended to give this information. The Haynes-Dinc extension addresses some of these issues.

Haynes-Dinc Model. Rigby and Anderson (1993) correctly argue that the traditional shift-share model measures the combined effects of output growth and productivity change on employment. In the traditional shift-share model, a region with above average employment growth either has a favorable industry mix or enjoys a competitive advantage over other regions. Therefore, as they point out, a positive (negative) shift may result from above (below) average output growth, and below (above) average productivity gains. Unless these effects are isolated, regional performance cannot be evaluated. To overcome this problem, they extend the basic shift-share method to separate the effects of changes in output and productivity on employment. Hence, they extend the shift-share model by incorporating both output and productivity changes and allocated all changes in productivity to labor.

Haynes and Dinc (1997) argue that Rigby and Anderson (1993) overestimate the production contributions of labor by ignoring the non-labor factors' contribution to productivity. To improve the Rigby and Anderson extension they separate contribution labor and capital to productivity. Following Kendrick (1961, 1973, 1983, 1984), they employ the total factor productivity (TFP) approach to separate labor and capital (or non-labor) productivity. In this approach, productivity is defined as the relationship between output of goods and services and the inputs of resources and is usually expressed in ratio of aggregate output to the sum of the inputs. Outputs are weighted by their costs per unit in constant prices. Inputs are combined in terms of their share of total costs in constant prices. Thus, to maintain the simplicity of the shift-share model this method is employed.

To investigate employment (L) change the shift-share equations took, therefore, the following forms:

$$TS_L \equiv NS_L + IM_L + RS_L \tag{18.5}$$

$$NS_L \equiv NS(a_L) + NS(b_L) = \sum E_{ir} \ (a_{nL} + b_{nL}) \tag{18.6}$$

$$IM_L \equiv IM(a_L) + IM(b_L) = \sum E_{ir} \ [(a_{inL} - a_{nL}) + (b_{inL} - b_{nL})] \tag{18.7}$$

$$RS_L \equiv RS(a_L) + RS(b_L) = \sum E_{ir} [(a_{irL}-a_{inL}) + (b_{irL}-b_{inL})] \quad (18.8)$$

where a's represent the rate of employment change in the region (and reference area) resulting from variations in output over the given time period with productivity constant; b's represent the rate of employment change resulting from variations in productivity over the given time period with output constant. For a full derivation of the above equations see Haynes and Dinc (1997).

In the Haynes-Dinc (1997) study, non-labor factors' contribution was not estimated directly, instead they calculated this contribution as a residual, the difference between actual employment change and employment change resulting from output and productivity variation. This can be formulated as:

$$\Delta E_P = \Delta E - \Delta E_L \quad (18.9)$$

where ΔE is the actual employment change over time in the region, ΔE_L is the employment change in the region or state resulting from the change in labor productivity and output. ΔE_P is the employment change resulting from the other production factors' contribution to total factor productivity.

This model was successfully evaluated and assessed for robustness with respect to alternative production function specifications (Haynes, Dinc and Paelinck, 1997).

Table 18.3 Evaluation guideline for the combined results

Position	Actual emp. change	S-S output effect	S-S productivity effect	S-S non-labor effect
1	+	+	+	+
2	+	+	+	-
3	+	+	-	+
4	+	-	+	+
5	-	+	+	+
6	-	-	-	-
7	-	-	-	+
8	-	-	+	-
9	-	+	-	-
10	+	-	-	-
11	+	+	-	-
12	-	-	+	+
13	+	-	-	+
14	-	+	+	-
15	-	+	-	+
16	+	-	+	-

Application. The application of the Haynes-Dinc extension of shift-share produces four outcomes for a given sector in a region: (a) employment change resulting from output growth (decline), (b) employment change resulting from productivity gain (loss), (c) employment change resulting from non-labor factors' contribution, and (d) actual employment change. Recall that the dynamic version of shift-share is employed so these outcomes reflect cumulative changes.

When we look at the causes of employment change (output growth or decline, variation in productivity and non-labor factors) and combine them with the actual change, a sector in a given state can be in one of the sixteen categories (18.3). Position 1 is the best position in which each of these components has a positive sign. We should note that here a positive sign for shift-share productivity affect means that the state in a given sector improved its productivity, even though productivity improvements in the short-term create employment decline.

Since we have four different outcomes from this application, we can utilize the Sectoral Development Diamond of Dinc and Haynes (1999). In this diamond, each dimension or axis represents one of the above outcomes, i.e., actual employment change of the sector, employment change resulting from output variation, etc. The distance from the center indicates a positive outcome for a given variable. If a given sector in a state is in the position 1, it will have a relatively large diamond because the better the position of a sector in any of these outcomes, the further away it will be from the center. On the other hand, if it is in the sixth position, it will be on the center. Depending on the position of a given sector the shape of the diamond could become asymmetric. This diamond provides a good summarizing and a quick visual comparison tool for policy makers and practitioners.

Diagram 18.1 Development diamond for individual sector

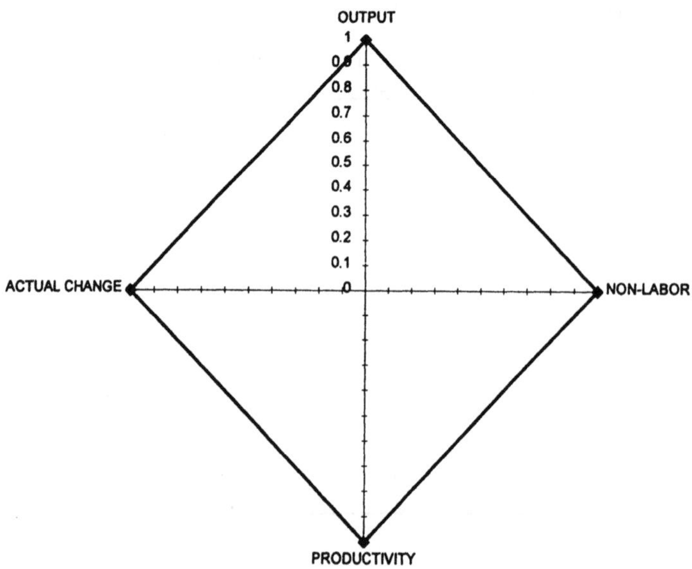

Further, individual manufacturing sectors are grouped in four categories based on product cycle and economic base theories. If a sector has reached its maturity and stopped growing, in some cases has started declining, it is called the old economy (i.e., apparel and textile products). On the other hand, if it has not reached the maturity level, it is labeled as the new or contemporary economy (printing and publishing, instruments). Some sectors may primarily serve the region or the state. These sectors are non-basic sectors and grouped as the local and regional economy (food and kindred products, paper and allied products and rubber and plastics products). Finally, some sectors' products are not for the most part consumed locally and these export oriented sectors are grouped as the core or export base economy (industrial machinery, electronic and electric products and transportation equipment).

The development diamond noted above can also be used to investigate the sectoral economic structure of a state. In this application, the new dimensions of the diamond are the new, old, local and export sectors.

Diagram 18.2 Sectoral development diamond

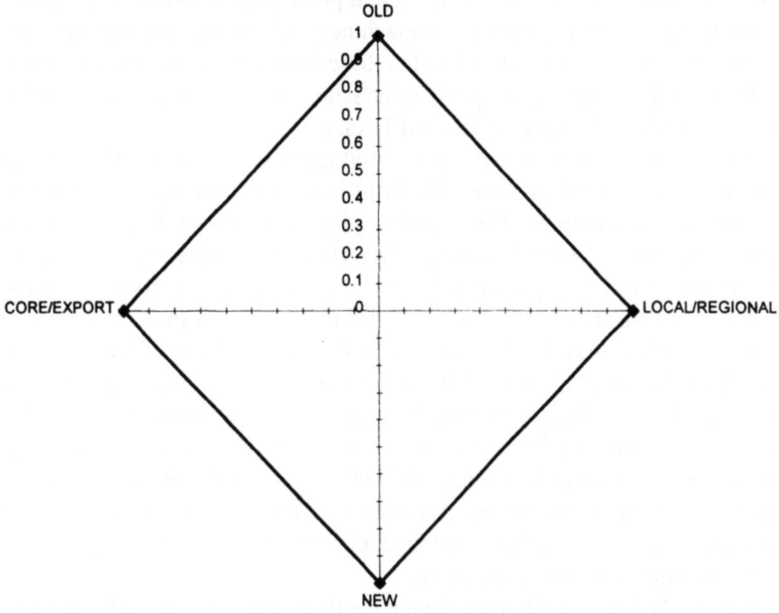

Data. The data employed are from the Annual Survey of Manufactures' Geographic Area Series, and the Census of Manufactures' Geographic Area Series. The study period is from 1987 to 1994.

Labor inputs were measured by hours worked rather than the number of employees, to take account of temporal, sectoral and spatial variations in the length of the unit of work. The labor inputs include only production workers in manufacturing sectors, which covers 70-75 percent of total manufacturing employees in the United States. Output is measured by value added and productivity is defined as output per hour worked according to the share of labor in the given sector's total value added. To determine the share of labor in income, the ratio of production wages to sector specific value added is used. Wages data of production workers are from the Annual Survey of Manufactures' and Census of Manufactures. Value added data by industry were deflated using the unpublished BLS industry price deflator. The total output of any state for a given year was deflated by a GDP price deflator.

18.4 Findings and Discussion

We have grouped our findings into three categories. The first category represents aggregate employment change in manufacturing based on available data at the sector level by state (Table 18. 4 and Figures 18.1 through 3). In the second category, individual manufacturing sectors are investigated as indicators of the old economy (i.e., apparel and textile products), the local and regional economy (food and kindred products, paper and allied products and rubber and plastics products), the new or contemporary economy (printing and publishing, instruments) and the core or export base economy (industrial machinery, electronic and electric products and transportation equipment) by state based on the general trend of these sectors. In the third category, a sample of contrasting states is examined by sector (Florida and Georgia vs. North Carolina and Texas).

Table 18.4 shows the aggregated results produced by the Haynes-Dinc model by state from 1987 to 1994. In this table, bold italic numbers represent the percentage share of each component in actual employment change. Figure 1 shows that with the exception of Florida, Georgia, Maryland, Louisiana and Virginia, all states in the south have outperformed their national counterparts and gained employment resulting from output growth. Texas had the largest employment growth (61,859,000 hours) due to output increases followed by Tennessee and North Carolina (36,839,000 and 35,465,000 hours respectively). On the other hand, Maryland had the largest decline in manufacturing employment resulting from a decline in outputs (4,191,000 hours). Other states that lost manufacturing employment are Louisiana and Virginia (508,000 and 170,000 hours respectively). Florida and Georgia performed worse than their national counterparts but they did not loose employment due to output growth (decline) rather their poor employment performance was due to productivity gains.

When we look at the employment change resulting from variation in productivity, six states (Alabama, Florida, Louisiana, Mississippi, Tennessee and Virginia) performed poorly and as a result of the decline in productivity their employment increased. Although, the remaining nine states have improved their productivity only two of them (Kentucky and Oklahoma) outperformed their national competitors. However, when all there components, NS, IM and RS, are combined all southern states (except Louisiana) lost employment because of improvement in productivity. In most states, employment losses from improvements in productivity outpaced employment gains from output growth. The best performers in terms of productivity improvement are North Carolina and Texas. They lost 62,965,000

Table 18.4 Employment change in southern manufacturing sectors, 1987-1994: H-D shift-share results

STATES	NS(a)	NS(b)	PS(a)	DS (b)	DS(a)	DS(b)	Total a	Total b	Total shift	Actual change	Other' factors	% change
Alabama	9600	-24515	-1799	3479	12298	14569	20099	-6466	13633	60300	46667	11.4
	0.16	-0.41	-0.03	0.06	0.20	0.24	0.33	-0.11	0.23	1.00	0.77	
Arkansas	6305	-15403	-260	3666	16978	-4201	23024	-15938	7086	59300	52214	18.1
	0.11	-0.26	0.00	0.06	0.29	-0.07	0.39	-0.27	0.12	1.00	0.88	
Florida	9600	-25689	-2603	6241	-4432	311	2565	-19137	-16572	-29400	-12828	-4.9
	0.33	-0.87	-0.09	0.21	-0.15	0.01	0.09	-0.65	-0.56	1.00	-0.44	
Georgia	10996	-28206	-3945	8721	4930	-66	11981	-19551	-7570	15100	22670	2.8
	0.73	-1.87	-0.26	0.58	0.33	0.00	0.79	-1.29	-0.50	1.00	1.50	
Kentucky	6281	-16415	-325	1823	22947	-21737	28904	-36328	-7425	62500	69925	18.9
	0.10	-0.26	-0.01	0.03	0.37	-0.35	0.46	-0.58	-0.12	1.00	1.12	
Louisiana	4465	-10843	85	1504	-5059	45457	-508	36119	35610	31900	-3710	18.0
	0.14	-0.34	0.00	0.05	-0.16	1.42	-0.02	1.13	1.112	1.00	-0.12	
Maryland	3796	-10588	-794	3357	-7193	-3328	-4191	-10559	-14750	-46100	-31350	-16.4
	-0.08	-0.23	-0.02	0.07	-0.16	-0.07	-0.09	-0.23	-0.32	1.00	-0.68	
Mississippi	5831	-14802	-1420	3692	16477	5964	20888	-5146	15742	78100	62358	23.1
	0.07	-0.19	-0.02	0.05	0.21	0.08	0.27	-0.07	0.20	1.00	0.80	
North Carolina	21192	-54273	-9597	8393	23870	-17085	35465	-62965	-27501	38900	66401	3.1
	0.54	-1.40	-0.25	0.22	0.61	-0.44	0.91	-1.62	-0.71	1.00	1.71	
Oklahoma	3813	-9541	-135	1978	17808	-13708	21486	-21272	215	19800	19585	9.8
	0.19	-0.48	-0.01	0.10	0.90	-0.69	1.09	-1.07	0.01	1.00	0.99	
South Carolina	9362	-24219	-4000	3775	14583	-23460	19945	-43904	-23959	-1800	22159	0.0
	5.20	-13.45	-2.22	2.10	8.10	-13.03	11.08	-24.39	-13.31	1.00	12.31	
Tennessee	126449	-32197	-3595	7597	27784	10102	36839	-14498	22341	109700	87359	16.9
	0.12	-0.29	-0.03	-0.07	0.25	0.09	0.34	-0.13	0.20	1.00	0.80	
Texas	20290	-51412	-1175	9925	42744	-17672	61859	-59159	2700	164800	162100	15.0
	0.12	-0.31	-0.01	0.06	0.26	-0.11	0.38	-0.36	0.02	1.00	0.98	
Virginia	8398	-21947	-2940	6494	-5628	3277	-170	-12176	-12346	-17300	-4954	-3.0
	0.49	-1.27	-0.17	0.38	-0.33	0.19	-0.01	-0.70	-0.71	1.00	-0.29	
West Virginia	1529	-4034	4	541	2922	-3878	4454	-7372	-2918	-3000	-82	-0.1
	0.51	-1.34	0.00	0.18	0.97	-1.29	1.48	-2.46	-0.97	1.00	-0.03	

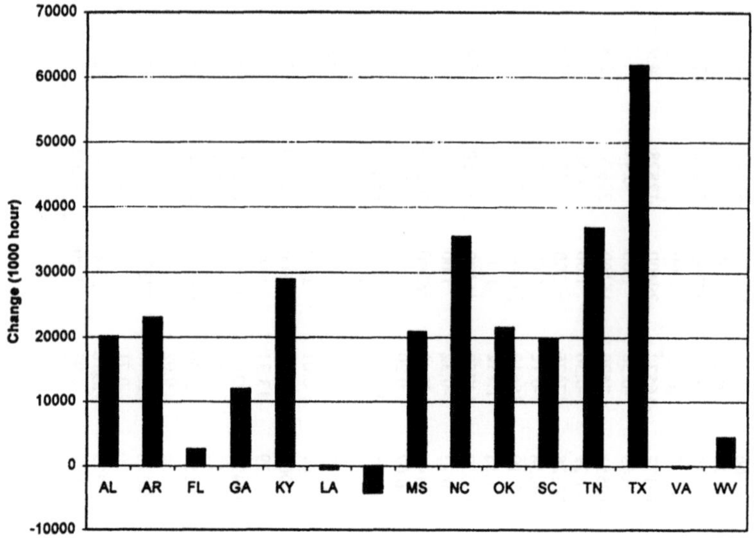

Figure 18.1 Employment change resulting from output change, 1987-1994. The states are presented in the following order: AL, AR, FL, GA, KY, LA, MD, MS, NC, OK, SC, TN, TX, VA, WV

and 59,159,000 hours of employment respectively. South Carolina and Kentucky followed these states with 43,904,000 and 36,328,000 hours employment losses.

Although output growth (decline) and productivity gains (losses) had an important impact on employment change in manufacturing, non-labor factors made a signifcant contribution to change in employment. In fact, in ten states non-labor factors were the driving forces behind this change (growth or decline). For example, non-labor factors (i.e., new investments, technological improvements) saved Texas, North Carolina, South Carolina, Georgia and Kentucky from heavy employment losses. On the other hand, non-labor factors in the form of exits and capital under-investment in such things as equipment and infrastructure deepened employment losses in Florida, Maryland, Louisiana and Virginia. These findings suggest that non-labor factors require a more detailed investigation in the region.

We have investigated individual sectors indicative of the old, new or contemporary, local and regional, and core or export base economy (Table 18.5). The apparel and textile products sector represents the old economy, which has been declining nationally and regionally for a long time. In this sector, with the exception of Tennessee and Texas all Southern states lost employment. In these two states, output growth and non-labor factors are the driving forces of this growth, although Texas improved its productivity and hence lost some employment. Among the remaining states only Kentucky, Louisiana and Oklahoma gained employment due to output growth. However, improvements in productivity and exits or under-investments (non-labor factors) caused a net decline in employment (Figure 18.4).

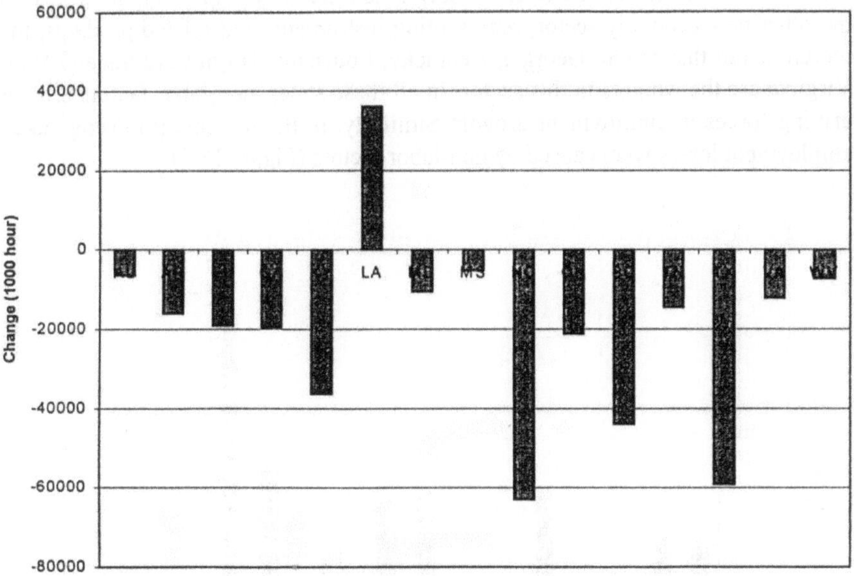

Figure 18.2 Employment change resulting from labor productivity gains, 1987-1994
The states are presented in the same order as in Figure 18.1.

Table 18.5 Sectors indicative of the different aspects of the Southern economy

The old economy	The new economy
Tobacco (SIC 21)*	Printing and publishing (SIC 27)
Textile mill products (SIC 22)*	Instruments and related products (SIC 38)
Apparel and textile products (SIC 23)	
Furniture and fixture (SIC 25)*	
The local and regional economy	**The core/export base economy**
Food and kindred products (SIC 20)	Industrial machinery and equipment (SIC 35)
Paper and allied products (SIC 26)	Electronic and electric equipment (SIC 36)
Rubber and plastics products (SIC 30)	Transportation equipment (SIC 37)

* Not examined due to missing sectors in some states

The printing and publishing, and instruments and related products sectors are assumed to be associated with the new or contemporary economy. In the printing and publishing sector, only Alabama and Maryland are the losers. Productivity gain and non-labor factors in these states are responsible for employment losses.

There was a slight increase in employment resulting from output growth, but it was not large enough to offset losses in these states. In the growing states, output growth and non-labor factors substantially contributed to an increase in employment in this sector. Tennessee and Georgia had the largest gain (Figure 18.5). In the other new economy sector, representing instruments and related products, the picture is not that bright. Georgia, Kentucky, Louisiana, North Carolina and West Virginia are the winners in this sector. In all these states, non-labor factors are the driving forces in employment growth. Similarly, in the remaining loosing states employment losses were caused by non-labor factors (Figure 18.6).

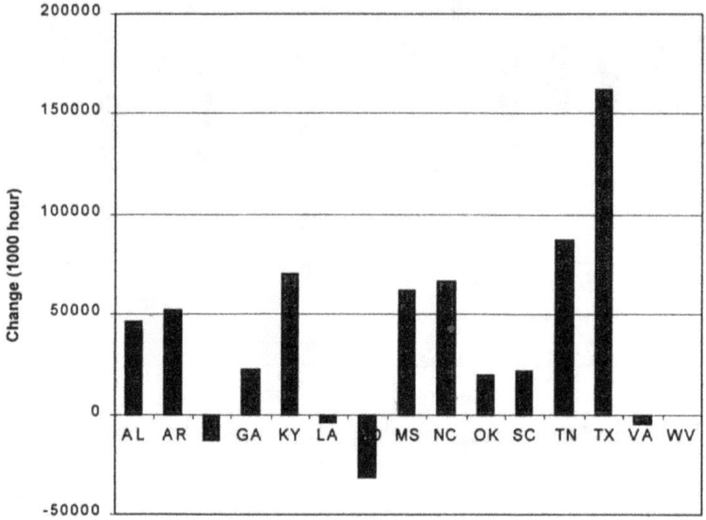

Figure 18.3 Employment change resulting from other (non-labor) factors, 1987-1994
The states are presented in the same order as in Figure 18.1.

Locally and regionally serving sectors are represented by the food and kindred products, paper and allied products, and the rubber and miscellaneous plastics sectors. In the food and kindred products sector, with the exception of Florida, Louisiana and Maryland all states had net employment growth. This gain resulted from output growth and non-labor factors' contribution. Except for a few states (AL, MS and WV) most improved their productivity and suffered employment losses. In this sector, Mississippi, Arkansas, Alabama and Texas are the leading states (Figure 18.7). The paper and allied products is another well-performing sector in the region. In this sector, only Florida lost employment resulting from productivity gains and non-labor factors. Although Oklahoma and West Virginia also lost employment, it was negligible. Other states with the contribution of non-labor factors and decline in productivity gained employment in this sector. Texas and Alabama are the leaders in this sector (Figure 18.8). In the rubber and miscellaneous plastics sector, only Alabama suffered a negligible employment loss. The other states enjoyed an increase in employment in which the big winners were Tennessee and Texas followed by North Carolina. The driving force in this growth was the contribution of non-labor factors to employment (Figure 18.9).

The industrial machinery and equipment, electronic and electric equipment and the transportation equipment sectors represent the core/export manufacturing activity. In the industrial machinery and equipment sector, Maryland is the only loosing state. Texas, North Carolina, South Carolina, Tennessee and Georgia gained substantial employment in this sector driven by non-labor factors' contribution. These states with the exception of North Carolina improved their productivity but employment gain from output growth almost canceled out the losses from the productivity gain. Hence, the remaining net change came from non-labor factors (Figure 18.10). The electronic and electric equipment sector did not enjoy the same growth in the South. Only four states, Texas, Alabama, South Carolina and North Carolina gained significant employment in this sector. Among seven loosing states, Maryland suffered the heaviest employment loss (Figure 18.11). Seven states in the transportation equipment sector lost employment due to all factors output decline, improvement of productivity and non-labor factors (exits, under-investment). Georgia, Texas, Florida and Virginia had the biggest lost. In this sector, Tennessee, Kentucky, South Carolina, North Carolina and Mississippi are the winners (Figure 18.12).

Finally, we have investigated four large manufacturing states of the South in detail: Florida, Georgia, North Carolina and Texas. Florida increased its employment in only four sectors: the printing and publishing (SIC 27), rubber and plastics (SIC 30), industrial machinery and equipment (SIC 35) and the miscellaneous manufacturing sectors (SIC 39). In three of these (SIC 27, 30 and 39), the driving force of employment gain was output growth and non-labor factors' contribution. On the other hand, in SIC 35 poor productivity saved Florida from employment losses, despite the decline in output and negative effects of non-labor factors (exits, under-investment). In other sectors, Florida generally improved its productivity but decline in demand, exits and under-investments caused employment losses. For example, in the transportation equipment sector, Florida enjoyed a significant productivity gain, and this gain and the negative impact of non-labor factors swapped all increases in employment resulting from output growth. Hence, Florida lost substantial employment in this sector (Figure 18.13).

Georgia relative to other successful Southern states enjoyed little employment growth in manufacturing. In the food and kindred products (SIC 20), industrial machinery and equipment (SIC 35) and rubber and plastics (SIC 30) sectors, Georgia performed well. Its overall productivity gain was modest in the lumber and wood products (SIC 24), transportation equipment (SIC 37), food and kindred products (SIC 20) and industrial machinery and equipment (SIC 35) sectors. In other sectors, it enjoyed either negligible productivity gains or decline in productivity (the electronic and electric equipment, SIC 36). In Georgia's manufacturing employment, non-labor factors played important roles both negatively and positively depending on the sector analyzed (Figure 18.14).

North Carolina is one of the well-positioned states in the south. Although it lost employment in five sectors, these were all part of the older and declining economy: tobacco products (SIC 21), textile mill products (SIC 22), apparel and textile products (SIC 23), furniture and fixtures (SIC 25) and stone, clay and glass (SIC 32) sectors. In all these sectors' productivity improved. On the other hand, North Carolina increased its manufacturing employment in fifteen sectors, which are core/export base, locally and regionally serving and new/or contemporary sectors. With two exceptions, North Carolina improved its productivity in all sectors. Similar to other Southern states non-labor factors contributed significantly to employment change in North Carolina (Figure 18.15).

Texas is the best performer among these states. It improved its productivity in all sectors we have investigated and lost employment in only three sectors: the stone, clay and glass (SIC 32). transportation equipment (SIC 37) and the instrument and related products (SIC 38). In these three sectors, employment losses resulted from non-labor factors, though productivity gains contributed to the losses. On the other hand, in the remaining sectors, increase in employment resulting from output growth and non-labor factors' contribution (new investments, new entries, and improvement of infrastructure) outpaced the losses resulting from productivity gains (Figure 18.16).

For demonstration purposes, we have applied the development diamond four states: Florida versus Georgia and North Carolina versus Texas. Recall that both North Carolina and Texas performed well and Florida was one of the poor performers. Georgia had a mixed employment change depending on the sectors. The below diamonds show the aggregate results for these states. In these diamonds, positive productivity, in fact, represents employment decline so it should be interpreted accordingly.

Diagram 18.3 Development diamond for FL and GA, 1987-1994

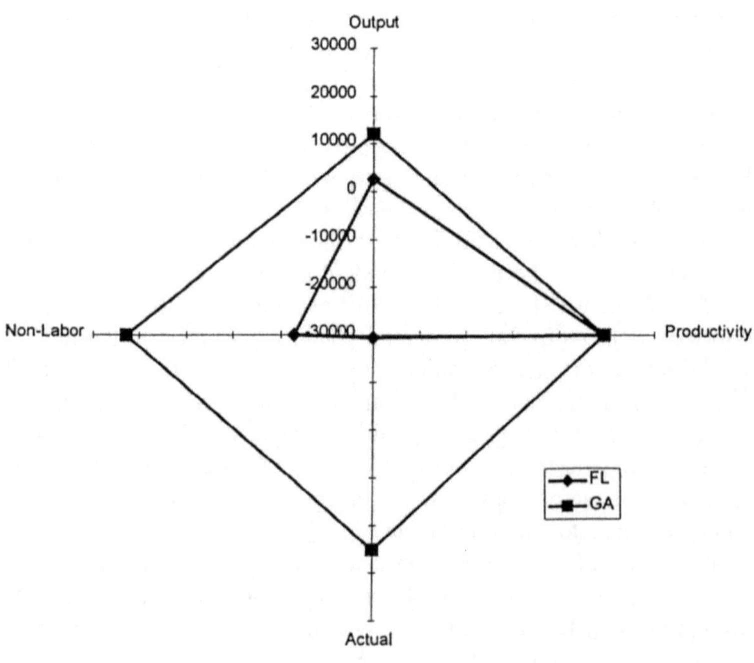

Diagram 18.4 Development diamond for NC and TX, 1987-1994

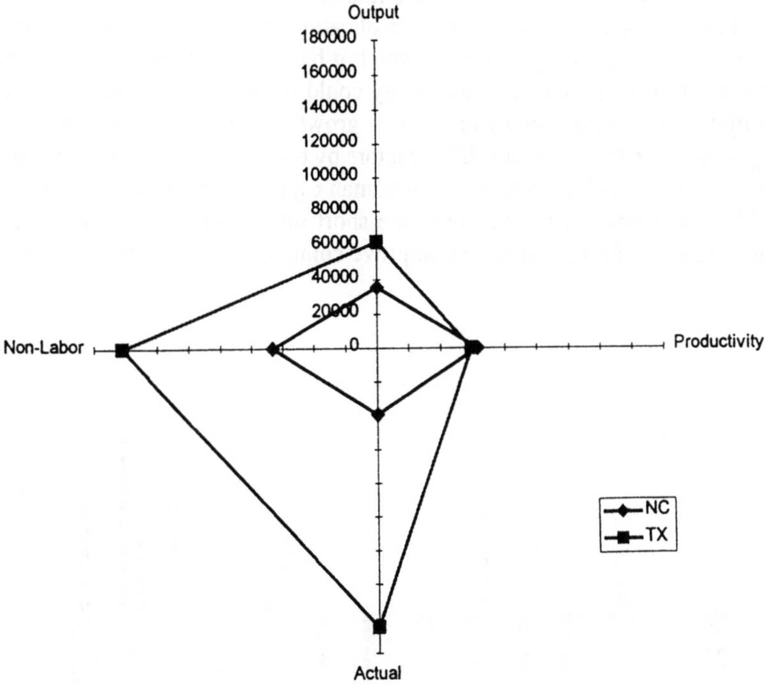

18.5 Conclusion

In this study, we have investigated the foundations of economic performance and employment change in manufacturing sectors in fifteen southern states by employing the Haynes and Dinc extension of shift-share. This enabled us to take account of industrial and regional differences in labor and capital (non-labor) productivity and the impacts of productivity differences on regional manufacturing employment change.

Findings of this study suggest that the change in manufacturing employment during the investigation period in the U.S. South has been driven by the differential effects of output growth (decline), new investments (underinvestments) in physical capital, and improvement in technology, though labor productivity has also played a crucial role in employment change. These findings are consistent with earlier studies (Haynes and Dinc, 1997; Ledebur and Moomaw, 1983; Hulten and Schwab, 1984; Moomaw and Williams, 1991).

Contrary to Malecki (1995) and Schaffer's (1993) concerns about the southern economy, our analysis also suggests that although there has been decline in manufacturing employment in some states and some sectors, the southern states have been performing relatively well in manufacturing. Such positive results are driven by output growth and capital (non-labor) investments. Labor productivity gains, except in few states and few sectors, have not been large enough to offset these components. However, the lack of sufficiently large gains in labor productivity could poses competitiveness problems in the future.

The findings of this study suggest some implications for regional growth theory. A state may focus on growing sectors in which it has a competitive edge or work to improve its declining sectors. One strategy could be marketing oriented which creates output growth and hence employment growth. A second strategy may be generating large investments in non-labor factors by providing incentives and other tools. The third one involves investment in human capital to improve productivity. This could cause employment decline in the short run but in the long run it can attract more business into the state and improve competitiveness of the sector and the state.

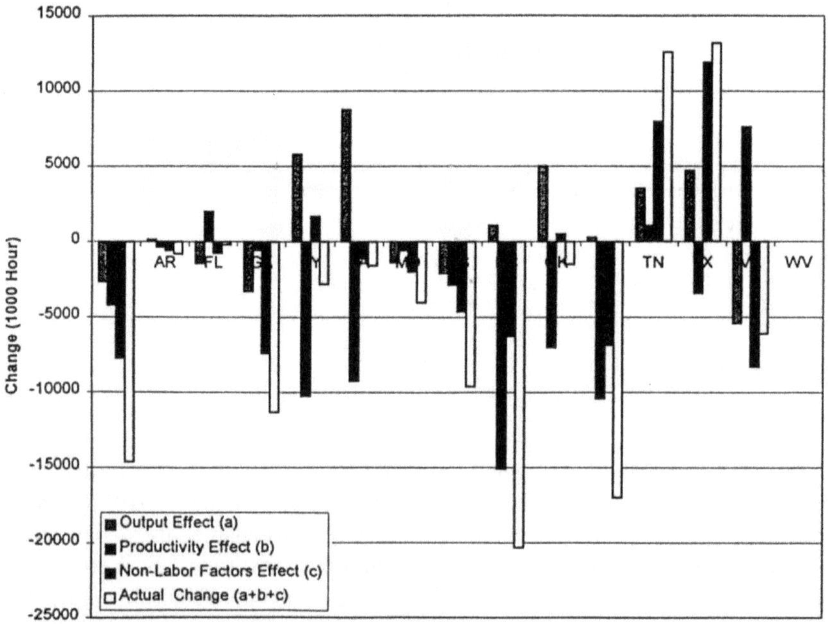

Figure 18.4 Employment change in apparel and textile products sector, 1987-1994
The states are presented in the same order as in Figure 18.1.

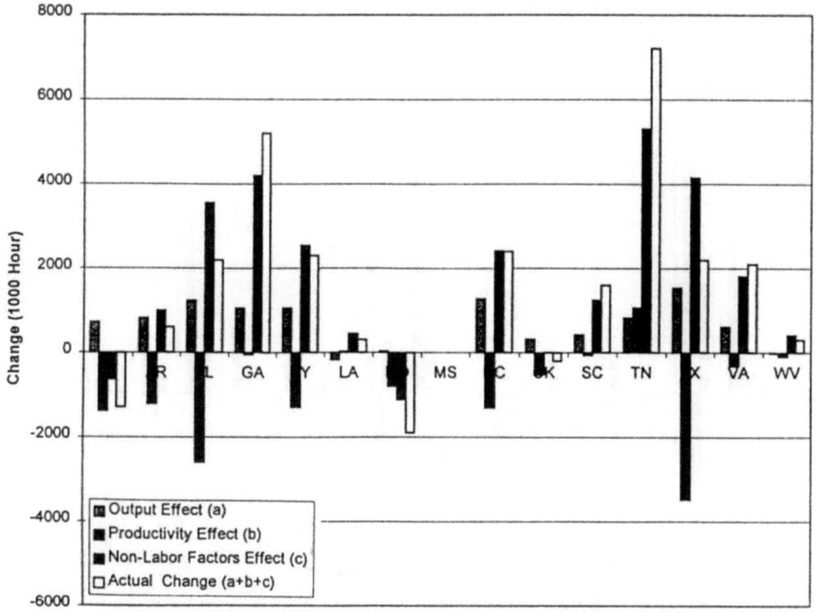

Figure 18.5 Employment change in printing and publishing sector, 1987-1994
The states are presented in the same order as in Figure 18.1.

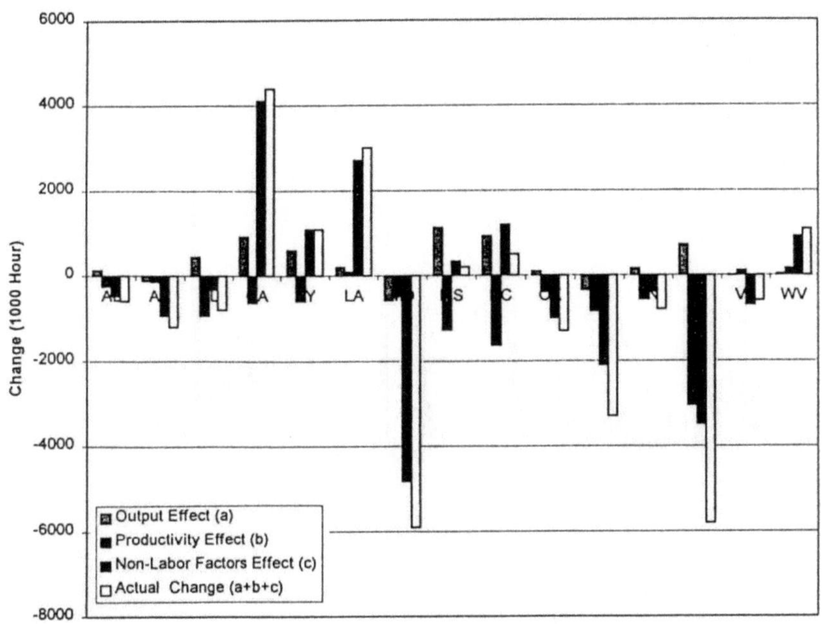

Figure 18.6 Employment change in instruments and related products sector, 1987-1994
The states are presented in the same order as in Figure 18.1.

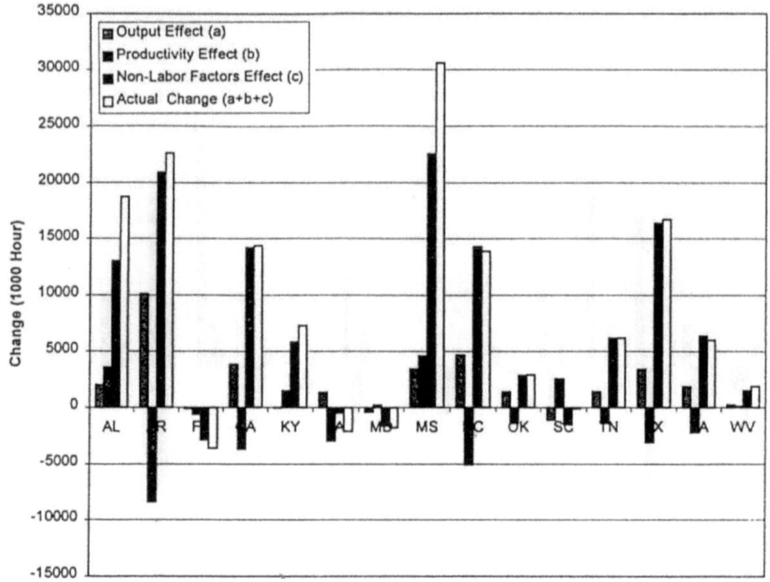

Figure 18.7 Employment change in food and kindred products sector, 1987-1994
The states are presented in the same order as in Figure 18.1.

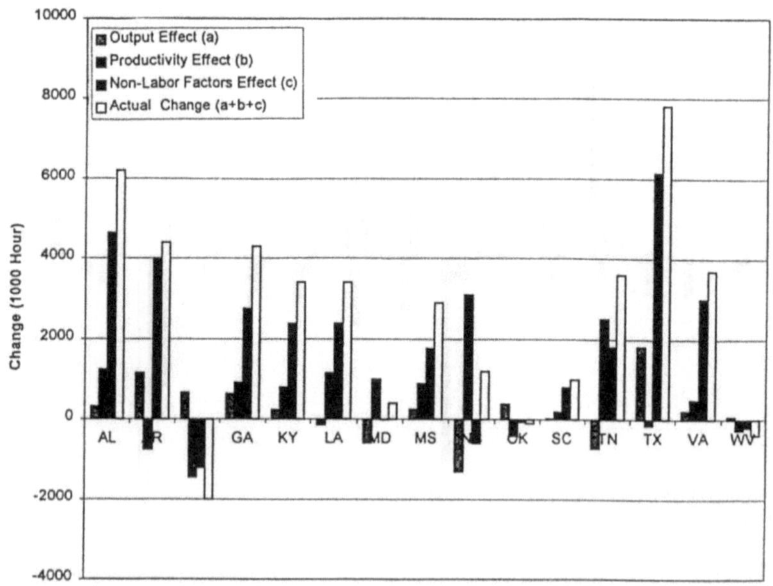

Figure 18.8 Employment change in paper and allied products sector, 1987-1994
The states are presented in the same order as in Figure 18.1.

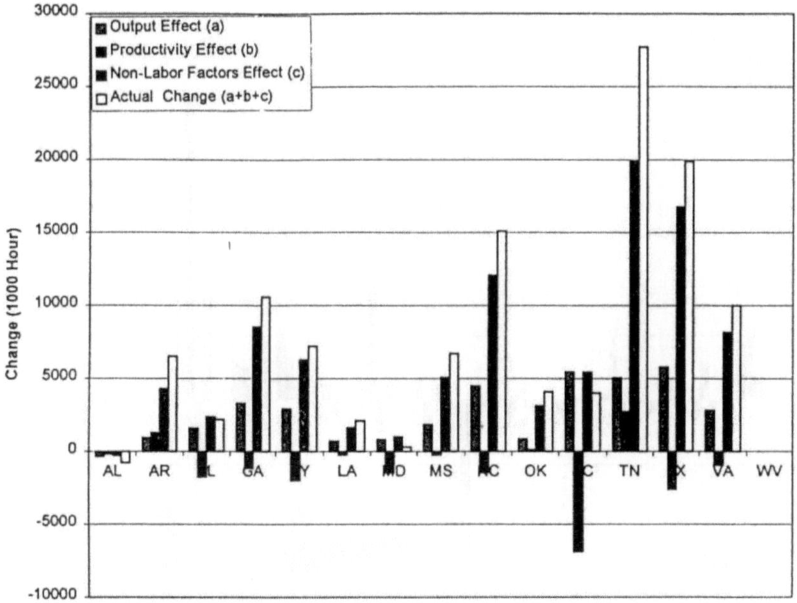

Figure 18.9 Employment change in rubber and misc. plastics sector, 1987-1994
The states are presented in the same order as in Figure 18.1.

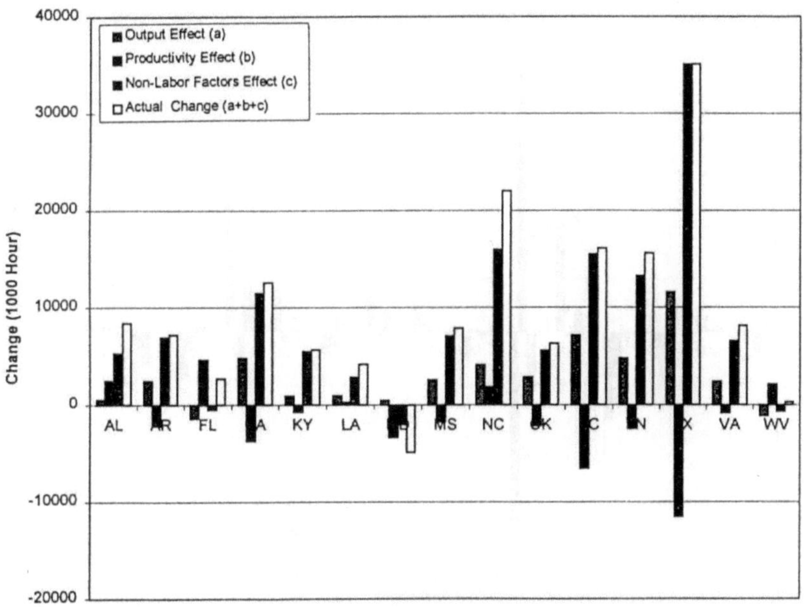

Figure 18.10 Employment change in industrial machinery and equipment sector, 1987-1994
The states are presented in the same order as in Figure 18.1.

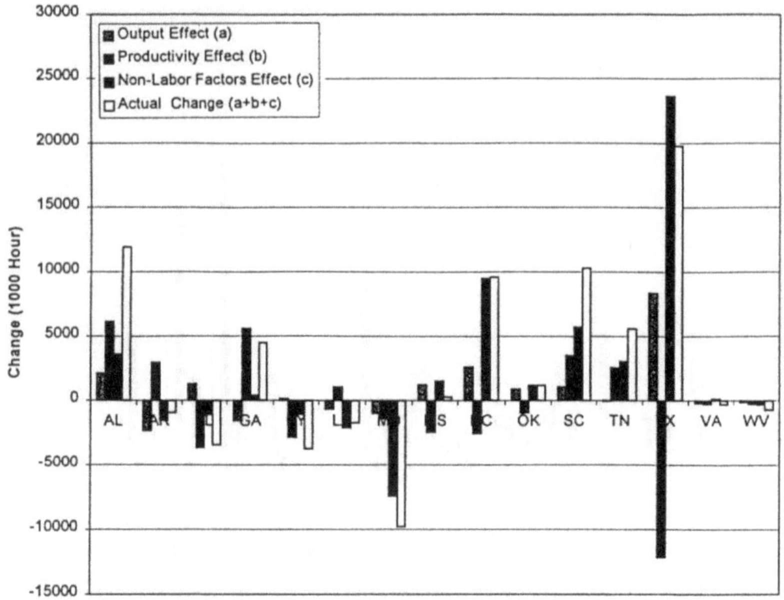

Figure 18.11 Employment change in electronic and electric equipment sector, 1987-1994
The states are presented in the same order as in Figure 18.1.

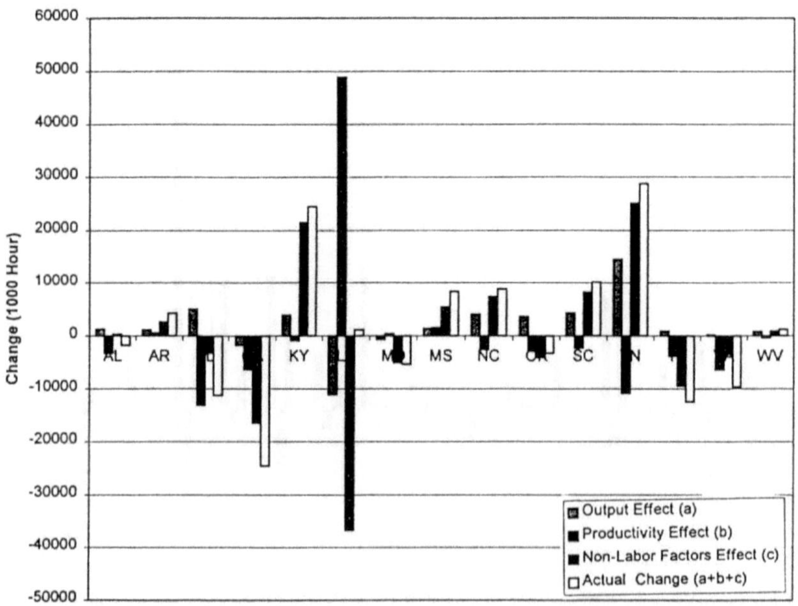

Figure 18.12 Employment change in transportation equipment sector, 1987-1994
The states are presented in the same order as in Figure 18.1.

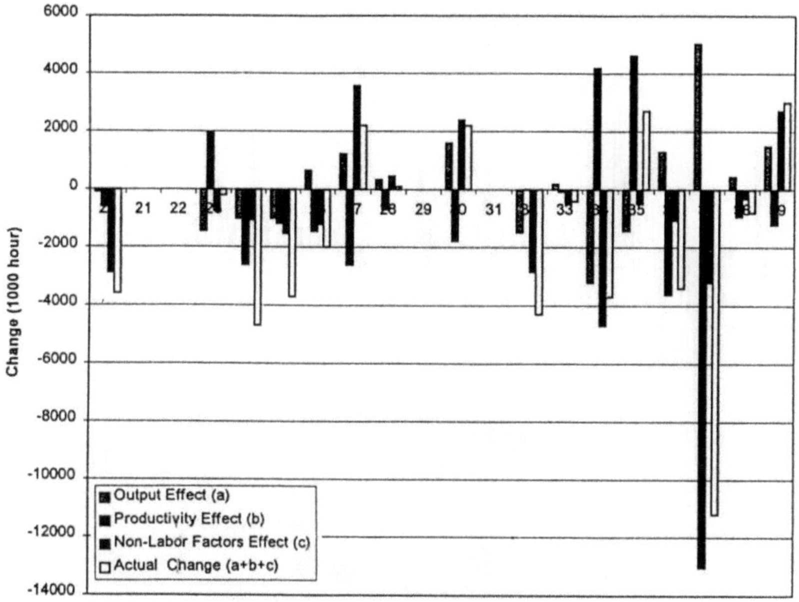

Figure 18.13 Manufacturing employment change in Florida by sector, 1987-1994
The sectors are presented in the following order: 20-39.

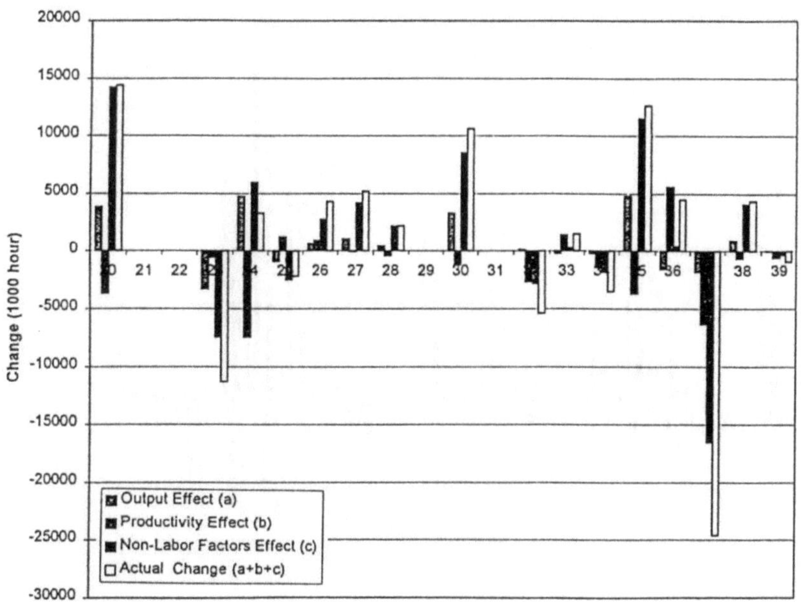

Figure 18.14 Manufacturing employment change in Georgia by sector, 1987-1994
The sectors are presented in the following order: 20-39.

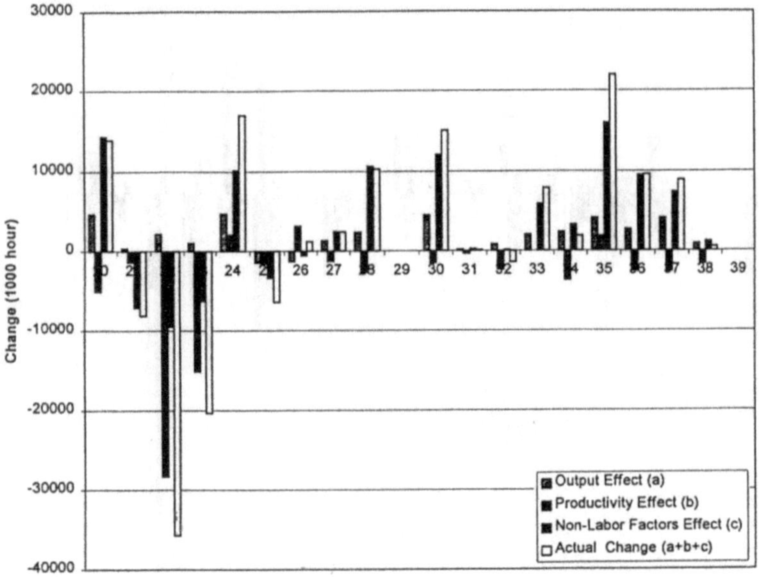

Figure 18.15 Manufacturing employment change in North Carolina by sector, 1987-1994 The sectors are presented in the following order: 20-39.

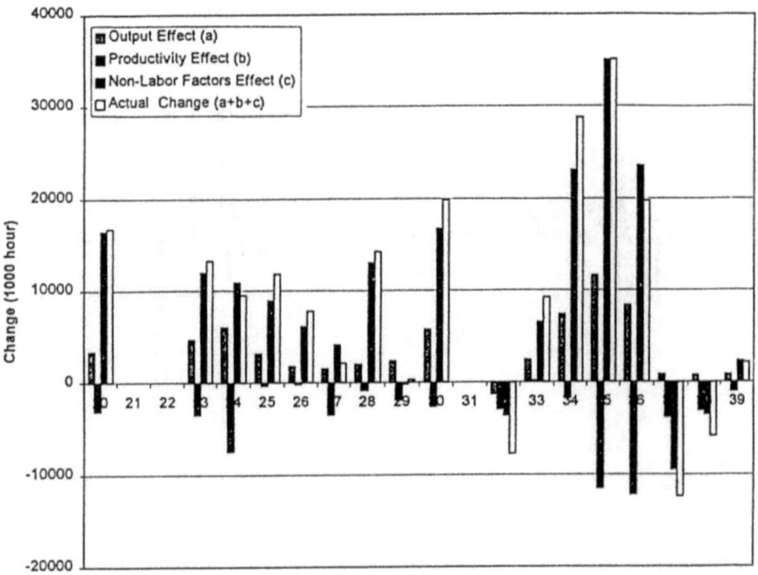

Figure 18.16 Manufacturing employment change in Texas by sector, 1987-1994 The sectors are presented in the following order: 20-39.

References

Barff, R.A. and P.L. Knight III, 1988, "Dynamic Shift-Share Analysis", *Growth and Change,* 15:1-10.

Casetti, E., 1984, "Manufacturing Productivity and Snowbelt-Sunbelt Shifts", *Economic Geography,* 60:313-324.

Connaughton, J.E. and R.A. Madsen, 1990, "The Changing Regional Structure of the U.S. Economy", *Growth and Change,* Vol. 21, 3:48-59.

Costrell, R.M., 1994, "Accounting for the Causes and Consequences of Industrial Employment Shift", *Industrial Relations,* Vol. 33, 3:346-364.

Crandall, R.W., 1993, "Manufacturing on the Move," The Brookings Institution, Washington, D.C.

Dinc, M., K.E. Haynes and L. Qiangsheng, 1998, "A Comparative Evaluation of Shift-Share Models and Their Extensions", *Australasian Journal of Regional Studies,* Vol. 4(2):275-302.

Dinc, M. and K.E. Haynes, 1999, "Sourced of Regional Inefficiency: An Integrated Shift-share, Data Envelopment and Input-output Approach", *The Annals of Regional Science* (in press).

Dunn, E.S., 1960, "A Statistical and Analytical Technique for Regional Analysis", *Papers of the Regional Science Association,* 6:97-112.

Erickson, R.A., 1976, "The Filtering-Down Process: Industrial Location in Non-Metropolitan Area," *Professional Geographer,* 28:254-260.

Erickson, R.A. and T.R. Leinbach, 1979, "Characteristics of Branch Plants Attracted to Non-Metropolitan Areas", in R.E. Lonsdale and H.L. Seyler (eds.), *Nonmetropolitan Industrialization,* V.H. Winston and Sons, Washington D.C.

Federal Reserve Bank of Chicago, 1996, "Assessing the Midwest Economy, No:2, Chicago.

Harrison, B. and B. Bluestone, 1988, *The Great U-Turn,* Basic Books, New York.

Haynes, K.E. and M. Dinc, 1997, "Productivity Change in Manufacturing Regions: A Multifactor/Shift-Share Approach", *Growth and Change,* 28:150-170.

Haynes, K.E., M. Dinc and J.H.P. Paelinck, 1997, "Identifying Sources of Regional Productivity Change in Manufacturing: Alternative Productivity Measurement Approach in a Shift-Share Framework", Paper presented at the 37th European Regional Science Association Annual meeting in Rome, Italy, August 26-29, 1997.

Hoover, E.M., 1948, *The Location of Economic Activity,* McGraw-Hill, New York.

Hulten, C.R. and R.M. Schwab, 1984, "Regional Productivity Growth in U.S. Manufacturing: 1951-78", *The American Economic Review,* Vol. 74, 1:152- 2.

Kendrick, J.W., 1961, *Productivity Trends in the United States*, Princeton University Press, Princeton

Kendrick, J.W., 1973, *Postwar Productivity Trends in the United States, 1948-1969,* National Bureau of Economic Research, New York.

Kendrick, J.W., 1983, *Interindustry Differences in Productuvity Growth,* American Enterprise Institute, Washington, D.C.

Kendrick, J.W., 1984, *Improving Company Productivity,* The Johns Hopkins University Press, Baltimore.

Knudsen, D. C., 1998, "Shift-Share Analysis: Further Examination of Models for the Description of Economic Change," Paper submitted to Socio-Economic Planning Sciences.

Ledebur, L.C. and R.L. Moomaw, 1983, "A Shift-Share Analysis of Regional Labor Productivity in Manufacturing", *Growth and Change,* Vol. 14, 1:2-9.

Loveridge, S. and A.C. Selting, 1997, "A Review and Comparison of Shift-Share Identities", *International Regional Science Review,* forthcoming.

Malecki, E.J., 1995, "Global Cities and Back Roads: Perspectives on the Southern Economy", *The Review of Regional Studies,* Vol. 25, 3:237-246.

Moomaw, R.L. and M. Williams, 1991, "Total Factor Productivity Growth in Manufacturing: Further Evidence from the States", *Journal of Regional Science,* Vol. 31, 1:17-34.

Moriarty, B.M., 1992, "The Manufacturing Employment Longitudinal Density Distribution in the USA", *The Review of Regional Studies,* Vol. 22, 1:1-24.

Moriarty, B.M., 1983, "Hierarchies of Cities and the Spatial Filtering of Industrial Development", *Papers of the Regional Science Association,* 52:59-82.

Norton, R.D., 1997, "Where are the New U.S. High-Tech Jobs?" Paper presented at the 37th European Regional Science Association Meeting in Rome, Italy, August 26-29, 1997.

Norton, R.D. and J. Rees, 1979, "The Product Cycle and Spatial Decentralization of American Manufacturing", *Regional Studies,* Vol. 13, 2:141-151.

Richardson, H.W. and J.H. Turek, (eds.), 1985, *Economic Prospects for the Northeast,* Temple University Press, Philadelphia.

Rigby, D.L. and W.P. Anderson, 1993, "Employment Change, Growth and Productivity in Canadian Manufacturing: An Extension and Application of Shift-Share Analysis", *Canadian Journal of Regional Science,* Vol. XVI, 1:69-88.

Rigby, D.L., 1992, "The Impact of Output and Productivity Changes on Manufacturing Employment", *Growth and Change,* Vol. 23, 4:405-427.

Sawers, L. and W.K. Tabb, (eds.), 1984, *Sunbelt/Snowbelt: Urban Development and Regional Restructuring,* Oxford University Press, New York.

Schaffer, W.A., 1993, "Stagnation, Decline and Development: A Trip Through the Southern Countryside", *The Review of Regional Studies,* Vol. 23, 3:213-218.

Stough, R.R., 1991, "Rise of the Southern Periphery in the United States: Understanding the Frostbelt-Sunbelt Shift", Working Paper No. 91:5, The Institute of Public Policy, George Mason University, Fairfax.

Thomson, W.R., 1969, "The Economic Base of Urban Problems," in H.W. Chamberlin (ed.), Richard D. Irwin Inc., Homewood Hills.

Vernon, R., 1966, "International Investment and International Trade in the Product Cycle", *Quarterly Journal of Economics,* 80:190-207.

Vernon, R., 1979, "The Product Cycle Hypothesis in a New International Environment", *Oxford Bulletin of Economics and Statistics,* 41:255-267.

19 Spatial Policy for Sparsely Populated Areas: A Forlorn Hope? Swedish Experiences

Hans Westlund
Swedish Institute for Regional Research (SIR)
Kyrkgatan 43B, S-831 34 Östersund, Sweden

19.1 Introduction

The regional policy in Sweden is usually divided in two parts. The specific regional policy comprises the responsible national ministry's means for regional development whereas the broad regional policy includes other policy areas with an important geographical dimension, e.g. education, infrastructure and subsidies to local and regional public sector bodies to equalise public service capacity all over the country. The specific Swedish regional policy was introduced in 1965 as a means to solve certain contradictory demands: on one hand enterprises' demand for labour and on the other the growing problems of depopulation of the sparsely populated areas, mainly in Norrland, the five northernmost counties of Sweden. A broad regional policy has in a sense always existed, but it was not until the introduction of the specific regional policy in 1965 that the regional dimensions of other policy areas were more systematically recognised and became objects of regional policy considerations.

Both the specific and the broad regional policy have been oriented towards central-places on different levels. The solution of the problems of the sparsely populated areas was for a long time considered to be to concentrate the population in central-places, which would guarantee public and commercial service and jobs to such an extent that further depopulation would be prevented.

From the perspective of getting people to migrate from the countryside to central-places, the policy seems to have been very successful. Between 1965 and 1995, the population in the sparsely populated areas[1] of Norrland decreased by 30.0%, or on average 1.18% per year. For the whole of Sweden, the figure was 19.1%. However, it should be stressed that this flight from the countryside started long before 1965.

The population development in the regional centres[2] of Norrland was the reverse. During the thirty-year period they grew by 18.9% or on average 0.58% per year, but for 1980-95 they grew by only 0.26% on average per year (Statistical

---

1 In the Nordic countries a "densely populated place" is defined as an agglomeration of houses with less than 200 meters between them and with more than 200 inhabitants. All other areas are defined as "sparsely populated areas".

2 Regional centres are here defined as the municipalities of Gävle-Sandviken, Sundsvall-Timrå-Härnösand, Östersund, Umeå and Luleå-Piteå-Boden.

Yearbook of Sweden). The population development of all densely populated places (here further, somewhat incorrect, called central-places) in Norrland was even less positive after 1980. Table 19.1 shows the total number of central-places and a division of them in increasing, decreasing or stable population development, 1980-90 and 1990-95, respectively.[3] Between 1980 and 1990, 54.6% of the central places lost population. 1990-95 this negative trend was reinforced and as a result 59% of the places experienced population losses. This negative trend was not only restricted to the smallest central-places. Of the 37 places with more than 4 000 inhabitants 1980, 25 lost population up to 1996 (Bylund, 1997). Thus, from the perspective of strengthening the central-places, the policy does not seem to have been particularly successful.

Table 19.1 Number of central-places in Norrland 1980 and 1990, divided in increasing, decreasing or stable population development.

	Total number	Increasing	Decreasing	Stable
1980-1990	362	147	198	17
1990-1995	359	129	212	18

Source: Bylund (1997)

In contrast to the concentration on central-places, a growth of the population on the countryside has been observed in certain areas in the countries of the western world. This counter-urbanisation has also been observed in Sweden, but seems to have been restricted to the countryside outside the biggest cities and around many of the regional centres. The more peripheral sparsely populated areas (among others, the main part of Norrland) have in general experienced continued population decline (Amcoff, 1997; Forsberg, 1997). An important observation is however the changes in the age structure of the countryside. While people over 65 years dominated the countryside in 1970, it was children under 15 years that constituted the biggest group 1990 (Amcoff, 1997).

The signs of declining smaller central-places and of rejuvenation of the countryside leads to some important questions about the effects of the broad regional policy and the place-policy (hereafter referred to under the concept *spatial policy*). Has the concentration on the central-places been founded on wrong assumptions? Were the comparative advantages of the countryside as residential areas neglected? Or is the big mistake of the spatial policy that it has spread out the resources on too many small central-places, instead of concentrating them on a few big ones?

This paper discusses these problems, but gives no definite answers. In Section 19.2, a brief summary of the spatial policy in Sweden is given. Section 19.3 discusses the ongoing counter-urbanisation and the forces behind it. Section 19.4 presents some empirical data on the changes in settlement patterns 1990-96 in four-

3 Between 1980 and 1990, three central-places had their population decreased under 200 inhabitants and thus disappeared out of the central-place statistics.

teen selected peripheral municipalities in southern and northern Sweden, all belonging to the European Unions Objective 5b or 6 support areas 1995-1999. The concluding section discusses the results and their implications on the future spatial policy for sparsely populated areas.

19.2 Spatial Policy in Sweden after 1945

1965 is usually regarded as the year when the modern Swedish regional policy was introduced. However, issues on industrial location and the problems of depopulation of sparsely populated areas had been dealt with far earlier. Already 1952 the Swedish *riksdag* (parliament) took a decision on location policy, containing trains of thought that would remain in the core of regional policy. One example was the idea of concentrations on commercial and cultural centres for the surrounding rural areas. The means for affecting enterprises' location were considered to be information and advisory service. Subsidies to individual firms were rejected.

A growing change in opinion could be traced in the official report on location 1963, when it as one of four aims for the location policy proposed that the structural changes and economic expansion should happen in such forms and pace protecting the individuals' safety. However, the report rejected any location efforts towards sparsely populated areas. According to the chairman of the report committee, MP Manfred Näslund, the efforts would be directed on constructing another one or a few centres with an attraction similar to the attraction that the three big cities in Sweden had: "The aim of the location policy is to support a concentration of the urban industries to such bigger urban places, which can function as central places for the population in the sparsely populated areas, while they simultaneously are attractive as location places for the manufacturing industry." (Quotation from Elander, 1978, p.62).

Even these modest attempts to support areas outside the three metropolitan areas were severely criticised by leading social democrats in the national trade union (LO) and by the director general of the national labour market board (AMS). In spite of this criticism, the report formed the base for the parliament's decision of location policy, which was introduced 1965. The big novelty was the possibility for individual firms of getting economic support for investments within a demarcated "support area", consisting of the areas most hit by depopulation i.e. mainly peripheral areas in northern and western Sweden.

During the first 22 years, economic support to enterprises was restricted to manufacturing industries. It was not until 1987 that enterprises in the service sector were able to receive regional economic support. Ironically enough, the Swedish manufacturing industry had the highest number employed ever in 1965, which meant that for the first 22 years, regional support was only given to enterprises in a declining sector. The regional policy is still, 1999 directed towards attracting enterprises. The growing awareness of well educated, high-competent people as a location factor has not yet had any impact on the official regional policy.

Parallel to the start of the regional policy, a big municipal reform was planned and introduced. With the growth of the welfare state the municipalities on the countryside were considered to be too small to carry on all the duties they should, according to the then prevalent ideology. From 1965 to 1972, about 2 000 municipalities, towns and cities were merged to about 275 municipalities. In the planning of this municipal reform, regional scientists took part as experts, using the central-place theory of Christaller (1933) as their theoretical tool. Their calculations came to the result that 2 200 inhabitants were considered as the minimum size for a place with tolerably good service functions (Bylund, 1997). The minimum size for the merged municipalities were set to 8 000 inhabitants.

In the more densely populated areas of southern Sweden, these mergers were merely a reflection of the expansion of the cities' and towns' labour markets and a confirmation that the countryside had been integrated as residential areas into the city-regions. However, in large parts of northern Sweden this was not the case. Most often, several areas, each one with their small central place, located at similar level in the central place-hierarchy, were merged and one of the centres were nominated as administrative centre.

This "place-policy", starting in two different policy areas and on quite different levels amalgamated during the 70s in the *place-classification policy*, which was one part of the parliament's regional policy decision 1973. One place in each municipality was given the position as "centre" and the centres were divided into four categories where the metropolitan centres were the highest and the municipal centres the lowest. For the specific regional policy, this place-classification never really came to function as an efficient tool and it was officially repealed 1982. However, for the counties' and municipalities' policies, the repealing of the place-classification did not mean an end of the policy of concentrating the development resources to the centres.

Simultaneous with the start of the centre-oriented spatial policy for the depopulating regions during the 60s, the labour market policy was strongly directed towards moving the labour from the countryside to the centres. In the beginning of the 70s, the official policy towards the sparsely populated areas was still that it was meaningless to make any efforts for redevelopment in them. However, the people who still were living there, would for equality reasons be guaranteed a certain level of public and commercial service. This view was shown very clearly in e.g. the directives for the official report on sparsely populated areas (SOU 1972:56). The report would not deal with location or agricultural policies, not with issues on general development or economic life, but solely with living conditions in the sparsely populated areas.

During the 70s, the official policy towards the sparsely populated areas slowly shifted, mainly due to four different reasons:

- The boom of the 60s was replaced by the oil crisis, which meant that the demand for labour in general in the metropolitan areas decreased. The new expanding sectors needed educated labour, which they could not fetch from the countryside. Moving low-educated people from the countryside to the biggest cities did not contribute very much to economic growth anymore.

- The opinion against "the depopulation policy" grew strong in the affected areas. For political reasons, it became necessary to acknowledge the right to development even to the sparsely populated areas.
- The "green wave" hit the western world and made the countryside popular in the public consciousness. In Sweden, the green wave coincided with a "wave" of solidarity, which favoured the depopulation areas as well.
- The above mentioned expansion of welfare policy demanded a forceful growth of the public sector. Extensive state subsidies to economically weak municipalities in the sparsely populated areas became necessary for the implementation of this equalisation policy. A great number of new jobs in the public sector were created.

In 1979, special means for supporting enterprises and commercial service in the sparsely populated areas became a part of the national regional policy. However, this policy did not make any difference between central-place and countryside in the municipalities which was classified as "sparsely populated". No analysis of the comparative advantages and disadvantages of the countryside was carried through. The result was that no particular policy was developed for the countryside that thus continued to be an appendage to the central-place. A reflection of this is that public investment in service, elderly-care, culture and sports buildings as well as blocks of flats were constructed in the central-places. During the 1990s, the number of empty public flats has increased sharply (SCB 1997) and with that the capital costs for many municipalities. Many municipalities have for economic reasons been forced to demolish parts of their stocks of public flats.

The policy depicted above has been the traditional top-down policy. This is no coincidence. The expansion of the public sector on local level – the municipalities – was a deliberate part of the social democrats' national equalisation programme. The tasks of the municipalities should be the same no matter what part of Sweden they were situated in nor if they were depopulating or expanding. The municipalities became well provided with service institutions for the inhabitants, but not very well equipped with institutions for local and regional development. Planning of public investments and urban residential areas were the most strategical tasks of the municipalities – not forming regional development strategies.

Not the regional authorities, but the state's county offices handled the specific regional policy. This was a clear marking of regional policy being primarily an equalisation policy from above and not a development policy from below. The labour market policy was conducted in the same way.

However, the local/regional initiatives was not wholly absent. The political bodies on regional level wanted to deal with development issues and not only with health-care, which the state had allotted to them. The bigger municipalities, i.e. the bigger cities, had offices for recruiting enterprises. The smaller municipalities gradually followed their example. It should of course be added that these limited bottom-up initiatives were concentrated on the central-places.

From around 1990, the hegemony of top-down policies has partly eroded. The main reason is that when policy shifted from *redistribution* to *development and economic growth*, it became obvious, even for the state that although redistribution must be handled from above, local and regional development and growth cannot. Therefore, local/regional development issues have increasingly become a task for the counties and municipalities.

This change of policy has not in itself implied any pronounced changes from the centre-oriented approach, described earlier. But bottom-up initiatives, such as the creation of thousands of village development groups all over Sweden from the latter half of the 80s and onwards, have at least made the distinction between countryside and central-places visible and stressed the need for a policy for *rural* development too.

In sum, development of the sparsely populated areas has been acknowledged as a national policy aim since about 1980, but the means for this development have mainly been the same as for developing towns and cities. No special national policy for the sparsely populated countryside has been designed. The policy for sparsely populated areas has therefore mainly been a policy for the small central-places and the policy for the small central-places has mainly been a copy of the policy for the bigger ones. The partial erosion of top-down policies during the 90s means an opportunity for developing a specified policy for the sparsely populated countryside, but the traditional centre-oriented policy is still dominating on local and regional level.

19.3 The Ongoing Counter-Urbanisation and the Comparative Advantages of the Countryside

Since the 1970s, a trend towards increasing population on the countryside has been noted in certain areas in European countries and in North America. The general process has been referred to under the concept counter-urbanisation. The pattern, speed, driving forces, etc. in the process differs between different countries and there are several theoretical approaches to explain the phenomenon (Kontuly, 1997). However, one of the most common denominators seems to be a migration of segments of the middle class from the cities to the city-close countryside – a process often designated as "gentrification" of the countryside.

Swedish research confirms the general trend of counter-urbanisation, but has also shown big geographical differences. The outmigration from the biggest cities to the city-close countryside is obvious, but the trend is similar for some other parts of Sweden too. But most parts of the traditional sparsely populated crisis-areas of Norrland and south-eastern Sweden, seem to continue to loose population (Forsberg, 1997). However, the development has been different over time and dependent on age groups. Those over 45 years had a more positive development in the central-places than on the countryside in almost every municipality during the 1970s. This pattern was almost as strong during the 1980s. The age group under 45 had a more positive development in the central-places in a majority of municipalities during the 1970s, but during the 1980s this group grew stronger on the countryside than in the central-places in most municipalities. Amcoff (1997) shows that the countryside has grown young again after 1970. In 1990, children under 15 years constituted the biggest share of the population on the countryside, when the population was divided in 15-year-groups.

Comparative national figures for the 1990s do not exist yet due to changes in the demarcations of the densely populated areas 1995. Studies of the parishes in the *Mälardalen* area[4] have shown a growing population in the "countryside parishes"[5] during the first half of the decade, but that the differences in population development between them and the "densely populated parishes"[6] were insignificant 1994 and 1995 (Amcoff, 1997).

Traditional migration theories are founded on gravity assumptions, claiming that people are attracted to big cities and that the attraction varies positively with the size of the city. The phenomenon of counter-urbanisation is left unexplained. As Forsberg (1997) points out, there is still not a theory that can be applied on the current situation of the countryside, but there are approaches. These approaches can be divided in attempts to explain outmigration from cities (push-factors) and inmigration to the countryside (pull-factors), respectively.

The first group of explanations is to a large extent focused on changes in relative incomes and costs in cities and countryside. These may be expressed in different types of crowding out tendencies from the cities. There are also explanations based on gravity assumptions, suggesting that the increase of the city-close settlements consists of inmovers from outside who cannot afford living in the centre.

The second group of explanations focuses more on changed set of values in the population as a whole or in certain age groups (Marsden et al., 1993). Technical development on several fields can be considered to have made it easier for people to let their sets of values govern their choice of residence. These factors should of course be expressed in economic terms, too. Choice of residence is based on rational considerations of individuals who want to maximise their utility. Counter-urbanisation in this perspective merely says that certain groups are maximising their utility by living on the countryside.

The central-place-policy carried on was based on the traditional assumption that work and dwelling were spatially united. In this perspective, the countryside had no comparative advantages as residential area except for farmers and woodsmen. Under the conditions of today, the countryside obviously has advantages as residential area compared with central-places. These advantages consist both of what is avoided and what is gained. Compared particularly with big cities, country dwellers avoid much crime, violence, heavy traffic in the neighbourhood, etc. The gains of the country are usually described in terms of space for and access to nature and outdoor activities of different kinds. In combination, these factors seems to make the countryside attractive, principally for families with children.

---

4 The counties of Stockholm, Uppsala, Södermanland, Västmanland and Örebro.

5 Defined as parishes with no central-place with 1000 inhabitants or more.

6 Defined as parishes with at least one central-place with more than 1000 inhabitants.

19.4 Settlement Changes in Sparsely Populated Municipalities During the 1990s

The conclusion of the above cited empirical research is that the counter-urbanisation has been limited to the areas around big and middle-sized cities, while the peripheral, sparsely populated parts of Norrland and south-eastern Sweden, have continued to loose population. However, the problem from the spatial policy point of view is that also the centres have been loosing population in these latter areas. Therefore, it would be of interest to investigate the population changes for the central-places and for the countryside respectively in the peripheral areas.

If the centres' share of the population in the peripheral areas is increasing, it would indicate that counter-urbanisation is restricted to the more densely populated regions of the country, i.e. mainly around the national and regional centres. Further, it would indicate that the centre-oriented policy still has had effects in strengthening the central-places relatively to the countryside, but that the resources have been spread out on too many small centres to strengthen their position versus higher ranked centres. The spatial policy carried on for development of sparsely populated regions would then be considered a forlorn hope and the likely conclusion would be to concentrate the resources on a smaller number of competitive centres.

On the other hand, if the countryside's share of the population in these areas is increasing, it is possible that this may be characterised as a form of counter-urbanisation, although the population is diminishing in absolute terms. But irrespective of how the phenomenon is described, it would indicate that the centre-oriented policy has not been adapted to the great changes in society which among other things have been expressed in the choice of living – both in central and peripheral regions. The spatial policy would then be characterised, not as a completely forlorn hope, but as a misdirected one, as it has been directed towards strengthening non-competitive centres, while "forgotten" the competitive aspects of the countryside as a residential location for households in the small centres or more distant, bigger centres.

To examine this problem, fourteen peripheral municipalities were selected in three parts of Sweden: in south-eastern Sweden, in western Sweden and in Norrland (Map 19.1). All the municipalities belong to EU:s Objective 5b (countryside) or 6 (remote) support areas. None of the municipalities have been hit by any quick, extensive industrial close down, but they have of course been subject to normal structural changes. As mentioned above, changes in the demarcations of the densely populated places were made in 1995, which make the data not comparable with earlier years. To create fixed boundaries, and thus avoid this problem, population data per square km were used for 1990 and 1996. Densely populated places comprised the places (villages and towns) with more than 200 inhabitants 1990 plus a "buffer zone", consisting of the part of the squares being outside the densely populated area demarcation. This means that the area of the central-places became somewhat larger compared with the official statistics, as well as number of inhabitants. However, for a comparison over a shorter time, this difference should not have any significant impact on the results.[7]

In order to check if the general pattern of rejuvenation of the countryside in Sweden applies also to the peripheral areas, the population has been divided in four age groups. Table 19.2 shows the results for each municipality. The total population exhibits a mixed pattern. The countryside has lost population shares in the four south-eastern and the three western municipalities, but has gained population shares in all eight Norrland municipalities. The countryside has lost its greatest shares in the three western municipalities.

Table 19.2 Changes in the countryside's share of the population in fourteen selected peripheral municipalities 1990-1996, in total and divided in age groups. Per cent units.

Municipality	Age groups					
	0-15	16-24	25-64	65-	0-65	Total
Ydre	1.03	3.63	0.13	-3.67	0.88	-0.12
Kinda	2.51	-3.84	0.68	-3.31	0.73	-0.13
Uppvidinge	-0.56	2.71	-0.63	-1.38	0.01	-0.33
Torsås	3.68	-1.15	-1.53	-1.36	0.05	-0.37
Tanum	0.14	-0.92	-0.57	-3.42	-0.27	-1.01
Torsby	0.99	-0.40	-0.29	-2.16	0.04	-0.51
Orsa	-1.53	-2.71	-1.30	-0.41	-1.46	-1.23
Ånge	1.58	-0.35	1.09	-1.80	1.11	0.38
Bräcke	3.35	2.88	3.06	-0.95	3.29	2.27
Strömsund	1.66	1.67	0.49	-2.74	1.00	0.05
Storuman	3.68	-0.68	2.22	-4.09	2.30	1.12
Arjeplog	3.15	6.41	1.60	-2.04	2.57	1.72
Jokkmokk	0.10	2.09	0.67	-1.79	0.78	0.31
Överkalix	1.26	5.36	-0.90	0.70	0.30	0.47
Total	1.34	0.61	0.29	-2.06	0.70	0.05

However, a look at two of the age groups gives us a more uniform pattern. In the age group up to fifteen years, the countryside increases its population share in all municipalities but two, whereas the group of 65 years and above is decreasing its shares in all municipalities but one. Thus the general trend of rejuvenation of the countryside seems to be confirmed even in the peripheral areas.

7 It may be argued that if the central-places change over time, so should their demarcations (cf. e.g. Keddie and Joseph 1991). From a long-term, macro perspective, this is no doubt correct. However, when investigating individual places and their surroundings this can create problematic marginal effects. If a village with just under 200 inhabitants, situated close to a central-place, increases its population with a few persons, it would be statistically incorporated with the densely populated place. Hence the central-place would in the statistics increase its population with 200 persons – and decrease the number of people living in the countryside just as much in the statistics.

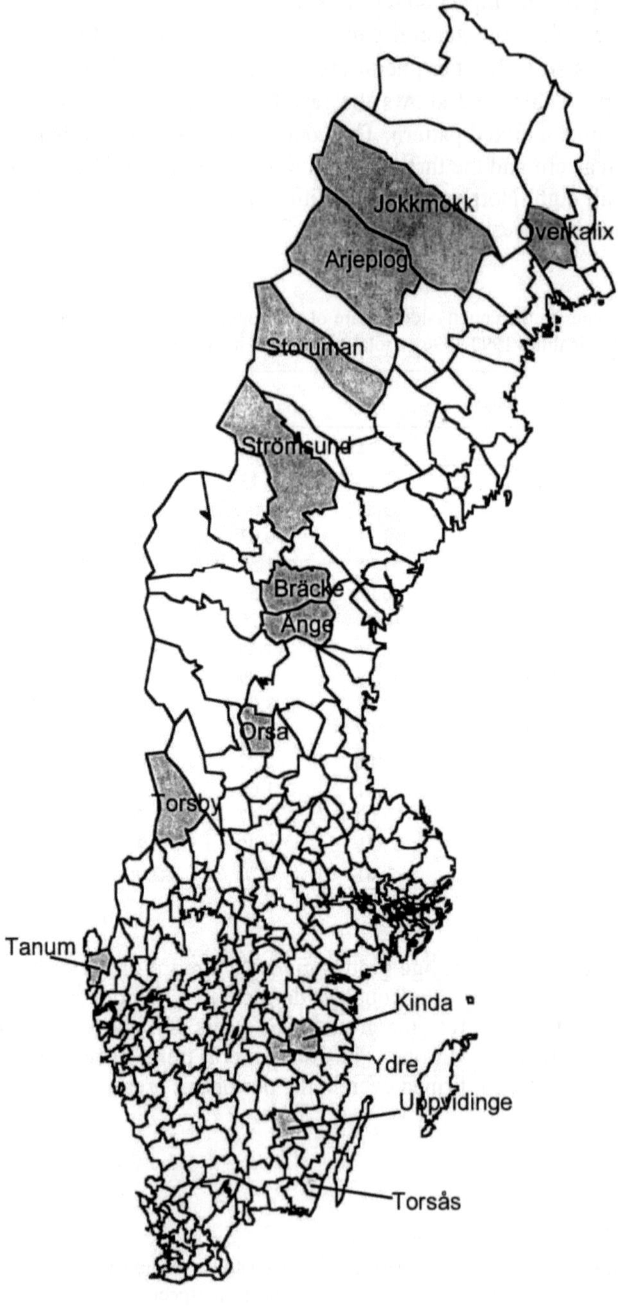

Figure 19.1 The fourteen selected municipalities

The changes in the shares of the other two age groups are less uniform. The age group where the biggest movements could be expected, i.e. 16-24 years, shows relatively great changes, but in different directions. The group increases its shares in many of the Norrland municipalities. It might be tempting to interpret this as a "flight from the (small) centres", but the probable explanation is that the young people in the central-places of Norrland are moving to regional and national centres to study and work to a greater extent than young people on the countryside. The most extensive age group, 25-64 years, increases its shares in eight of the fourteen municipalities. Here too, the figures for the Norrland municipalities are the most positive for the countryside.

In sum, the population development in the sparsely populated areas seems to have a pattern similar to the more densely populated regions. There is a trend of rejuvenation and particularly if elderly people are excluded, the countryside seems to increase its shares of the population of the municipalities. The big difference is that this is going on in areas with decreasing populations in absolute numbers.

There seems to be geographical differences as well. The studied Norrland municipalities had the most positive results for the countryside, while the three western municipalities had the most negative results. There is no evident explanation to these differences. A possible hypothesis for further research might be the existence of a higher level of rural development activities in the Norrland municipalities.

These results must of course be subjects for further research. Principally, the connections between the changes and structural and individual socio-economic variables should be investigated. A bigger sample is also needed.

19.5 Spatial Policy Interpretations

The results indicate that the population in the peripheral municipalities has exhibited another pattern of change than was observed during the 1960s and 1970s. The sparsely populated areas have lost in total population but gained in young people and at least during the 1990s, their share in the municipalities' population less than 65 years have in general increased. There has been a flight from the small central-places, mainly to the regional and national central-places, but to a certain extent to the countryside too.

A likely conclusion is that living in a flat in a small central-place is probably the least attractive form of living at all – with the exception of certain age groups like young, single people who do not want to leave their home-municipality and elderly people who do not manage to live on the countryside anymore. For most people, the alternative to living in the countryside is living in the city – not in a small hole. The many unlet flats in the Norrland municipalities will probably continue to be unlet.

With this in mind, do the peripheral municipalities have any opportunities to carry on a new spatial policy for economic renewal, in spite of the depopulation? Do they have any comparative or competitive advantages?

From an employment perspective, the advantage of these areas should be in growing, space-using activities. The almost only such activity is tourism. All over Europe, tourism has been regarded as the new hope for regions in crisis. In

Sweden, the most positive example of development through tourism is the alpine ski centre Åre, with a very high share of new enterprises during the 1990s. In most other cases, tourism has created new jobs, but not enough to counteract the loss of jobs in other sectors. Even if e.g. wildlife, hunting and fishing tourism will experience growth, this conclusion will probably be the general one with few exceptions.

The other advantage of the sparsely populated areas is space for living and for the residents' leisure activities. Compared with the city-close countryside, these areas have the disadvantage of being peripheral. Today the sparsely populated areas mainly function as residential areas for the declining small centres. But what are their opportunities for developing into residential areas for more distant centres?

Their only likely possibility is probably to reinforce some of the current tendencies in the post-industrial society, foremost telecommuting combined with the growing desires among certain groups to spend an active leisure outdoors. Like improved communications and transport technology have facilitated living in the city-close countryside and getting the incomes from the city, their continued development will presumably facilitate living in the periphery and earning the incomes somewhere else.

A large share of the Swedish population has a city-residence where they stay during the workingweek and a country residence for vacations and holidays. With the ongoing changes in working life, a growing number of people are prolonging their stay in the countryside with a couple of days of telecommuting. So far, this development has been spontaneous and not deliberately supported by policy measures. If the peripheral areas were able to develop a policy to reinforce these tendencies and get the weekend commuters to take up residence, it would probably be their best way to promote economic renewal.

If this conclusion is correct, it means that a policy for economic renewal of sparsely populated areas mainly must be formed from below and that it should be directed more towards attracting individual as inhabitants, rather than traditional attempts in attracting enterprises.[8] If anything, this would be a shift of the Swedish spatial policy for sparsely populated areas – from a top-down to a bottom-up approach, from concentrations on production localities for enterprises in the centres to quality of life for individuals on the countryside.

Acknowledgements

The author is indebted to Ms. Gunilla Eriksson, BA, at the National Rural Area Development Agency, Östersund, Sweden, who has compiled the statistical material for the study.

8 According to Nelson (1997) strategies for recruiting individuals with "footloose" incomes, i.e. incomes not generated in the local community, is used by several rural areas in the United States.

References

Amcoff, J., 1997, "Kontraurbanisering i Sverige?", in H. Westlund (ed.), *Lokal utveckling för regional omvandling*, SIR report No. 100, Fritzes, Stockholm.

Bylund, E., 1997, "Geografisk strukturpolitik för glesbygdsutveckling", in H. Westlund (ed.), *Lokal utveckling för regional omvandling*, SIR report No. 100, Fritzes, Stockholm.

Christaller, W., 1933, *Die Zentralen Orte in Süddeutschland*, Jena.

Elander, I., 1978, *Det nödvändiga och det önskvärda. Socialdemokratisk ideologi och regeringspolitik 1940-72,* Arkiv, Lund.

Forsberg, G., 1997, "Att förstå den nya landsbygdens mosaik", in H. Westlund (ed.), *Lokal utveckling för regional omvandling,* SIR report No. 100, Fritzes, Stockholm.

Keddie P.D. and A.E. Joseph, 1991, "The Turnaround of the Tournaround. Rural Population Changes in Canada; 1976 to 1986", *Canadian Geographer*, Vol. 35, No. 4.

Kontuly, T., 1997, "Contrasts of the Counterurbanization Experience in European Nations", Paper presented at the 36th Western Regional Science Association Congress, The Big Island of Hawaii, February 23-27, 1997.

Marsden, T., J. Murdoch, P. Lowe, R. Munton and A. Flynn, 1993, *Constructing the Countryside*, UCL Press, London.

Nelson, P. B., 1997, "Migration, Sources of Income, and Community Change in the Nonmetropolitan Northwest", *Professional Geographer*, Vol 49, No 4.

SCB, 1997, Bo 35, SM9701, Stockholm.

SCB, *Statistical Yearbook of Sweden* 1965-1997, Stockholm.

SOU 1972:56, *Glesbygder och glesbygdspolitik,* Stockholm.

Westlund, H., (ed.), 1997, *Lokal utveckling för regional omvandling*, SIR report No 100. Fritzes, Stockholm.

20 Theories of Endogenous Regional Growth – Lessons for Regional Policies

Börje Johansson[a], Charlie Karlsson[a] and Roger Stough[b]

[a] Jönköping International Business School, Box 1026, SE-551 11 Sweden

[b] The Institute of Public Policy, The Mason Enterprise Center for Regional Analyis and Entrepreneurship, and The Transport Policy and Logistics Center, George Mason University, Fairfax VA, 22030-4444 USA

This concluding chapter creates a synthesis of the major aspects and lessons for regional policy presented in the preceding chapters. Common elements are found in the identification of a critical group of decision makers and their objectives, and in the specification of policies that give support to change processes that are essential in the development of knowledge resources, the implementation of new technology and the formation of clusters. *Conclusions* regarding these issues are embedded in the analysis of infrastructure, institutions and associated local policies supporting endogenous regional growth. This also means that local or regional advantage is understood as the result of territorial competition based on competitive advantages that evolve in a dynamic process. It is also recognised, in major parts, that policy itself is endogenous.

20.1 Who are the Policy Makers and which are the Objectives?

Contributions to the new growth theory show that human capital and R&D output are essential elements in a regional growth process. Agglomeration effects comprise an additional element of similar prominence, and this means that the concept of functional region must be placed in focus. It is in this spatial context that we can understand externalities and cluster phenomena, increasing returns from factor accessibility and knowledge spillover and, in particular, accessibility to knowledge workers. We may then contemplate how to identify the pertinent policy makers. Moreover, how can policy decisions in administrative regions influence individual functional regions as well as networks of such regions?

Jacobs (1984) observes that generically "systems confusion" arises when general policy measures are exercised for and across large and heterogeneous territories. The argument is that with heterogeneity the measures may be inappropriate for every individual region inside the territory, and hence inefficient. A contrasting observation is made in Cheshire and Gordon (1998). They observe that the boundaries of an administrative region (municipality, town etc) may not coincide with those of the corresponding functional region. The two types of territories may not be congruent, and when the administrative regions are fragmented, the functional region may be much larger than each of the associated administrative regions in which policies are executed. This form of mismatch makes development policies less effective, imprecise and perhaps even counter productive. The tentative message from Cheshire and Gordon is that if there is a choice, one should prefer to have an administrative region encompassing the relevant functional region rather than to have a case where several administrative regions combine to a functional region or even worse, a case where different parts of a functional region belong to different administrative regions. The latter two cases, which may be more frequent, require that policy makers develop insightful cooperation.

Who are the policy makers? Following Stough (1990), politicians and political parties constitute only one of several categories of decision makers that may participate in the attempts to form processes that foster economic development in a region. He identifies a set of critical properties of successful policies of regional economic development. Local initiative is vital and tends consistently to be carried out by nongovernment community organisations, which demonstrate a record of effective economic development planning. Such plans have the nature of a platform for cross-section collaboration. This in turn facilitates the creation of access to a broad range of local and extra local resources.

A contrasting insight that adds perspective is that regions are not isolated islands. Indeed, in the globalised economy of the 21st century it is more important than ever to identify and understand the linkages a region's economy has to other parts of the world economy (including the national economy). Some of the associated modelling problems and opportunities that arise from these observations are discussed in chapters 9 and 18.

20.2 Knowledge, Technology and Clusters – Policies Supporting Change Processes

Referring to Harrington and Ferguson (Chapter 3), endogenous growth theory offers considerably enlarged possibilities to study growth processes and still retain the neoclassical foundation for analysis. First, endogenous growth theory allows for models with increasing returns, which means that these models refer to empirical observations in a more adequate way than earlier models. Second, the models become policy relevant, because in endogenous growth models policy matters. Such policies comprise efforts to stimulate the growth of knowledge intensity of the labour supply and knowledge production in the form of R&D, and to facilitate cluster formation. In a regional perspective knowledge-oriented policies also have to consider measures that can attract knowledge that is embodied in labour as well as capital input resources. As a consequence, infrastructure policies will also influence endogenous growth. It can do so by improving accessibility and the milieu of a region.

During the 1970's and the subsequent decade much attention was given to technology-based theories of regional development, depicting technology as the prime motor of regional economic development. On the one hand, many contributions in this period made use of spatial product cycle models. On the other hand, the focus on Schumpeterian processes later gave birth to *economics of entrepreneurship* as special research field. In both cases the economic milieu of a region is looked upon as a slowly changing factor, which in this way can be analysed as a partly given arena for processes of product development and renewal of production methods through innovation and imitation. Typically, policy analysis then will examine how the economic milieu can form a fertile ground for innovations and technology implementation. In this context Rees (Chapter 5) places four factors side by side as technology support elements. These factors are agglomeration, milieu, learning and leadership. Such factors can be thought of as preconditions when a region attempts to change its role from being a recipient of innovations via branch plants to become a generator of innovation via indigenous growth. We may also observe that *economic milieu* in some contributions is treated as a broad concept including factors such as

- Agglomeration, size and density of the region or size of the region's market potential
- Production factors and accessibility to industry input suppliers
- Residential environment or the entire household milieu
- Business milieu, entrepreneurial tradition, composition of sectors and sectoral clusters in the region
- Leadership and learning capacities of the economic system in the region
- General accessibility

This broad concept of economic milieu can also be related to Maillat 's idea of a learning region, which emphasises created resources such as skills, know-how, qualifications as well as methods solving problems (Chapter 12). In this view a region maintains its competitive advantages through learning processes. Markets of such learning economies are considered to be embedded in habits, rules and norms, which form an arena for communication and exchange of qualitative non-price information.

In the paradigm of a learning regional economy, the purpose of interaction is to find a solution to a problem through cooperation and collaboration. This requires conversations among actors that create new knowledge, which in turn presupposes trust and confidence among participants in the learning process. These features are accentuated when technology and economic conditions change rapidly. In this context geographic proximity and established culture become salient attributes.

Work on the concept of economic milieu has brought renewed attention to agglomeration and localisation economies. In this re-discovery of old concepts, it is also clear that benefits of agglomeration are not just economic but also cultural in nature. The latter is related to concepts such as social capital and entrepreneurial tradition. Entrepreneurship emphasises aspects of risk taking, uncertainty and social acceptance of failure as a natural part of innovation ambitions. This perspective also puts the local context of the entrepreneur into focus. In summary, the regional environment of active entrepreneurs is assumed to have the following characteristics:

- A high frequency of family or small businesses
- A diversified pattern of firms and activities also in smaller regions, in which the sector composition is specialised
- An infrastructure that supports the activity pattern
- Accessibility to skilled resources that are adjusted to the specialisation of the region
- A solid financial community
- Presence of government and community incentives to start new business.

The above factors can help to understand how the responsiveness to innovation varies between regions. Responsiveness is analysed in Chapter 6. A major conclusion is that in order to analyse innovation responsiveness the associated increasing returns must be carefully specified.

With the background presented in this Section, can anything be said about the role of universities in a region's economic development? The location of university education is analysed in Chapter 4. It is clear that the location has an impact on the regional supply of labour, but does it affect innovation processes and economic renewal in any other ways. The empirical evidence suggests that R&D aspects of a local university are more important for small than for large regions. This conclusion should be combined with other types of learning processes that are related to cluster phenomena as stressed in Part 4 of this volume.

20.3 Policies for Building a Regional Arena – Infrastructure and Institutions

Various contributions in the past make a distinction between tangible and intangible infrastructure. The latter is sometimes referred to as soft infrastructure or social capital and may include knowledge as well as institutions. How do we distinguish between infrastructure and institutions? As a start we may recognise that institutions are defined as collectively held beliefs, values, mores and rules that condition individual actions. In early contributions, such as Veblen (1899), a distinction is made between ceremonial and instrumental institutions, where the latter is associated with systems that legitimate and motivate problem-solving action or skills acquisition. These aspects relate to so-called neo-institutional economics, which defines and interprets institutions in relation to their effects on transaction costs (e.g. North, 1990). In this framework, institutions are reflected in the constraints that human beings impose on themselves. In this way institutions can be related to the theory of choice in neoclassical economic theory as they influence the formation of efficient and reliable transaction procedures as well as the ways of making decisions.

How does an institutional setting affect transaction costs? Which attributes are essential? Both formal and informal rules of interactive behaviour have to be considered, and the interplay between formal and informal institutions is of critical importance. Another issue is the extension of institutions. Are they society-wide or are only certain groups effectively influenced? Moreover, how large is the territory of a ruling institutional system? In this context, can a territory encompass a whole set of functional regions? Finally, how durable or lasting is such a system, and how does it evolve?

An institutional setting may influence transaction costs by reducing the costs of carrying out market activities such as negotiations, contract formation, search and adjustment. The role of institutions has a special function in processes of change, in which novelty by combination and structural adjustments increase uncertainties. In this way the analysis of institutions can be brought into close contact with models of endogenous growth, which emphasise leadership, learning, social capital and innovation.

In view of the above, a clue in provided as to how statements like the following should be interpreted: Institutions are social rule structures with associated standing patterns of behaviour and procedure. These standing patterns form a basis for decision under uncertainty in attempts to combine activities in new ways and introduce new products on the market.

Why and how is leadership important? In Chapter 3 Stough associates leadership in a region with the region's institutional system. Leadership is characterised as a collaborative relationship between local institutional actors and is based on mutual trust and cooperation. It requires the capability among actors to mobilise institutional, psychological, political and economic resources and to satisfy criteria and motives of followers. Leaders in this sense persuade and guide followers by example. In a down-to-the earth terminology, leadership involves the capacity to detect opportunities, to find out how local resources can be combined, and to stimulate other actors to engage in the development process.

In the described sense, leadership is catalytic. How does this relate to infrastructure, which can also be described as catalytic? We have already hinted at the connection between institutions and intangible infrastructure, but what is the role of physical infrastructure? The latter may to a large extent be thought of as networks for interaction and economic exchange. Just like institutions, network infrastructure facilitates economic interaction both socially and economically. In this way transaction costs may be reduced due to infrastructure formation.

However, infrastructure is also related to the size and density of a region. The local market potential of a region can expand only if infrastructure is allowed to match such an expansion. Moreover, both the local and external market potentials of a region set the limits for how firms in a region can make use of both internal and external scale economies. In a short-term perspective this means that the size of a region's market potential has the character of a slowly changing resource factor, and hence something that affects a region's comparative advantage. Arguments along this line point in the direction that regions should strive to develop a large local market. Does this mean that medium-sized and small regions are forced to give up?

In Chapter 8 it is suggested that there is indeed a way out of this dilemma. The option for small and medium-sized regions is to foster the development of niches characterised by external economies and cluster interaction. In this way the firms in a region will get sufficiently large market potentials within their fields of specialisation. Contact-intensive inputs to specialised industries can cluster around these industries in the region and thereby reduce transaction costs, benefiting from a sufficiently large demand from the localised industries. The latter may export products, which due to low distance sensitivity can capture demand in external markets.

What does this type of recipe look like? Well, formation of specialised external economies in small and medium-sized regions cannot be found everywhere. Rather we find it in regions with well-developed local institutions and leadership structures that help to maintain and develop the necessary positive externalities, that are vital for successful development in the not so large urban regions. By forming this conclusive circle of theory and empirical observations, we have synthesised the messages about how to build a regional arena for economic development in both large and small regions.

These conclusions may be interpreted in view of the results reported on in Chapter 10, in which it is recognised that the size of each urban region is determined in a system of regions, where agglomeration economies and increasing returns are basic factors. On the basis of empirical observations from Japan, the following conclusions emerges:

- There is a threshold level, regarding city size. Below this level agglomeration economies cannot be confirmed for most sectors. This threshold level could be approximated by 200,000 inhabitants.
- As for the effect of transport network improvements (which increase the external market potential), there are two opposing forces. The first is a concentration effect in large regions, due to agglomeration economies, and the second is a dispersion effect, due to lower factor prices in peripheral regions. The dispersion response to lowered transport cost came out as the stronger tendency.

Our conclusion is that below the threshold level, establishment of localisation economies based on cluster-like externalities is of major importance. This conclusion is empirically assessed in Chapter 11 and also in other parts of the book.

20.4 Dynamic Formation of Local Advantages – Territorial Competition

Under influence of models of dynamic complexity in the 1980's and early 1990's, it became popular among regional researchers to make a distinction between fast and slow processes, where variables such as prices and outputs change on a much faster scale than institutions, infrastructure and values. In Andersson (1993) the slowly adjusting regional structure plays the role of an arena for the faster processes of economic and policy games. A major conclusion from these types of distinctions is that regional policies should put more effort into the building of an arena than in participating in the fast adjustments. To a large extent the same conclusions come from the contributions in this book. However, the emphasis on regional leadership may have an intermediate role by involving itself also in processes of a less slow pace. Still, the importance of leadership is in the rearranging of structures and making way for new resource combinations.

The theory of comparative advantage, as introduced by Ricardo, is one of the most robust ideas in economic thought. Using established jargon, the theory is based on the existence of trapped resources, and these are nothing but slowly adjusting characteristics or capacities of a region. Then what does this theory say? The answer is: A region has a comparative advantage in producing a commodity z, if the number of other commodities, y, that has to be given up in order to produce one extra unit of z, is smaller in the region than in other competing regions. This embeds the more special case of absolute advantages, which obtain generically when scale economies prevail.

With this background one may contemplate dynamic comparative advantage, which relates to processes in which the advantages are created gradually over time. We may now recall that regional or local advantages can be based on (i) accessibility to a labour supply with special skills and high education, (ii) physical infrastructure and the size of the local market potential, (iii) entrepreneurial tradition and the presence of regional leadership, local supply of risk capital and similar factors discussed and analysed in this book. Obviously, the development and maintenance of an entrepreneurship tradition and of leadership are definitely endogenous processes. The attraction of labour with university education and the growth of the region's market potential are also endogenous processes that can be supported and stimulated by regional policies. In addition, regional growth generates new resources for improving infrastructure and education programmes, etc.

Without making the list of these examples longer, we may conclude that the idea of dynamic comparative advantages can be incorporated in a theory of regional endogenous growth. To a large extent the regional supporting policies will themselves be part of the pertinent self-reinforcing mechanism. Empirical recognition of this form of policy endogeneity in the dynamic creation of regional advantages can be found in Chapter 13-18.

20.5 Is Policy itself Endogenous? A Chart of Instruments

It has been claimed that regional competition started to increase as a consequence of a gradually stronger economic integration within the European union (Dicken and Tickel,.1992). Who are then the agents of regional competition and what are their objectives? One way of structuring these questions is to first divide the competition actors into the following four groups in a region: (i) firms, (ii) policy makers, (iii) interaction between policy makers and firms, and (iv) the region as a system of many actors.

Firms compete by substituting imports from other regions, by exporting to other regions, by making direct investments and by developing cooperative links with firms and subsidiaries in other regions. On a general level the objectives of firms comprise a desire to survive, to grow, to make profits and to satisfy return expectations from owners (that may be non-local).

In order to improve competitiveness, policy makers use taxation, organise of public services, make investment in infrastructure and the design of the built environment, and stimulate in-migration of warranted labour categories and tax payers. Schematically the objectives are complex and partly conflicting, with a desire to be re-elected, ambitions to improve the welfare of inhabitants but also to increase the tax base, and to reduce unemployment.

As an interacting group, firms stimulate policy makers to create a desired economic milieu with competitive advantages, and policy makers try to attract (and keep) firms which bring income and employment to the region. In favourable cases this interaction may involve the regional society as a whole.

In the headline of this section we ask if policy itself is endogenous. Partly this question has already been given a positive answer in the preceding section. It is evident that one of the messages of endogenous growth theory is that policy matters, which does mean that the competitive advantages develop in a creative process, and that policy support is an important component is this process. Which are then instruments and areas of influence. The following policy areas have been emphasised in many parts of the book:

- Communication and transport networks
- Household milieu, housing environment
- Higher education and training services
- Support to technology development
- Local regulation and taxes
- Inter-business interaction, cluster formation
- Conditions for the supply of venture capita

References

Andersson, Å.E., 1993, Economic Structure of the 21st Century, in Å.E. Andersson, D.F. Batten , K. Kobayashi and Y. Yoshikawa (eds), *The Cosmo-Creative Society - Logistical Networks in a Dynamic Economy,* Springer-Verlag, Berlin.

Cheshire , P.C. and I.R. Gordon , 1998, Territorial Competition: Some Lessons for Policy, *Annals of Regional Science*, 32:321-346.

Dicken, P. and A. Tickel, 1992, Competitors or Collaborators? The Structure of Inward Investment Promotion in the Northern England, *Regional Studies*, 26:99-106.

Jacobs, J., 1984, *Cities and the Wealth of Nations*, Random House, New York.

North , D., 1990, *Institutions, Institutional Change, and Economic Performance*, Cambridge University Press, Cambridge, Mass.

Stough , R., 1990, Potentially Irreversible Global Trends and Changes: Local and Regional Strategies for Survival, Paper prepared for presentation at the meetings of the American Association for the Advancement of Science, New Orleans, Louisiana, Febr. 17-20, 1990.

Veblen T.B., 1899, The Theory of the Leisure Class: An Economic Study of Institutions, Macmillan, New York.

List of Figures

List of Tables

Authors Index

Subject Index

Titles in the Series

C. S. Bertuglia, M. M. Fischer
and *G. Preto* (Eds.)
Technological Change,
Economic Development and Space
XVI, 354 pages. 1995. ISBN 3-540-59288-1
(out of print)

H. Coccossis and *P. Nijkamp* (Eds.)
Overcoming Isolation
VIII, 272 pages. 1995. ISBN 3-540-59423-X

L. Anselin and *R. J. G. M. Florax* (Eds.)
New Directions in Spatial Econometrics
XIX, 420 pages. 1995. ISBN 3-540-60020-5
(out of print)

H. Eskelinen and *F. Snickars* (Eds.)
Competitive European Peripheries
VIII, 271 pages. 1995. ISBN 3-540-60211-9

J. C. J. M. van den Bergh, P. Nijkamp
and *P. Rietveld* (Eds.)
Recent Advances in
Spatial Equilibrium Modelling
VIII, 392 pages. 1996. ISBN 3-540-60708-0

P. Nijkamp, G. Pepping and *D. Banister*
Telematics and Transport Behaviour
XII, 227 pages. 1996. ISBN 3-540-60919-9

D. F. Batten and *C. Karlsson* (Eds.)
Infrastructure and the Complexity
of Economic Development
VIII, 298 pages. 1996. ISBN 3-540-61333-1

T. Puu
Mathematical Location and
Land Use Theory
IX, 294 pages. 1997. ISBN 3-540-61819-8

Y. Leung
Intelligent Spatial Decision Support Systems
XV, 470 pages. 1997. ISBN 3-540-62518-6

C. S. Bertuglia, S. Lombardo
and *P. Nijkamp* (Eds.)
Innovative Behaviour in Space and Time
X, 437 pages. 1997. ISBN 3-540-62542-9

A. Nagurney and *S. Siokos*
Financial Networks
XVI, 492 pages. 1997. ISBN 3-540-63116-X

M. M. Fischer and *A. Getis* (Eds.)
Recent Developments in Spatial Analysis
X, 434 pages. 1997. ISBN 3-540-63180-1

R. H. M. Emmerink
Information and Pricing
in Road Transportation
XVI, 294 pages. 1998. ISBN 3-540-64088-6

F. Rietveld and *F. Bruinsma*
Is Transport Infrastructure Effective?
XIV, 384 pages. 1998. ISBN 3-540-64542-X

P. McCann
The Economics of Industrial Location
XII, 228 pages. 1998. ISBN 3-540-64586-1

L. Lundqvist, L.-G. Mattsson
and *T. J. Kim* (Eds.)
Network Infrastructure
and the Urban Environment
IX, 414 pages. 1998. ISBN 3-540-64585-3

R. Capello, P. Nijkamp and *G. Pepping*
Sustainable Cities and Energy Policies
XI, 282 pages. 1999. ISBN 3-540-64805-4

M. M. Fischer and *P. Nijkamp* (Eds.)
Spatial Dynamics of European Integration
XII, 367 pages. 1999. ISBN 3-540-65817-3

J. Stillwell, S. Geertman
and *S. Openshaw* (Eds.)
Geographical Information and Planning
X, 454 pages. 1999. ISBN 3-540-65902-1

G. J. D. Hewings, M. Sonis, M. Madden
and *Y. Kimura* (Eds.)
Understanding and Interpreting Economic
Structure
X, 365 pages. 1999. ISBN 3-540-66045-3

A. Reggiani (Ed.)
Spatial Economic Science
XII, 457 pages. 2000. ISBN 3-540-67493-4

D. G. Janelle and *D. C. Hodge* (Eds.)
Information, Place, and Cyberspace
XII, 381 pages. 2000. ISBN 3-540-67492-6

P. W. J. Batey and *P. Friedrich*
Regional Competition
VIII, 290 pages. 2000. ISBN 3-540-67548-5

MIX
Papier aus verantwortungsvollen Quellen
Paper from responsible sources
FSC® C105338

If you have any concerns about our products,
you can contact us on
ProductSafety@springernature.com

In case Publisher is established outside the EU,
the EU authorized representative is:
Springer Nature Customer Service Center GmbH
Europaplatz 3, 69115 Heidelberg, Germany

Printed by Libri Plureos GmbH
in Hamburg, Germany